Hydrologic Time Series Analysis: Theory and Practice

Hydrologic Time Series Analysis: Theory and Practice

By

Deepesh Machiwal

Central Arid Zone Research Institute
Regional Research Station, Bhuj, Gujarat, India

and

Madan Kumar Jha

Indian Institute of Technology Kharagpur, West Bengal, India

With contributions of

Zhenxing Zhang, Robert D. Pody, Andrew D. Dehoff and John W. Balay, USA; M.N. Khaliq and L. Sushama, Canada; Dileep K. Panda and Ashwani Kumar, India; and Philip G. Oguntunde and Babatunde J. Abiodun, South Africa

Springer

CP

A C.I.P. Catalogue record for this book is available from the Library of Congress.

ISBN 978-94-007-1860-9 (HB)
ISBN 978-94-007-1861-6 (e-book)

Copublished by Springer,
P.O. Box 17, 3300 AA Dordrecht, The Netherlands
with Capital Publishing Company, New Delhi, India.

Sold and distributed in North, Central and South America by Springer,
233 Spring Street, New York 10013, USA.

In all other countries, except SAARC countries—Afghanistan, Bangladesh, Bhutan, India, Maldives, Nepal, Pakistan and Sri Lanka—sold and distributed by Springer, Haberstrasse 7, D-69126 Heidelberg, Germany.

In SAARC countries—Afghanistan, Bangladesh, Bhutan, India, Maldives, Nepal, Pakistan and Sri Lanka—sold and distributed by Capital Publishing Company, 7/28, Mahaveer Street, Ansari Road, Daryaganj, New Delhi, 110 002, India.

www.springer.com

Printed on acid-free paper

Printed in India.

Preface

In the 21st century, the entire world is suffering from freshwater scarcity due to the ever-increasing water demand in different sectors, worldwide population growth and increasing pollution of vital freshwater resources. Therefore, efficient planning and management of water resources is of utmost importance to ensure sustainable development on the earth. Statistical analyses of hydrologic time series play a central role in the planning and management of water resources. In fact, statistical analyses of every hydrologic time series must always be carried out for determining fundamental time series characteristics, i.e. normality, homogeneity, stationarity, presence of trends and shifts, periodicity, persistence and stochastic component. However, such a practice is currently missing among the hydrologists and hydrogeologists. As a result, hydrologic time series analysis has received less attention even in the era of information technology, especially in developing countries. A comprehensive review of literature by the authors revealed that the past studies on time series analysis mostly focus on a specific time series characteristic only and that the current application domain of time series is limited. Based on the research experience of the authors, it was found that a suitable book dealing with both theory and application of time series analysis techniques is lacking, particularly in the field of water resources engineering. Therefore, many hydrologists and hydrogeologists face difficulties in adopting time series analysis as one of the tools for their research. Thus, there is a need to have a book with a proper blend of theoretical and practical aspects of time series analysis, and its use in the field of water resources engineering.

The present book is an attempt to fill the above gap by providing adequate theoretical background as well as practical applications of various tools/ techniques for analyzing time series data. This book is divided into two parts: Part I describes theoretical aspects of tools/techniques available for time series analysis, and Part II presents applications of various time series tests through selected case studies. Chapter 1 deals with an overview of water problems and challenges, fundamentals of time series analysis, and the significance of time series analysis in hydrology. Chapters 2 to 6 constitute Part I, which present an overview of time series characteristics in hydrology/water resources engineering, statistical measures for summarizing time series data and evaluating system performance, methods for checking the normality of time series data, theoretical details of 31 available statistical tests along with detailed procedures for applying them to real-world time series data, theory and methodology of stochastic modelling, and current status of time series analysis in hydrological sciences. Chapters 7 to 12 constitute Part II, which demonstrate the application of most time series tests through a case study in India (authors' own work) and present a comparative evaluation of various time series tests (also authors' own work). In addition, four invited case studies are included as Chapters 9, 10, 11 and 12 from India and abroad (USA, Canada and South Africa). The contributors of the invited case studies were chosen based on their proven knowledge in the specific area of their contribution and these chapters were meticulously reviewed and edited by the authors. Thus, Chapters 9, 10, 11 and 12 have been revised at least twice.

This book will not only serve as a textbook for the students and teachers of water resources engineering field but will also solve the purpose of reference book to educate researchers/scientists about the theory and practice of time series analyses in hydrological sciences. This book will be very useful to a wide range of students, researchers, teachers and professionals such as undergraduate and postgraduate students, teachers and researchers of civil, environmental, agricultural and ecological engineering fields as well as to the practising hydrologists and hydrogeologists.

Deepesh Machiwal

Madan Kumar Jha

About the Authors

Deepesh Machiwal obtained his Bachelor of Engineering from Rajasthan Agricultural University, Bikaner, India in 1999, and Master of Engineering from Maharana Pratap University of Agriculture and Technology (MPUAT), Udaipur, Rajasthan, India in 2001. He obtained his PhD. from Indian Institute of Technology (IIT) Kharagpur, West Bengal, India in 2009. Between June 2001 and October 2001 he worked in the capacity of Senior Research Fellow in Department of Soil and Water Engineering, College of Technology and Engineering (CTAE), MPUAT. Between November 2001 and June 2005 he served as a Senior Research Fellow in Department of Agricultural & Food Engineering, IIT Kharagpur. He served from July 2005 to January 2011 as an Assistant Professor in Department of Soil and Water Engineering, CTAE, MPUAT, Udaipur, India. In February 2011, he joined as Senior Scientist in one Regional Research Station of Central Arid Zone Research Institute, situated at Bhuj, India. Dr. Machiwal has to his credit 10 publications in international refereed journals and six publications in national refereed journals, two technical reports and 17 publications in conference proceedings. He has also contributed four book chapters. He is a reviewer of national journal and international journals related to water resources engineering. He is also a life member of eight national professional societies or associations.

Madan Kumar Jha is Professor of Groundwater Hydrology at the Department of Agricultural & Food Engineering, Indian Institute of Technology (IIT) Kharagpur, West Bengal, India. He obtained his Bachelor in Technology (B.Tech., Agril. Engineering) from Rajendra Agricultural University, Bihar, India in 1990 and was the recipient of 'University Gold Medal' for outstanding academic performance. He obtained his M.Eng. from Asian Institute of Technology, Bangkok in 1992 availing full postgraduate scholarship, and PhD. from Ehime University, Japan in 1996 availing Monbusho Scholarship. Before joining IIT Kharagpur in 1999, Prof. Jha also worked as Water Resources Engineer at Panya Consultants Co. Ltd., Bangkok and as Postdoctoral Fellow at Kochi University, Japan availing 'JSPS Research Fellowship'. He has authored/co-authored three books and has contributed nine book chapters. To date, he has to his credit 50 publications in international refereed journals and six publications in national refereed journals, six technical reports and more than 60 publications in conference proceedings. He is serving as 'Assistant Managing Editor' of *Journal of Spatial Hydrology*, USA since January 2010; 'Editor' of *Journal of Water Resource and Protection*, USA since September 2010; 'Associate Editor' of *International Agricultural Engineering Journal,* China since July 2008; 'Associate Editor' of *Journal of Agricultural Engineering,* India since 2010 and 'Editor' of *Research Journal of Chemistry and Environment*, India since 2004. He is also serving as a reviewer of several international journals of water resources engineering field. He has been awarded 'Fellow' of the Institution of Engineers India (IEI), Kolkata in July 2011 as well as 'Fellow' of the Indian Water Resources Society, Roorkee in July 2008. In addition, Prof. Jha has to his credit salient scholastic awards and international fellowships, including international 'AMA-Shin-Norinsha-AAAE Young Researcher Award' by the Asian Association for Agricultural Engineering, Bangkok in 2005; 'Distinguished Services Award' in the field of Soil and Water Engineering by the Indian Society of Agricultural Engineers, New Delhi in 2002; 'JSPS Research Fellowship' (1997-1999); 'Alexander von Humboldt Fellowship' (2004-2005); and 'JSPS Invitation Fellowship (Long-Term)' in 2009. Prof. Jha is a regular or life member of several international and national professional societies/associations.

Acknowledgements

The completion of a book with a proper blending of theoretical and practical aspects of time series analysis, and its use in water resources engineering certainly involves help and support from others besides authors' own exertion. This book is a part of constant efforts of the two authors, who extensively explored literature and did research on the application of time series analysis in hydrology and climatology for last several years. This book includes four invited case studies and the authors are grateful to the contributors who accepted our invitation gladly and extended full cooperation during revision of the manuscripts according to our detailed comments and suggestions. Each chapter has been meticulously revised at least twice. Chapters 2, 3, 4 and 5 have been greatly benefitted from Helsel and Hirsch (2002), Loucks and van Beek (2005), USEPA (2006) and Shahin et al. (1993), which is gratefully acknowledged. We also offer our gratitude to other individuals who helped indirectly in the preparation of this book.

Thanks are also due to Capital Publishing Company, New Delhi, India and to Springer, The Netherlands for their kind cooperation and timely efforts in publishing this book to make it available to the scientific community. We also thank the editorial team of the publishers for careful editing of the manuscripts.

Finally, the first author (Deepesh Machiwal) would be failing in his duties if he does not record special thanks to his parents, wife Savita and daughter Mahi who have been a constant source of encouragement during the entire course of this hard task. The second author (Madan Kumar Jha) is especially grateful to his wife Manisha and son Piyush for their constant love, forbearance and encouragement, and sincere support during the writing of this book.

Contents

1

Introduction

1.1 Water Problems and Challenges: An Overview

Water is the most precious resource of the earth because no life is possible without water. It is essential for the survival and livelihood of every human. It also regulates ecosystems, grows our food and powers our industry. Hardly any economic activity can be sustained without water. Undoubtedly, water plays a vital role in our life. Different dimensions of water functions in society and nature are (Falkenmark and Rockstrom, 2004): (i) water as *life-support* and hence as a basic need and as a human and animal right; (ii) water as an *economic* commodity in some uses; (iii) water as an *integral part of ecosystem* (sustaining it and being sustained by it); (iv) water as a *sacred resource*; and (v) water as *an inevitable component of cultures and civilizations*. Thus, water is the key resource for the human/animal health, socio-economic development, and the survival of earth's ecosystems. On the other hand, natural ecosystems also play a crucial role in the availability and quality of water through their purifying and regulating services, thereby sustaining human development on the earth. In other words, water has social, economic and environmental values and is essential for sustainable development (Falkenmark and Rockstrom, 2004; UNESCO, 2003, 2009). In contrast with many other vital resources of the earth, there is no substitute for water in most activities and processes where it is needed!

At present, about 10% of the world's freshwater supplies are used for maintaining health and sanitation, whereas agriculture accounts for about 70% and industries about 20% of the world's freshwater supplies (Shiklomanov, 1997; Shiklomanov and Rodda, 2003). Food production is the most water-intensive sector. It has been estimated that about one litre of liquid water gets converted to water vapour to produce one calorie of food. Every person is responsible for consuming 2000 to 5000 litres of water every day depending on one's diet and the method of food production, which is far more than 2 to 5 litres we drink every day (Rodriguez and Molden, 2007). A meat-based diet requires much more water than a vegetarian diet; for example, we need about

1000 litres of water to produce one kilogram of wheat, whereas we need about 5000 to 13,500 litres of water to produce one kilogram of meat. The demand for water is gradually increasing with growing population as well as rapid urbanization and industrialization in different parts of the world (Postel, 1998; Shiklomanov and Rodda, 2003; UNESCO, 2003, 2009; Grafton and Hussey, 2011). As a result, water demand is surpassing the available freshwater resource. On top of it, in future, more people will need more water not only for food and sanitation but also for fibre, livestock and industrial crops (bio-energy).

Unfortunately, the excessive use and continued mismanagement of freshwater resources for human development (to supply ever-increasing water demands for food, feed, fibre and fuel) have led to water shortages, increasing pollution of freshwater, loss of biodiversity, and degraded ecosystems across the world (e.g., Postel, 1998; de Villiers, 2001; Steffen et al., 2002; UNESCO, 2003; UN Water, 2007; Vörösmarty et al., 2010; Grafton and Hussey, 2011). As a result, freshwater scarcity has emerged as one of the most pressing problems in the 21st century. According to Molden (2007), one in three people at present face water shortages, around 1.2 billion people (almost one-fifth of the world's population) live in areas of '*physical water scarcity*' (i.e., where the available water resources cannot meet the demands of the population), and 500 million people are approaching this situation. Another 1.6 billion people (almost one quarter of the world's population) face '*economic water scarcity*' (i.e., where countries lack the necessary infrastructure to harness water from rivers and aquifers). Furthermore, about 2.5 billion people lack adequate sanitation, and 884 million people are without access to safe water (UNICEF and WHO, 2008). It has been estimated that half of the population of the developing world is exposed to polluted sources of water that increase disease incidence. Between 1991 and 2000, over 665,000 people died in 2557 natural disasters, of which 90% were water-related disasters and a vast majority of victims (97%) were from developing countries (IFRC, 2001).

If the present trend continues, based on the widely used Falkenmark indicator for water scarcity, nearly 1.4 billion people will experience '*chronic water scarcity*' (i.e., water supply less than 1000 m^3/capita/annum) within the first 25 years of this century, mostly in semi-arid regions of Asia, North Africa and Sub-Saharan Africa. Also, 1.8 billion people will be living in countries or regions with '*absolute water scarcity*' (i.e., water supply less than 500 m^3/capita/annum), and two thirds of the world's population could be under '*water stress*' (i.e., water supply less than 1700 m^3/capita/annum) conditions by 2025 (UN Water, 2007). Urban and industrial water use in the world is projected to double by 2050. With increasing evidence of unsustainable water use in several parts of the world, particularly in developing nations, India is under '*water stress*' conditions today and will face '*chronic water scarcity*' by 2025. The problem of water management in general and water shortages in particular will worsen in many parts of the world due to global climate change. Higher temperatures and changes in extreme weather conditions are projected to affect the availability and distribution of rainfall, snowmelt, river flows and groundwater, and deteriorate water quality, which in turn can have severe impacts on both urban and rural regions of the world (IPCC, 2007). Climate change is considered as a major challenge to the efficient management of natural resources and a barrier to the transition from poverty to prosperity (UNDP, 2007). Thus, in the beginning of the 21st century, we are bound to face the stark reality that the current patterns of water development and consumption are not sustainable in several countries of the world. Therefore, there is an urgent need for widespread realization that freshwater is a finite and vulnerable resource, which must be used efficiently, equitably and in an ecologically sound manner for present and future generations to ensure sustainable development on the earth.

Inadequate water resource systems reflect failures in planning, management and decision making not only in the water sector but also in other sectors of society directly or indirectly dependent on water. It is the need of the hour for scientists/engineers as well as for planners and decision makers to efficiently plan, develop, operate and manage water resource systems so as to ensure adequate, cost-effective, good-quality and sustainable supply of water for humans and nature (Falkenmark and Rockstrom, 2004; Loucks and van Beek, 2005; Grafton and Hussey, 2011). The complex and deep interactions that have existed between humans and water systems throughout the human history need to be understood by modern scientists/engineers, planners and decision makers (Postel and Richter, 2003). It is also essential to recognize that unlike much basic economic theory, the goods and services provided by ecosystems are not at all substitutable and ecosystems cannot easily be replaced by technology (Kaufmann, 1995). At this point in human evolution, it is vital that people understand the crucial link between human welfare and ecosystem well-being (Arrow et al., 1995; UNESCO, 2003, 2009), and institutions must be strengthened to support effective water governance (Walker, 2009). Natural scientists and social scientists need to work together to better understand human-environment interactions (IPCC, 2001) as well as to bridge the growing knowledge gap between water management and ecology. More and more research is needed to predict how potential ecosystem perturbations may affect short- and long-term ecosystem functionality. Given the dynamic and evolving nature of ecosystems, a major technical challenge is quantifying how much the ability of ecosystems to meet human needs is changing over time.

Today, one of the biggest challenges is how we can effectively balance freshwater for human development and ecosystems welfare in achieving equity, environmental sustainability, and economic efficiency in the face of looming global climate change. Quantitative analysis using statistical and mathematical modelling tools as well as modern information technologies such as remote sensing, GIS, decision support system, expert system, etc. can support and improve water resources planning and management (Loucks and van Beek,

2005; Jha and Peiffer, 2006; Jha, 2010). The "think globally, act locally" slogan of the late 1980s reminds us of our professional attitudes to and scientific responsibilities for environment (nature) in general and freshwater resources in particular, which must not be forgotten. Following holistic and multidisciplinary approaches as well as using modern concepts and tools/ techniques, water scientists and engineers must make sincere and sustained efforts to improve their understanding about hydrologic/hydrogeologic processes and their linkage with our ecosystems; thereby improving both the process and product. It is worth mentioning that the existing tools and technologies, irrespective of their sophistication, will not eliminate the need to reach conclusions and make decisions on the basis of incomplete and uncertain data, and scientific knowledge (Loucks and van Beek, 2005). In other words, the importance of professional judgement and that of research, development and education in the planning and management of water resources must not be undermined amidst increasing popularity and reliance on new/ emerging tools and technologies in the 21st century.

1.2 What is Time Series?

The term *time series* is defined as "a sequence of values collected over time on a particular variable" (Haan, 1977). A time series can consist of the values of a variable observed at discrete times, averaged over a given time interval, or recorded continuously with time. It may consist of only deterministic events, only stochastic events, or a combination of deterministic and stochastic events. Generally, a hydrologic time series is composed of a stochastic component superimposed on a deterministic component (Haan, 1977; Shahin et al., 1993). The deterministic component can be classified as a trend, a jump, a periodic component, or a combination of these (Haan, 1977). The time intervals for most hydrologic time series are hour, day, week, month, season or year. Data in business, economics, engineering, environment, medicine, earth sciences, hydrology, climatology, meteorology and other areas are often collected in the form of time series. Some examples of the general time series are share prices on successive days, company profits in successive years, and sales figures in successive weeks/months/years; while the examples of hydrologic time series are hourly/daily/monthly/annual temperature (air or water) readings, precipitation in successive days/weeks/months/years, hourly/daily/monthly/ annual evaporation or evapotranspiration readings, hourly/daily/monthly/annual soil moisture, hourly/daily/weekly/monthly/annual streamflow or river-stage readings, hourly/daily/weekly/monthly/annual groundwater-level readings, hourly/daily/weekly/monthly/annual tide-level readings, daily water consumption in domestic, industrial or agricultural sectors, etc.

Furthermore, the *realization* of a process is the outcome of an experiment in which the process is observed, and hence a single time series is known as a *realization* (Shahin et al., 1993). The term *ensemble* denotes a collection of all possible realizations of a process, and it is used in the theory of stochastic processes and time series analysis in lieu of the well-known statistical term 'population' (Haan, 1977; Shahin et al., 1993). The properties of a time series can be obtained based on a single realization over a time interval or based on several realizations at a given time. The properties based on a single realization are known as *time average properties*, whereas those based on several realizations at a particular time are known as *ensemble properties* (Haan, 1977). If the *time average properties* and the *ensemble properties* of a time series are same, the series is said to be *ergodic* (Haan, 1977). *Ergodicity* is the property by which each realization of a given process is a complete and independent representative of all possible realizations of the process (Shahin et al., 1993). Thus, the *ergodicity* allows the scientists/researchers to determine the statistical properties of a process from a single realization.

1.3 Time Series Analysis

Time series analysis is the investigation of a temporally distributed sequence of data or the synthesis of a model for prediction wherein time is an independent variable. Sometimes, time is not actually used to predict the magnitude of a random variable such as peak runoff rate, but the data are ordered by time. The main intent of time series analysis is to detect and describe quantitatively each of the generating processes underlying a given sequence of observations (Shahin et al., 1993). Hydrologic time series are analyzed for several reasons. The main reason as reported in the literature is to detect a trend due to another random hydrologic variable. Secondly, time series may be analyzed to develop and calibrate a model that would describe the time-dependent characteristics of a hydrologic variable. Thirdly, time series models may be used to predict future values of a time-dependent variable. Besides the time-dependent data series, there are space-dependent data series of hydrologic systems, which are known as 'spatial data series'. Thus, in the spatial data series, the data are location specific instead of depending on time as in the time series. The examples of spatial data series are: the variability of groundwater levels over a groundwater basin, spatial variation of aquifer or soil properties, spatial variation of rainfall in a catchment/basin, and so on. Most of the time series analysis methods can equally be applied to spatial data series (Shahin et al., 1993). Therefore, spatial data series is sometimes referred to as time series.

There are four major steps involved in a time series analysis (McCuen, 2003): (i) detection, (ii) analysis, (iii) synthesis, and (iv) verification. In the detection step, systematic components of the time series such as trends or periodicity are identified. It is also necessary to decide in this step whether the systematic effects are physically and statistically significant. In the analysis step, the systematic components are analyzed to identify their characteristics including magnitudes, form and their duration over which the effects exist. In the synthesis step, information from the analysis step is accumulated to develop a time series model and to evaluate goodness-of-fit of the developed model. Finally, in the verification step, the developed time series model is evaluated using independent sets of data. For further details of the time series analysis, the readers are referred to the specialized books on time series analysis such as Yevjevich (1972), Salas et al. (1980), Bras and Rodriguez-Iturbe (1985), Cryer (1986), and Clarke (1998).

1.4 Classification of Time Series

A time series can be classified in many ways according to different criteria. Three widely used classifications of the time series are described below. The details about the classification of hydrologic time series can be found in Salas (1993).

1.4.1 Discrete or Continuous Time Series

Time series can be either continuous or discrete. A time series is called 'discrete' if the observations are recorded at different time instants or at different points in space (Haan, 1977; Shahin et al., 1993). On the other hand, if the observations are recorded continuously in time or space, then the series is known as a 'continuous time series'. 'Discrete time series' is often derived from a 'continuous time series'. Usually in hydrology, a time series is of the discrete type. As a result, the case studies presented in this book are restricted to the discrete time series. A continuous plot of a 'discrete time series' should not be confused with a 'continuous time series'.

1.4.2 Full or Partial Duration Series

A 'full time series' is the one which contains all the recorded observations over time or space (Haan, 1977; Shahin et al., 1993). As the name suggests, a 'partial duration series' contains only selected observations which are extracted from the full time series. For instance, daily rainfall recorded at a specific location over a given period of time constitutes a full time series of rainfall. A time series of one-day maximum rainfall can be extracted from the full rainfall time series by arranging the maximum rainfall occurring in a day for each year in the order of occurrence. Note that the maximum rainfall time series contains less information than the original full rainfall time series. That is, a 'partial duration series' always contains less information than the 'full time series'. In addition, the observation points in a partial duration series may not be equidistant.

1.4.3 Univariate or Multivariate

If only one variable is observed at each time, the time series is known as 'univariate time series'. However, if two variables are observed at the same time (simultaneously), the series is known as 'bivariate time series'. If more than two variables are observed simultaneously at a time, the series is known as a 'multivariate time series'. This book deals with univariate time series only.

1.5 Structure of Time Series

A time series is often adequately described as a function of four components: *trend, seasonality, dependent stochastic component* and *independent residual component*. In general, a time series can be mathematically expressed as (Shahin et al., 1993):

$$x_t = T_t + S_t + \varepsilon_t + \eta_t \qquad (1)$$

where T_t = trend component, S_t = seasonality, ε_t = dependent stochastic component, and η_t = independent residual component.

In the time series analysis, it is assumed that the data (observations) consist of a *systematic pattern* and *random noise* (error); the latter usually makes the pattern difficult to be identified. The *systematic pattern* is represented by the first two components of Eqn. (1), which are deterministic in nature, whereas the stochastic component accounts for the random error. Generally, the stochastic component contains a dependent part which may be represented by an ARMA(p,q) model, where 'p' and 'q' are the orders of the autoregressive and moving-average models, respectively, and an independent part that can only be described by some sort of probability distribution function. When $p = 0$, the ARMA(p,q) represents an MA(q) model, and when $q = 0$, it represents an AR(p) model.

Thus, the process of hydrologic time series analysis should be viewed as a process of identifying and separating the total variation in measured data into above-mentioned four components. When a time series has been analyzed and the components accurately characterized, each component can then be modelled. Methods for identifying trends in time series are described in Chapter 4 and the methods for identifying stochastic component are described in Chapter 5.

1.6 Salient Characteristics of Time Series

Most statistical analyses of hydrologic time series at the usual time scale encountered in water resources studies are based on a set of fundamental assumptions, which are: the series is homogenous, stationary, free from trends and shifts, non-periodic with no persistence (Adeloye and Montaseri, 2002). The term 'homogeneity' implies that the data in the series belong to one population, and therefore have a time invariant mean. Non-homogeneity arises due to changes in the method of data collection and/or the environment in which it is done (Fernando and Jayawardena, 1994). On the other hand, 'stationarity' implies that the statistical parameters of the series computed from different samples do not change except due to sampling variations. A time series is said to be *strictly stationary* if its statistical properties do not vary with changes of time origin. A less strict type of stationarity, called *weak stationarity* or *second-order stationarity*, is that in which the first- and second-order moments depend only on time differences (Chen and Rao, 2002). In nature, *strictly stationary* time series does not exist, and *weakly stationary* time series is practically considered as stationary time series.

There are many ways by which changes in the hydro-meteorological series can take place. A change can occur gradually (known as 'trend') or abruptly (known as 'step change' or 'jump'), or may take more complex form (Shahin et al., 1993). A 'trend' is defined as "a unidirectional and gradual change (falling or rising) in the mean value of a variable" (Shahin et al., 1993). A time series is said to have trends, if there is a significant correlation (positive or negative) between the observed values and time. Trends and shifts in a hydrologic time series are usually introduced due to gradual natural or human-induced changes in the hydrologic environment producing the time series (Haan, 1977; Salas, 1993). Gradual or natural changes in hydrologic variables could be caused by a global or regional climate change, which would be a representative of changes occurring over the study area. Changes in the observed variables that may not be able to be extrapolated over a study area could be caused by a gradual urbanization of the area surrounding the monitoring site, changes in the method of measurement at the monitoring site, or by moving the monitoring site even a short distance away. 'Step changes' or 'jumps' in a time series usually result from catastrophic natural events such as earthquakes, tsunami, cyclones, or large forest fires which quickly and considerably alter the hydrologic regime of an area. The man-made changes such as the closure of a new dam, the beginning or termination of groundwater pumping, or other such developmental activities may also cause jumps in some hydrologic time series (Haan, 1977). Jumps can be either positive or negative. The 'jump' or 'step change' is usually noted in the overall record at a monitoring site, but this information is not always presented with the site's data series. Thus, the variables that appear to have a trend may actually just represent a change in climatological or hydrological conditions near the monitoring site. Under such conditions, the affected climatological data should be modified so that the values are better representative of the study area as a whole (Hameed et al., 1997). A key element in this process is the ability to demonstrate whether a change or trend is present in the climatological data series and to quantify this trend, if it is present.

'Periodicity' is another characteristic of time series (natural hydrologic time series), which represents a regular or oscillatory form of movement that is recurring over a fixed interval of time (Shahin et al., 1993). It generally occurs due to astronomic cycles such as earth's rotation around the sun (Haan, 1977; Kite, 1989). Annual cycles are often apparent in rainfall, evapotranspiration, streamflow, groundwater level, soil moisture and other types of hydrologic data (Haan, 1977). Weekly cycles may be present in the water-use data of domestic, industrial, or agricultural sectors; many times the water-use time series contain both annual and weekly periodicities (Haan, 1977). In order to identify and quantify the periodicity in a hydrologic or climatologic time series, the time scale should be considered less than a year (i.e., month or six-month). The periodicity effect is not discernible in an annual time series, and hence half-annual or monthly time series normally encountered in hydrology can be used for analyzing the periodicity.

Lastly, the phenomenon of 'persistence' is highly relevant to the hydrologic time series, which means that the successive members of a time series are linked in some dependent manner (Shahin et al., 1993). In other words, 'persistence' denotes the tendency for the magnitude of an event to be dependent on the magnitude of previous event(s), i.e., a memory effect. For example, the tendency for low streamflows to follow low streamflows and that for high streamflows to follow high streamflows. Thus, 'persistence' can be considered synonymous with autocorrelation (O'Connel, 1977). Hurst (1951, 1957) was the first person to describe 'persistence' comprehensively in his studies on a reservoir design across the Nile River. The phenomenon was defined in terms of a parameter called "Hurst's coefficient", the average value of which is approximately 0.73 for very large samples. However, its theoretical value for an independent Gaussian process to which hydrologic series are assimilated should be 0.5 (Capodaglio and Moisello, 1990). If the theoretical and the observed values of Hurst's coefficient do not correspond, it is known as "Hurst's phenomenon". All the stochastic models that have been proposed to represent hydrologic time series have attempted to include the persistence phenomenon. However, with the time series records commonly available in hydrology, it is virtually impossible to identify any long-term persistence in the hydrologic time series (Capodaglio and Moisello, 1990). Chapter 4 deals with various methods/tests used for identifying the above characteristics of a time series.

1.7 Time Series Analysis vis-a-vis Hydrology

In early days, the application of statistics in hydrology was restricted to only surface water problems, especially for analyzing the hydrologic extremes such as floods and droughts. However, during the past three decades or so, the application of statistics in hydrology has expanded considerably to encompass the problems of both surface water and groundwater systems, including atmospheric systems. With such a broad domain coupled with the rapid advancement in computer and data management technologies, statistics has emerged as a powerful tool for analyzing hydrologic problems. Particularly, time series analysis has become a major tool in hydrology in the era of information technology. Today, besides the basic statistical analysis of hydrologic time series, the applications of time series analysis in hydrological sciences include development of mathematical models to generate synthetic hydrologic data, to forecast hydrologic events, to identify trends and shifts in hydrologic data, to fill in missing observations, and to extend short hydrologic records (Salas, 1993). Certainly, time series analysis has become a vital tool in hydrological sciences and its importance has dramatically enhanced in the recent past due to ever-increasing interest in the scientific understanding of climate change.

In epilogue, statistics is just one of the several tools available for application in hydrological sciences. Like other tools and techniques of hydrology/water resources engineering, statistical models and methods can serve as valuable tools in the analysis and solution of several real-world water problems. It should be noted that the usefulness of any tool or technique, and hence the reliability of a hydrologic analysis/estimate depends squarely on the proficiency and knowledge of the hydrologists/water resources engineers. Unfortunately, the time and energy associated with the development of a model and the complexity involved in modelling or analysis often so focus the modellers, especially novice modellers, that they believe that the model is indeed a full representation of reality/natural systems. However, in reality, no model whether statistical or mathematical or some combination of the two can describe the actual and complete hydrology of any natural system (Haan, 2002); it is always simpler than the prototype/natural system. We should never forget that a model is a simplified form of reality and that it is simply a tool to assist in decision making, not a replacement for it! No models or techniques, no matter how complex they are, can replace the vital role of hydrologists' competency and their in-depth knowledge of water systems in making efficient decisions for solving water problems.

References

Adeloye, A.J. and Montaseri, M. (2002). Preliminary streamflow data analyses prior to water resources planning study. *Hydrological Sciences Journal*, **47(5):** 679-692.

Arrow, K., Bolin, B., Costanza, R., Dasgupta, P., Folke, C., Holling, C., Jansson, B.-O., Levin, S., Maler, K.-G., Perrings, C. and Pimentel, D. (1995). Economic growth, carrying capacity and the environment. *Science*, **268:** 520-521.

Bras, R.L. and Rodriguez-Iturbe, I. (1985). Random Functions and Hydrology. Addison-Wesley, Reading, M.A.

Capodaglio, A.G. and Moisello, U. (1990). Simple stochastic model for annual flows. *Journal of Water Resources Planning and Management, ASCE*, **116(2):** 220-232.

Chen, H.-L. and Rao, A.R. (2002). Testing hydrologic time series for stationarity. *Journal of Hydrologic Engineering, ASCE*, **7(2):** 129-136.

Clarke, R.T. (1998). Stochastic Processes for Water Scientists: Development and Applications. John Wiley and Sons, New York.

Cryer, J.D. (1986). Time Series Analysis. PWS Publishers, Duxbury Press, Boston, MA.

de Villiers, M. (2001). Water: The Fate of Our Most Precious Resource. Mariner Books, Mifflin, Boston, 368 pp.

Falkenmark, M. and Rockstrom, J. (2004). Balancing Water for Humans and Nature: The New Approach in Ecohydrology. Earthscan, London.

Fernando, D.A.K. and Jayawardena, A.W. (1994). Generation and forecasting of monsoon rainfall data. Proceedings of the 20th WEDC Conference on Affordable Water Supply and Sanitation, Colombo, Sri Lanka, pp. 310-313.

Grafton, R.Q. and Hussey, K. (editors) (2011). Water Resources Planning and Management. Cambridge University Press, Cambridge, U.K., 777 pp.

Haan, C.T. (1977). Statistical Methods in Hydrology. Iowa State University Press, Iowa, 378 pp.

Haan, C.T. (2002). Statistical Methods in Hydrology. 2nd edition, Iowa State University Press, Iowa, 496 pp.

Hameed, T., Marino, M.A., DeVries, J.J. and Tracy, J.C. (1997). Method for trend detection in climatological variables. *Journal of Hydrologic Engineering, ASCE*, **2(4):** 154-160.

Hurst, H.E. (1951). Long term storage capacity of reservoirs. *Transactions, ASCE*, **116:** 770-800.

Hurst, H.E. (1957). A suggested statistical model of some time series which occur in nature. *Nature*, **180(4584):** 494.

IFRC (2001). World Disasters Report 2001. International Federation of Red Cross and Red Crescent Societies, Geneva, Switzerland.

IPCC (2001). Third Assessment Report: Climate Change 2001. Intergovernmental Panel on Climate Change (IPCC), Geneva, Switzerland, http://www.ipcc.ch (accessed in March 2005).

IPCC (2007). Climate Change 2007: Impacts, Adaptation and Vulnerability. Contribution of Working Group II to the Fourth Assessment Report of the Intergovernmental Panel on Climate Change (IPCC), Cambridge University Press, Cambridge, U.K.

Jha, M.K. (editor) (2010). Natural and Anthropogenic Disasters: Vulnerability, Preparedness and Mitigation. Springer, Berlin, Germany and Capital Publishing Company, New Delhi, India, 615 pp.

Jha, M.K. and Peiffer, S. (2006). Applications of Remote Sensing and GIS Technologies in Groundwater Hydrology: Past, Present and Future. BayCEER, Bayreuth, Germany, 201 pp.

Kaufmann, R.K. (1995). The economic multiplier of environmental life support: Can capital substitute for a degraded environment? *Ecological Economics*, **12:** 67-79.

Kite, G. (1989). Use of time series analyses to detect climatic change. *Journal of Hydrology*, **111:** 259-279.

Loucks, D.P. and van Beek, E. (2005). Water Resources Systems Planning and Management: An Introduction to Methods, Models and Applications. Studies and Reports in Hydrology, UNESCO Publishing, UNESCO, Paris, 680 pp.

McCuen, R.H. (2003). Modeling Hydrologic Change: Statistical Methods. Lewis Publishers, CRC Press LLC, Florida, 433 pp.

Molden, D. (editor) (2007). Summary of Water for Food, Water for Life: A Comprehensive Assessment of Water Management in Agriculture. Earthscan, London, U.K., 40 pp.

O'Connel, P.E. (1977). ARIMA models in synthetic hydrology. *In:* T.A. Ciriani, U. Maione and J.R. Wallis (editors), Mathematical Models for Surface Water Hydrology. John Wiley and Sons, Inc., New York.

Postel, S.L. (1998). Water for food production: Will there be enough in 2025? *BioScience*, **48:** 629-638.

Postel, S. and Richter, B. (2003). Rivers for Life: Managing Water for People and Nature. Island Press, Washington D.C., USA, 220 pp.

Rodriguez, D. and Molden, D. (2007). Water for food, water for life: Influencing what happens next. Water Front, Stockholm International Water Institute (SIWI), No. 2, pp. 12-13.

Salas, J.D. (1993). Analysis and modeling of hydrologic time series. *In:* D.R. Maidment (editor-in-chief), Handbook of Hydrology. McGraw-Hill, Inc., USA, pp. 19.1-19.72.

Salas, J.D., Delleur, J.W., Yevjevich, V. and Lane, W.L. (1980). Applied Modeling of Hydrologic Time Series. Water Resources Publications, Littleton, CO.

Shahin, M., Van Oorschot, H.J.L. and De Lange, S.J. (1993). Statistical Analysis in Water Resources Engineering. A.A. Balkema, Rotterdam, the Netherlands, 394 pp.

Shiklomanov, I.A. (1997). Comprehensive Assessment of the Freshwater Resources of the World: Assessment of Water Resources and Water Availability in the World. Comprehensive Assessment of the Freshwater Resources, World Meteorological Organization (WMO), Geneva, Switzerland, 88 pp.

Shiklomanov, I.A. and Rodda, J.C. (editors) (2003). World Water Resources at the Beginning of the 21st Century. UNESCO and Cambridge University Press, Cambridge, U.K.

Steffen, W., Jaeger, J., Carson, D.J. and Bradshaw, C. (editors) (2002). Challenges of a Changing Earth. Springer Verlag, Berlin.

UN Water (2007). Coping with water scarcity: Challenge of the twenty-first century. Report for World Water Day 2007, http://www.unwater.org/wwd07/downloads/documents/escarcity.pdf (accessed on 23 March 2007).

UNDP (2007). Human Development Report 2007/2008. United Nations Development Program (UNDP), United Nations Plaza, New York, 384 pp.

UNESCO (2003). The United Nations World Water Development Report: Water for People, Water for Life. World Water Assessment Program, the United Nations Educational, Scientific and Cultural Organization (UNESCO), Paris, France.

UNESCO (2009). The 3rd United Nations World Water Development Report: Water in a Changing World. World Water Assessment Program, UNESCO, Paris, France.

UNICEF and WHO (2008). Progress on Drinking Water and Sanitation: Special Focus on Sanitation. UNICEF, New York and WHO, Geneva, 54 pp.

Vörösmarty, C.J., McIntyre, P.B., Gessner, M.O., Dudgeon, D., Prusevich, A., Green, P., Glidden, S., Bunn, S.E., Sullivan, C.A., Reidy Liermann, C. and Davies, P.M. (2010). Global threats to human water security and river biodiversity. *Nature*, **467(7315):** 555-561.

Walker, B.H., Barrett, S., Polasky, S., Galaz, V., Folke, C., Engström, G., Ackerman, F., Arrow, K., Carpenter, S., Chopra, K., Daily, G., Ehrlich, P., Hughes, T., Kautsky, N., Levin, S., Mäler, K.-G., Shogren, J., Vincent, J., Xepapadeas, T. and de Zeeuw, A. (2009). Looming global-scale failures and missing institutions. *Science*, **325(5946):** 1345-1346.

Yevjevich, V.M. (1972). Stochastic Processes in Hydrology. Water Resources Publications, Fort Collins, CO.

PART I

Tools/Techniques for Time Series Analysis

2

Statistical Characteristics of Hydrologic Time Series

Any hydrologic time series can be appropriately analyzed when knowledge about the basic statistical characteristics of the data series itself is first considered. Many time series analysis procedures are based on the assumptions that the time series possess certain characteristics which, in fact, are not true (Adeloye and Montaseri, 2002; Helsel and Hirsch, 2002; Rao et al., 2003). The results of such analyses based on false assumptions may provide incorrect and unreliable interpretations, or unnecessarily inconclusive. Therefore, it is essential to know about the common characteristics of hydrologic time series, which can help in selecting appropriate data analysis procedures for a given hydrologic time series.

One of the most important tasks while analyzing any time series is to describe and summarize the time series data in forms, which easily convey their important characteristics. If we want to know expected chloride concentration in rainfall at a location or spatial variability of infiltration rate or expected flood for 100-year recurrence period, then it is required to understand summary statistics of the underlying hydrologic data series. Key statistical characteristics often described include: a measure of the central tendency of the data, a measure of spread or variability, a measure of the symmetry of the data distribution, and perhaps estimates of extremes such as some large or small percentile (Snedecor and Cochran, 1980; Upchurch and Edmonds, 1991). This chapter discusses various methods for analyzing hydrologic time series data. In this chapter, important statistical characteristics of hydrologic time series are discussed, together with the salient measures for evaluating the performance of water resources systems.

It is worth explaining the concept behind two basic terms of statistics called 'population' and 'sample' in order to refresh the memory of the readers. According to Helsel and Hirsch (2002), the data about which a statement or summary is to be made are called 'population' or sometimes 'target population'. Examples of population might be major ion concentrations in all waters of an aquifer or stream reach, or all streamflows over some time at a particular site. All such data are seldom available to us. It may be impossible both physically and economically to collect all data of interest (all the groundwater in an aquifer over the study period). Alternatively, a subset of the entire data called 'sample' is selected and measured in such a way that conclusions about the sample may be extended to the entire population. Statistical characteristics computed from the sample are only inferences or estimates about characteristics of the entire population, such as location or central tendency, spread or dispersion, skewness and kurtosis. Measures of location are usually the sample mean and sample median. Measures of spread include the sample standard deviation and sample interquartile range. Use of the term 'sample' before each statistic explicitly demonstrates that they only estimate the population value, the population mean or median, etc. As sample estimates are far more common than the measures based on the entire population, the term 'mean' used in this book should be interpreted as the 'sample mean', and similarly other statistics should be interpreted.

2.1 Measures of Location

Out of six measures of location (mean, median, mode, geometric mean, harmonic mean, and trimmed mean), the 'mean' and 'median' are two commonly used measures of location.

2.1.1 Classical Measure: Arithmetic Mean

The arithmetic mean ($\bar{x}$) is calculated by summing up of all data values, x_i and dividing the sum by the sample size n:

$$\bar{x} = \sum_{i=1}^{n} \frac{x_i}{n} \tag{1}$$

For data which are in one of n groups, Eqn. (1) can be rewritten to show that the overall mean ($\bar{x}$) depends on the mean for each group, weighted by the number of observations n_i in each group (Shahin et al., 1993; Helsel and Hirsch, 2002):

$$\bar{x} = \sum_{i=1}^{n} \bar{x}_i \frac{n_i}{n} \tag{2}$$

where $\bar{x}_i$ is the mean for i^{th} group. The influence of any single observation $x_{(j)}$ on the mean can be seen by placing all but that single observation in one 'group', or

$$\bar{x} = \bar{x}_{(j)} \frac{(n-1)}{n} + x_{(j)} \frac{1}{n} \tag{3}$$

or $$\bar{x} = \bar{x}_{(j)} + (x_{(j)} - \bar{x}_{(j)})\frac{1}{n} \quad (4)$$

where $\bar{x}_{(j)}$ is the mean of all data values excluding $x_{(j)}$. Each observation's influence on the overall mean $\bar{x}$ is the distance between the observation and the mean excluding that observation. Hence, all observations do not have the same influence on the mean. An extreme/outlier observation, either high or low, has a much greater influence on the overall mean $\bar{x}$ than does a more 'typical' observation, one closer to its $\bar{x}_{(j)}$.

The influence of extreme or outlier can also be illustrated by realizing that the mean is the balance point of the entire data values, when each point is arranged on a number line (Fig. 2.1). Data points far from the central location apply a stronger downward force than those closer to the centre. If one data point nearby the central location on number line is removed, the balance point would only need a little adjustment to keep the whole dataset in balance. On the contrary, if one outlier value very far from central location is removed, the balance point would shift considerably (Fig. 2.2). This sensitivity to the magnitudes of a small number of points in the dataset defines why the mean is not a robust/resistant measure of location. It is not resistant to changes in the presence of, or to changes in the magnitudes of, a few outlier observations. When this strong influence of a small number of observations in a dataset is desirable, the mean is an appropriate measure of central location. This usually occurs when computing units of mass, such as the average precipitation from a number of sites in a raingauge network. High rainfall amounts represented by a raingauge would exert more influence (due to greater mass of rainfall) on the final average rainfall amount than low rainfall amounts.

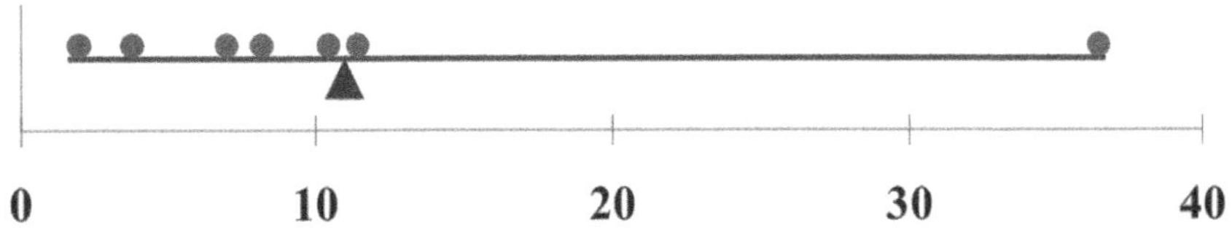

Fig. 2.1. Mean shown by triangle acting as a balance point of time series data (Helsel and Hirsch, 2002).

Fig. 2.2. Shift of mean in the left direction after removal of outlier.

2.1.2 Robust Measure: Median

The median is the middle value of data series when the data are ranked in their order of magnitude. It is 50th percentile (P_{50}) of the dataset. For a data series with an odd number of observations, the median is the mid data point which has an equal number of observations both above and below it. For a data series with an even number of observations, it is the average of the two central observations. In order to compute the median, first rank the observations from the smallest (x_1) observation to the largest (x_n) observation and then use one of the following equations depending on the number of observations (n):

$$P_{50} = \frac{x_{n+1}}{2} \text{ when } n \text{ is odd} \tag{5}$$

and

$$P_{50} = \frac{1}{2}(x_{(n/2)} + x_{(n/2+1)}) \text{ when } n \text{ is even.} \tag{6}$$

Opposite to the mean, the median is highly resistant and slightly affected by the magnitude of a single observation in a data series, being determined solely by the relative order of observations. This robustness to the effect of a change in value or presence of outlier observations is often a desirable property.

The median is always preferred over the mean in case a robust summary statistics is desired that is not strongly influenced by a few extremely low or high observations. One such example is the expected daily rainfall to occur across a network of raingauge stations for a given day. Suppose one of the raingauge stations recorded unusually higher daily rainfall than that recorded by the other raingauge stations. Using the median, one raingauge station with unusually high daily rainfall will not have a greater effect on the expected daily rainfall than raingauge stations with low daily rainfalls. However, if the mean is used then the expected daily rainfall may be pulled towards the outlier, and be higher than daily rainfalls recorded by most of the raingauge stations.

2.1.3 Additional Measures of Location

In addition to classical and robust measures of location, four additional measures of location are 'mode', 'geometric mean', 'harmonic mean' and 'trimmed mean', which are less frequently used. Mode is defined as the most frequently observed value in a given data series. It is the value having the highest bar in a histogram. The mode is more applicable for the grouped data, data which are recorded only as falling into a finite number of categories, than for the continuous data. Although it is very easy to obtain, it is a poor measure of location for the continuous data because its value often depends on the arbitrary grouping of the data (Helsel and Hirsch, 2002). The geometric mean (GM) is often used to compute summary statistic for positively skewed datasets. It is the mean of the logarithms, transformed back to their original units:

$$\text{GM} = \exp\left[\sum_{i=1}^{n} \frac{\ln(x_i)}{n}\right] \tag{7}$$

For the positively skewed data series, the GM is usually fairly close to the median of the series. In fact, the GM is an unbiased estimate of the median when the logarithms of the datasets are symmetric (Helsel and Hirsch, 2002). This is because the median and mean logarithms are equal. When transformed back to original units, the GM continues to be an estimate for the median, but is not an estimate for the mean.

In mathematics, the harmonic mean (sometimes also called 'sub-contrary mean') is one of several kinds of averages. Typically, it is appropriate for situations when the average of rates is desired (Shahin et al., 1993). The harmonic mean (HM) of positive real numbers of a time series, $x_1, x_2, ..., x_n > 0$ is defined as:

$$HM = \frac{n}{\frac{1}{x_1} + \frac{1}{x_2} + ... + \frac{1}{x_n}} = \frac{n}{\sum_{i=1}^{n} \frac{1}{x_i}} \tag{8}$$

The harmonic mean is related to the arithmetic and geometric means. Equivalently, the harmonic mean is the reciprocal of the arithmetic mean of reciprocals. For all positive datasets containing at least one pair of non-equal values, the harmonic mean is always the least of the three means, while the arithmetic mean is always the greatest of the three and the geometric mean is always in between. Since the harmonic mean of a list of numbers tends strongly toward the least elements of the list, it tends (compared to the arithmetic mean) to mitigate the impact of large outliers and aggravate the impact of small ones.

Moreover, compromises between the median and mean can be made by trimming off several of the lowest and highest observations in the time series, and then calculating the mean of remaining data. Unlike the mean, such an estimate of location is not influenced by the most extreme (or abnormal) tails of the sample. Nevertheless, unlike the median, it allows the magnitudes of most data values to affect the location estimate (Helsel and Hirsch, 2002). This estimate of location is known as 'trimmed mean' because it is computed after trimming away a desirable percentage of the data. The most common trimming is to remove 25% of the data on each tail—the resulting mean of the central 50% of data is commonly called 'trimmed mean', but it is more precisely 25 percent trimmed mean. A zero percent trimmed mean results in the arithmetic mean itself, whereas trimming all but 1 or 2 mid data points produces the median. Percentage of trimming should be explicitly stated when 'trimmed mean' is used. The trimmed mean is a robust estimator of location because it is not strongly influenced by outliers, and works well for a wide variety of distributional shapes such as normal, lognormal, etc. (Helsel and Hirsch, 2002). It may be considered as a weighted mean, where data beyond the cutoff 'window' are given a weight of zero, and those within the window a weight of one (Fig. 2.3).

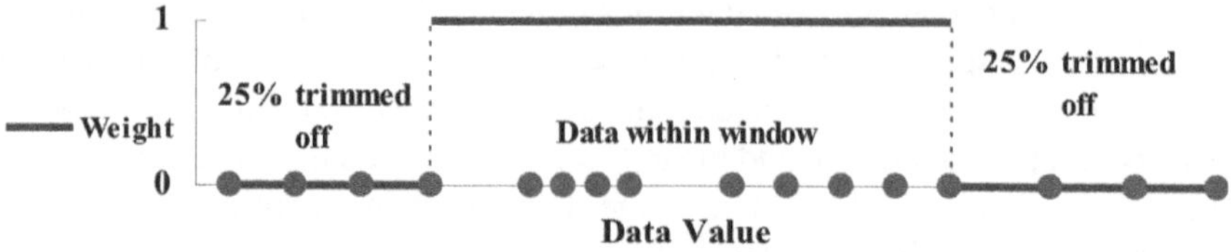

Fig. 2.3. Window diagram for the trimmed mean (Helsel and Hirsch, 2002).

2.2 Measures of Spread/Dispersion

It is very important to know the statistical dispersion or the variability of time series data, which can be quantified by the measures of spread. Two widely used measures of spread are described in subsequent sections.

2.2.1 Classical Measures

The 'sample variance' and 'sample standard deviation' (square root of sample variance) are classical measures of spread (dispersion), which are the most common measures of dispersion. Similar to the mean, the classical measures of spread are strongly influenced by outlier values. The sample variance (s^2) and the sample standard deviation (s) for a time series x_1, x_2, ..., x_n are mathematically expressed as follows:

$$s^2 = \sum_{i=1}^{n} \frac{(x_i - \bar{x})^2}{(n-1)} \tag{9}$$

$$s = \sqrt{\sum_{i=1}^{n} \frac{(x_i - \bar{x})^2}{(n-1)}} \tag{10}$$

Both the classical measures (s and s^2) of dispersion are computed using the squares of deviations of data values from the mean of the time series, so that magnitudes of the measures are even more influenced by outliers than that for the mean. In presence of outliers in the time series, the classical measures of dispersion become unstable and inflated. Under such condition, the classical measures may indicate much greater spread than is indicated by a majority of the hydrologic time series data.

2.2.2 Robust Measures

Robust measures of spreading about the mean include 'range', 'interquartile range', 'coefficient of variation' and 'median absolute deviation'. As the value of the range, standard deviation and coefficient of variation increases, the population variability also increases (Helsel and Hirsch, 2002). The interquartile range (IQR) is the most commonly used resistant measure of spread that measures the range of the central 50% of the data in a time series, and is not influenced at all by the 25% data on either tail. It is, therefore, the width of the non-zero weight window for the trimmed mean of Fig. 2.3.

The IQR is computed by subtracting the 25th percentile values from the 75th percentile value. The 75th (upper quartile), 50th (median) and 25th (lower quartile) percentiles split the entire time series data into four equal-sized quarters. These three quartiles help in depicting graphical distribution of time series data in the form of box and whisker plot (see Section 3.1.3 of Chapter 3 for details). The 75th percentile (P_{75}), which is also called the 'upper quartile', is a value which exceeds no more than 75% of the data and is exceeded by no more than 25% of the data in a time series. The 25th percentile (P_{25}) or 'lower quartile' is a value which exceeds no more than 25% of the data and is exceeded by no more than 75% of the data in a time series. Consider a time series arranged in chronological order of magnitudes of data: x_i, $i = 1$ to n. The percentiles (P_j) are computed using the following formula (Helsel and Hirsch, 2002):

$$P_j = x_{(n+1)\bullet j} \tag{11}$$

where n is the sample size of x_i, and j is the fraction of data less than or equal to the percentile value (for the 25th, 50th and 75th percentiles, $j = 0.25$, 0.50 and 0.75, respectively).

The range is the length of the smallest interval which contains all the data of time series. It is calculated by taking difference between the maximum and minimum values of the time series. Since it only depends on two of the observations, it is a poor and weak measure of dispersion except when the sample size is large.

The coefficient of variation (CV) gives a normalized measure of spreading about the mean, and is estimated as:

$$\text{CV}(\%) = \frac{s}{\bar{x}} \times 100 \tag{12}$$

The standard deviation of data series must always be understood in the context of the mean of the data series. Thus, the CV being a dimensionless number is advantageous over the standard deviation. Therefore, when comparing between datasets with different units or widely different means, one should use the coefficient of variation instead of the standard deviation. On the contrary, consideration of the CV also has limitations in certain cases. For example, when the mean of the data series is close to zero, the CV value will approach infinity and hence it is sensitive to small changes in the mean. Also, unlike the standard deviation, it cannot be used to construct confidence intervals for the mean.

Hydrologic variables with larger CV values are more variable than those with smaller values. Wilding (1985) suggested a classification scheme for identifying the extent of variability for soil properties based on their CV values, where CV values of 0-15, 16-35 and >36 indicate little, moderate and high variability, respectively. Typical ranges of CV values of salient soil properties are reported in the literature (Jury, 1986; Jury et al., 1987; Beven et al., 1993; Wollenhaupt et al., 1997).

One robust estimator of spread other than the IQR, being more resilient to outliers in a dataset than the standard deviation, is the median absolute deviation (MAD). In the standard deviation, the distances from the mean are squared, so on an average, large deviations are weighted more heavily, and thus outliers can heavily influence it. In the MAD, the magnitude of the distances of a small number of outliers is irrelevant. The MAD is computed by first creating a new difference time series by listing the absolute value of differences $| d |$ between each data value and median of a time series.

$$|d_i| = x_i - P_{50} \tag{13}$$

where P_{50} = median of the original time series.

Thereafter, the MAD is computed as the median of the absolute difference time series as follows:

$$\text{MAD} = P_{50}\,|d| \tag{14}$$

Quartile coefficient (qc) of dispersion is another descriptive statistic which measures dispersion and is used to make comparison within and between datasets. The test-statistic is computed using the first (P_{25}) and third (P_{75}) quartiles for each data set. The quartile coefficient of dispersion (qc) is given as:

$$qc = \frac{P_{75} - P_{25}}{P_{75} + P_{25}} \tag{15}$$

2.3 Measures of Skewness

Hydrologic time series data are usually skewed, which means that data in the time series are not symmetric around the mean or median, with extreme values extending out longer in one direction. The probability density function for a lognormal distribution shown in Fig. 2.4 demonstrates this skewness in the data. When extreme values extend the right tail of the distribution (as shown in Fig. 2.4), the distribution of time series data is said to be skewed to the right, or positively skewed. Whereas, when extreme values extend the left tail of the distribution, the time series data are said to be skewed to the left, or negatively skewed. For the skewed data values, the mean is not expected to be equal to the median, but is pulled toward the tail of the distribution. Thus, for the positively skewed data, the mean exceeds more than 50% of the data (Fig. 2.4). The standard deviation is also inflated by data in the tail. In hydrology, all kinds (e.g., rainfall, streamflow, groundwater levels, etc.) of time series data often have positive skewness. Therefore, statistical properties including only the mean and standard deviation or variance are not sufficient for the studies related to water resources development and management. This is because of the fact that the mean and standard deviation alone may not describe the properties of the majority of the data values very well when the data are skewed. Also, both the mean and the standard deviation are inflated by outlier observations. Robust summary statistics, such as the median and other percentile values have greater applicability to the skewed hydrologic data. The skewed data are questionable regarding the applicability of hypothesis (parametric) tests, which are based on the assumptions that the time series data follow a normal distribution. These parametric tests may be of questionable value when applied to hydrologic time series, as the time series are often neither normal nor even symmetric (Helsel and Hirsch, 2002).

2.3.1 Classical Measure of Skewness

The 'coefficient of skewness (g)' is the most common measure of skewness. It is defined as the adjusted third moment about the mean divided by the cube of the standard deviation (s), and is mathematically expressed as follows:

$$g = \frac{n}{(n-1)\ (n-2)} \sum_{i=1}^{n} \frac{(x_i - \bar{x})^3}{s^3} \tag{16}$$

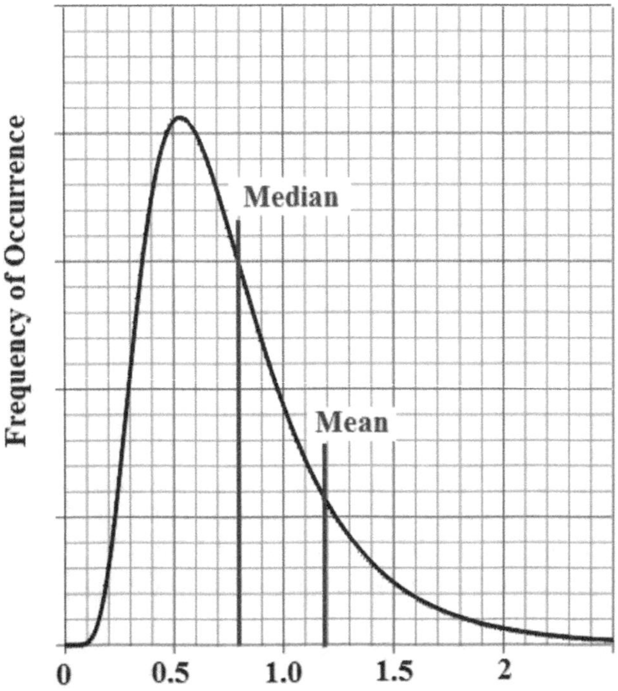

Fig. 2.4. Probability density function of a lognormal distribution.

A positively skewed distribution of hydrologic time series with right extended tail has a positive coefficient of skewness, whereas a time series with negative-skewed distribution with left extended tail has a negative coefficient of skewness. The presence of outliers has a significant influence on the coefficient of skewness (g). For instance, an otherwise symmetric distribution having one outlier will produce a large (and possibly misleading) measure of skewness.

2.3.2 Robust Measure of Skewness

A robust measure of skewness is the 'quartile skew coefficient (qs)', which is defined as the difference in distances of the upper and lower quartiles from the median, divided by the IQR (Kenney and Keeping, 1954). Mathematically, it is expressed as:

$$qs = \frac{(P_{75} - P_{50}) - (P_{50} - P_{25})}{P_{75} - P_{25}} \tag{17}$$

Similar to the coefficient of skewness, a right-skewed distribution has a positive quartile skew coefficient and a left-skewed distribution has a negative quartile skew coefficient. Also, similar to the trimmed mean and IQR, the quartile skew coefficient uses the central 50% of the data.

2.4 Additional Robust Measures

As an additional robust measure, percentile values other than three quartiles may be used to produce a series of robust measures of location, spread and skewness. For example, the 15% trimmed mean can be coupled with the range between the 15th and 85th percentiles as a measure of spread, and a corresponding measure of skewness to produce a consistent series of robust statistics. The robust measure of skewness for 15% trimmed mean (qs_{15}) is mathematically expressed as:

$$qs_{15} = \frac{(P_{85} - P_{50})\ (P_{50} - P_{15})}{P_{85} - P_{15}} \tag{18}$$

Geologists have used the 16th and 84th percentiles for many years to compute a similar series of robust measures of the distributions of sediment particles (Inman, 1952). However, the measures based on quartiles have become generally standard, and additional measures should be clearly defined prior to their use (Helsel and Hirsch, 2002). The median, IQR and quartile skew can be easily summarized graphically using box and whisker plots (see Section 3.1.3 of Chapter 3), which are widely used by scientists and researchers.

2.5 Measures of Peakedness or Flatness

'Kurtosis' is a measure of peakedness or flatness of a data series distribution relative to the normal distribution. That is, datasets with high kurtosis tend to have a distinct peak near the mean, decline rather rapidly, and have heavy tails. For a time series data, $x_1, x_2, \ldots, x_n$, the formula for kurtosis can be written as follows:

$$\gamma = \frac{\sum_{i=1}^{n}(x_i - \bar{x})^4}{(n-1)\, s^4} \tag{19}$$

The kurtosis for a standard normal distribution is three, and hence some authors provide the following definition of kurtosis, which is often referred to as 'excess kurtosis' (Shahin et al., 1993):

$$\gamma = \frac{\sum_{i=1}^{n}(x_i - \bar{x})^4}{(n-1)\, s^4} - 3 \tag{20}$$

A high kurtosis distribution has a sharper *peak* and longer, fatter *tails*, while a low kurtosis distribution has a more rounded *peak* and shorter, thinner *tails*. Distributions with a zero excess kurtosis are called 'mesokurtic' or 'mesokurtotic'. On the other hand, distributions with a positive excess kurtosis are called 'leptokurtic' or 'leptokurtotic'. In terms of shape, a leptokurtic distribution has a more acute *peak* around the mean (i.e., a lower probability than a normally distributed variable of values near the mean) and *fatter tails* (i.e., a higher probability than a normally distributed variable of extreme values). A distribution with a negative excess kurtosis is called 'platykurtic' or 'platykurtotic'. In terms of shape, a platykurtic distribution has a lower, wider *peak* around the mean (i.e., a higher probability than a normally distributed variable of values near the mean) and *thinner tails* (if viewed as the height of the probability density—that is, a lower probability than a normally distributed variable of extreme values).

Alternatively, peakedness of a data distribution may be described by another measure known as 'percentile coefficient of kurtosis' (PCK). PCK is mathematically expressed as follows (Shahin et al., 1993):

$$\mathrm{PCK} = \frac{1}{2}\left(\frac{P_{75} - P_{25}}{P_{90} - P_{10}}\right) \tag{21}$$

2.6 Statistical Measures for System Performance Evaluation

Many theoretical and practical approaches have been proposed in the literature for identifying and quantifying objectives and for considering multiple criteria/objectives in water resources planning and management. Over the years, various tools have been designed to provide information which is of great importance to the planning and decision-making in the field of water resources engineering. The information used in decision-making process is generally derived from the analyses of hydrologic time series. In the era of information technology, time series modelling and/or simulation analysis (using numerical, empirical and analytical models) is very frequently carried out. The time series modelling may include stochastic models such as autoregressive (AR) models, moving average (MA) models, autoregressive moving average (ARMA) models, autoregressive integrated moving average (ARIMA) models, seasonal autoregressive integrated moving average (SARIMA) models, etc. (Box and Jenkins, 1976; Shahin et al., 1993; Hipel and McLeod, 1994). Performance of individual time series models may be compared to decide the best model so as to ensure appropriate simulation of the hydrologic time series. Water resources planners and managers, coupled with other decision makers typically involved in decision-making often face a challenge in selecting one out of many alternatives, each characterized by different values provided by multiple performance criteria.

In statistics, there are many methods for summarizing time-series data resulting from field observations or simulation analyses. The weighted arithmetic mean and the geometric mean are two common methods of summarizing multiple time series data (Loucks and van Beek, 2005). As mentioned earlier, the overall mean itself usually provides little information about a dynamic process. Multiple time series plots are normally difficult to compare. Another approach to summarize and compare hydrologic time-series data is the variance (Section 2.3.1). For example, let us consider annual rainfall time series consisting of 200, 675, 475, 175, 780, 890, 945, 875, 400 and 300 mm rainfall for ten years. The mean of these values is 571.5 mm and the variance can be calculated as:

$$\text{Variance} = \sum_{i=1}^{n}(x_i - \bar{x})^2/(n-1) \qquad (22)$$

$$= [(200-571.5)^2+(675-571.5)^2+...+(300-571.5)^2]/10$$

$$= 89893.3\,\text{mm}^2$$

The plot of above rainfall data and their mean is shown in Fig. 2.5. The mean and variance for the time series shown in Fig. 2.5, however, are the same for its upside-down image, as shown in Fig. 2.6. They do not even depend on the order of the time-series data. Consider these two sets of time series again (Fig. 2.7), each having the same mean and variance. Assume that any value equal to or less than the dashed line (just above 300 mm) is considered unsatisfactory. This rainfall value is called a *threshold value*, dividing the rainfall time series data into satisfactory and unsatisfactory values (Hashimoto et al., 1982a, b). It is apparent from Fig. 2.7 that the impact of these two rainfall time series may differ. The original time series shown in blue colour remained in an unsatisfactory condition for a shorter time than did the 'rotated' time series shown in red colour. However, its maximum extent of failure when it failed was more than the rotated time series. These characteristics of hydrologic time series can be captured by the measures/criteria namely 'reliability', 'resilience' and 'vulnerability' as suggested by Hashimoto et al. (1982 a, b). These measures/criteria are described in subsequent sections.

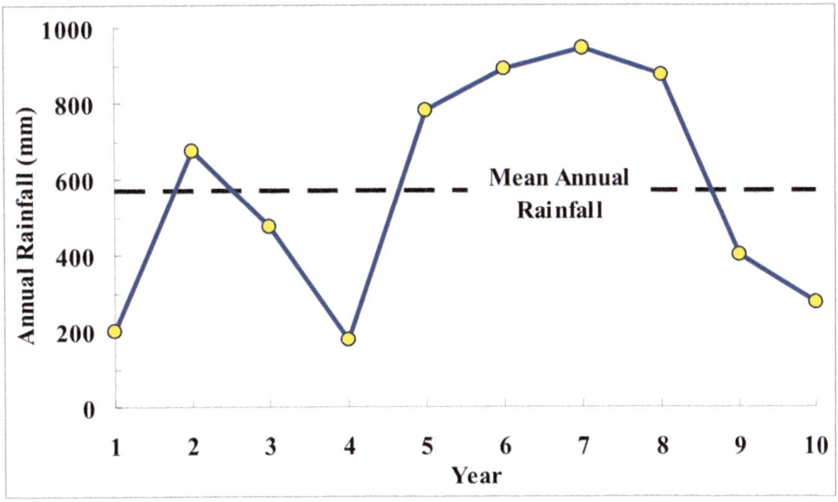

Fig. 2.5. Plot of annual rainfall time series with a mean of 571.5 mm and a variance of 89893.3 mm^2.

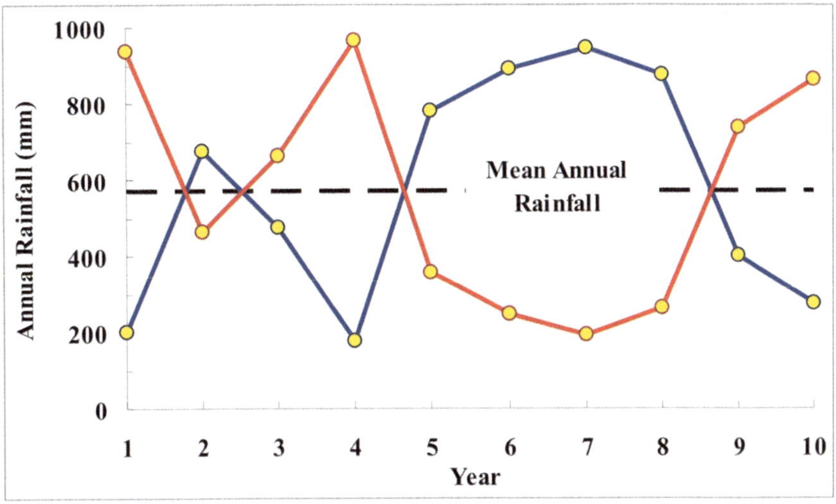

Fig. 2.6. A plot of two different time series of annual rainfall having the same mean and variance.

2.6.1 Reliability

The 'reliability' of a system is defined as the number of data in a satisfactory state divided by the total number of data in the time series. Assuming satisfactory values in the hydrologic time series x_n containing n values are those equal to or greater than some threshold x^T, the reliability of the system can be expressed as:

$$\text{Reliability } (x) = [\text{Number of time periods } t \text{ when } x_t \geq x^T]/n \tag{23}$$

The reliability of the original time series shown in blue colour (Fig. 2.7) is 0.7, which suggests that it failed three times in ten. The reliability of the rotated time series shown in red colour (Fig. 2.7) is also 0.7, indicating three times failure in ten. In general, a more reliable system is better than a less reliable system, but it is not always true. The reliability measure does not tell anything about how fast a system recovers and returns to a satisfactory value, nor does it indicate how bad an unsatisfactory value might be if it occurs. It may be fine that a system that fails relatively often, but by insignificant amounts and for short durations, will be much preferable to the system whose reliability is much higher but when a failure does occur, it is likely to be much more severe. 'Resilience' and 'vulnerability' measures, which are discussed below, can quantify these system characteristics.

2.6.2 Resilience

The 'resilience' of a system is defined as the probability that if a system is in an unsatisfactory state, the next state will be satisfactory. In other words, it is the probability of having a satisfactory value in time period $t - 1$, given an unsatisfactory value in any time period t. It can be expressed as:

$$\text{Resilience } (x) = \frac{\left[\begin{array}{c}\text{Number of times a satisfactory value} \\ \text{follows an unsatisfactory value}\end{array}\right]}{\left[\begin{array}{c}\text{Number of times an unsatisfactory} \\ \text{value occurred}\end{array}\right]} \tag{24}$$

Note that 'resilience' cannot be defined if no unsatisfactory values occur in the time series. For the original time series shown in blue colour (Fig. 2.7), the resilience is 2/2 = 1, again assuming the value of 300 mm or less is considered a failure. We cannot judge the resilience of the blue time series on the basis of the last failure in period 10 because we do not have an observation in period 11. For the rotated time series shown in red colour (Fig. 2.7), the resilience is 1/3 = 0.33.

2.6.3 Vulnerability

The term 'vulnerability' is a measure of the extent of the differences between the threshold value and the unsatisfactory time series values. Obviously, this is a probabilistic measure. Some use expected values, some use maximum observed values, and others may assign a probability of exceedance to their vulnerability measures. Assuming an expected value measure of vulnerability is to be used, vulnerability can be expressed as follows:

$$\text{Vulnerability}(x) = \frac{[\text{Sum of positive values of } (x^{\mathrm{T}} - x_t)]}{\left[\begin{array}{c}\text{Number of times an unsatisfactory}\\ \text{value occurred}\end{array}\right]} \tag{25}$$

The expected vulnerability of the original blue time series (Fig. 2.7) is [(300 – 200) + (300 – 175)]/2 = 125. Similarly, the expected vulnerability of the time series shown by the red line in Fig. 2.7 is [(300 – 248) + (300 – 193) + (300 – 263)]/3 = 65.33.

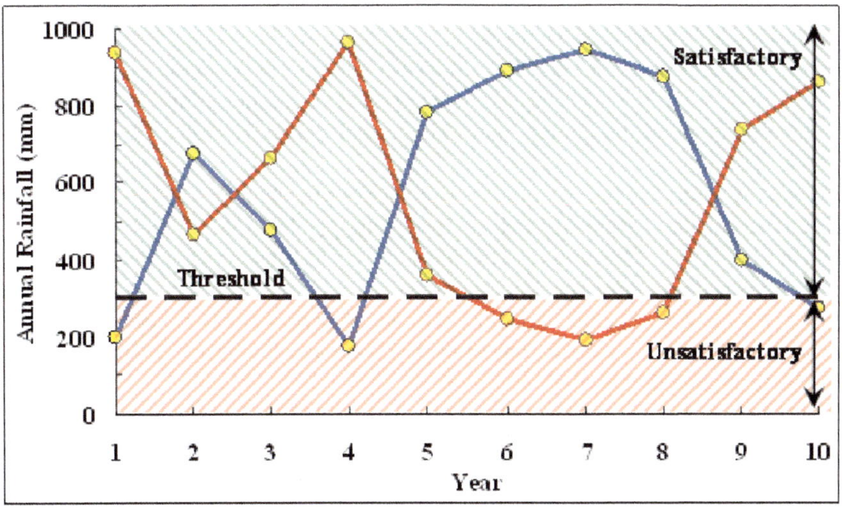

Fig. 2.7. Threshold value distinguishing values considered satisfactory and unsatisfactory.

Thus, depending on whether a threshold value is considered a failure or not in the above example, the 'reliability' and 'resilience' of original time series (blue line) is equal or more than the rotated time series (red line). However, the expected vulnerability of the original time series is more than that of the rotated time series. It shows the typical tradeoffs researchers/ scientists or decision makers can identify using these three measures of system performance. Note that the above-mentioned three measures of system performance 'reliability', 'resilience' and 'vulnerability' (R-R-V) in a combined form are used as a sustainability criterion for assessing the sustainability of existing water resources systems (Loucks, 1997; Kay, 2000; Kjeldsen and Rosbjerg, 2001). The interested readers are referred to Kjeldsen and Rosbjerg (2004) for the application of 'reliability', 'resilience' and 'vulnerability' (R-R-V) measures to real-world hydrologic time series and their comparative evaluation. Recently, a cohesive approach for considering and expressing various aspects of system resilience has been proposed by Wang and Blackmore (2009) focussing on water resources systems.

References

Adeloye, A.J. and Montaseri, M. (2002). Preliminary streamflow data analyses prior to water resources planning study. *Hydrological Sciences Journal*, **47(5):** 679-692.

Beven, K.J., Henderson, D.E. and Reeves, A.D. (1993). Dispersion parameters for undisturbed partially saturated soil. *Journal of Hydrology*, **143:** 19-43.

Box, G.E.P. and Jenkins, G.M. (1976). Time Series Analysis: Forecasting and Control. Holden-Day, San Francisco, 575 pp.

Hashimoto, T., Loucks, D.P. and Stedinger, J.R. (1982a). Robustness of water resource systems. *Water Resources Research*, **18(1):** 21-26.

Hashimoto, T., Stedinger, J.R. and Loucks, D.P. (1982b). Reliability, resiliency and vulnerability criteria for water resource system performance evaluation. *Water Resources Research*, **18(1):** 14-20.

Helsel, D.R. and Hirsch, R.M. (2002). Statistical Methods in Water Resources. Chapter A3, Book 4, Hydrologic Analysis and Interpretation, Techniques of Water-Resources Investigations of the United States Geological Survey (USGS), USGS, Reston, VA, 510 pp.

Hipel, K.W. and McLeod, A.I. (1994). Time Series Modeling of Water Resources and Environmental Systems. Elsevier, Amsterdam, pp. 463-465.

Inman, D.L. (1952). Measures for describing the size distribution of sediments. *Journal of Sedimentary Petrology*, **22:** 125-145.

Jury, W.A. (1986). Spatial variability of soil properties. *In:* S.C. Hern and S.M. Melancon (editors), Vadose Zone Modeling of Organic Pollutants. Lewis Publishers, Chelsea, pp. 245-269.

Jury, W.A., Russo, D., Sposito, G. and Elaod, H. (1987). The spatial variability of water and solute transport properties in unsaturated soil: I. Analysis of property variation and spatial structure and statistical models. *Hilarda*, **55:** 1-32.

Kay, P.A. (2000). Measuring sustainability in Israel's water system. *Water International*, **25(4):** 617-623.

Kenney, J.F. and Keeping, E.S. (1954). Mathematics of Statistics—Part One. D. Van Nostrand Company, Inc., New York, 102 pp.

Kjeldsen, T.R. and Rosbjerg, D. (2001). A framework for assessing water resources system sustainability. *In:* A.H. Schumann, M.C. Acreman, R. Davis, M.A. Marino, D. Rosbjerg and Xia Jun (editors), Regional Management of Water Resources. IAHS Publ. 268, IAHS Press, Wallingford, U.K., pp. 107-113.

Kjeldsen, T.R. and Rosbjerg, D. (2004). Choice of reliability, resilience and vulnerability estimators for risk assessments of water resources systems. *Hydrological Sciences Journal*, **49(5):** 755-767.

Loucks, D. P. (1997). Quantifying trends in system sustainability. *Hydrological Sciences Journal*, **42(4):** 513-530.

Loucks, D.P. and van Beek, E. (2005). Water Resources Systems Planning and Management: An Introduction to Methods, Models and Applications. Studies and Reports in Hydrology, UNESCO Publishing, UNESCO, Paris.

Maier, H.R., Lence, B.J., Tolson, B.A. and Foschi, R.O. (2001). First-order reliability method for estimating reliability, vulnerability, and resilience. *Water Resources Research*, **37(3):** 779-790.

Moy, W.-S., Cohon, J.L. and ReVele, C.S. (1986). A programming model for analysis of the reliability, resilience, and vulnerability of a water supply reservoir. *Water Resources Research*, **22(4):** 489-498.

Rao, A.R., Hamed, K.H. and Chen, H.-L. (2003). Nonstationarities in Hydrologic and Environmental Time Series. *Water Science and Technology Library*, **45**, 392 pp.

Shahin, M., Van Oorschot, H.J.L. and De Lange, S.J. (1993). Statistical Analysis in Water Resources Engineering. A.A. Balkema, Rotterdam, the Netherlands, 394 pp.

Snedecor, G.W. and Cochran, W.G. (1980). Statistical Methods. The Iowa State University Press, Ames, Iowa, USA.

Upchurch, D.R. and Edmonds, W.J. (1991). Statistical procedures for specific objectives. *In:* M.J. Mausbach and L.P. Wilding (editors), Spatial Variabilities of Soils and Landforms. Special Publication Number 28, Soil Science Society of America, Inc., Madison, WI, pp. 49-71.

Wang, C. and Blackmore, J.M. (2009). Resilience concepts for water resource systems. *Journal of Water Resources Planning and Management, ASCE*, **135(6):** 528-536.

Wilding, L.P. (1985). Spatial variability: Its documentation, accommodation, and implication to soil surveys. *In:* D.R. Nielsen and J. Bouma (editors), Soil Spatial Variability, Proceedings of Workshop of the ISSS and the SSSA, Las Vegas, November 30-December 1, 1984, Pudoc. Wageningen, the Netherlands, pp. 166-194.

Wollenhaupt, N.C, Mulla, D.J. and Gotway Crawford, C.A. (1997). Soil sampling and interpolation techniques for mapping spatial variability of soil properties. *In:* F.J. Pierce and E.J. Sadler (editors), The State of Site Specific Management for Agriculture. American Society of Agronomy, Madison, Wisconsin, pp. 19-54.

3

Methods for Testing Normality of Hydrologic Time Series

Statistical methods are applied in all the phases of time series analysis from collecting data to evaluating results in hydrologic studies. Advances in computer technology has enabled most of the scientists/researchers to apply statistical analyses effectively; however, some of the researchers do not check parametric test assumptions, especially the normality assumption (Adeloye and Montaseri, 2002). Many methods of time series analysis depend on the basic assumption that data were sampled from a normal distribution (Madansky, 1988; USEPA, 1996; Thode, 2002). This assumption is very crucial for the reliability of results especially for parametric tests. These days many statistical software packages are available, which include several tests for checking the normality of time series data. However, the important point is to judge which test should be used under what condition (USEPA, 1996).

In general, the normality assumption can be evaluated by graphical and statistical methods (USEPA, 1996; Thode, 2002). The graphical methods provide us with some information about the shape of the distribution, but do not guarantee that the distribution is normal and do not test whether the difference between the normal distribution and the sample distribution is significant. On the other hand, numerical methods provide only quantitative information. Major statistical methods to assess the assumption of normality are (USEPA, 1996; Thode, 2002): Chi-square goodness-of-fit test, Kolmogorov-Smirnov (KS) test, Lilliefors corrected Kolmogorov-Smirnov test, Anderson-Darling test, Cramer-von Mises test, Shapiro-Wilk test, Jarque-Bera test, and D'Agostino-Pearson omnibus test. It is worth mentioning that there is an inherent problem with normality tests. Because of a small sample size, normality tests have little power to reject the null hypothesis that the data come from a normal distribution. Hence, small samples always pass normality tests. However, with large samples, minor deviations from normality may be treated as statistically significant, even though small deviations from a normal distribution may not affect the results of a parametric test (GraphPad, 2007).

Thus, the best way to decide whether time series data are normal or not is to apply both graphical and statistical methods for examining normality.

This chapter presents an overview of commonly used graphical and statistical methods for checking the normality of time series data. The step-by-step procedures for applying these methods are also included in this chapter. This chapter can serve as a guideline for the researchers and scientists as well as for practising engineers in selecting an appropriate normality test for their time series data.

3.1 Graphical Methods

Graphical methods provide detailed information about a hydrologic time series that may not be apparent from statistical methods. Histograms, stem-and-leaf plots, and normal probability plots are some of the graphical methods which are useful for determining whether or not a given set of time series data follow a normal distribution curve (USEPA, 1996). Both the histogram and stem-and-leaf plot of a normal distribution are bell-shaped. The normal probability plot of a time series having normal distribution follows a straight line. However, for the non-normally distributed data, there are large deviations from the straight line in the tail or middle of a normal probability plot. The subsequent section deals with six graphical methods used for checking normality of time series data.

3.1.1 Frequency Plots/Histogram

Two classical methods for summarizing hydrologic time series data are 'frequency plot' (Fig. 3.1) and 'histogram' (Fig. 3.2). The basic principles used by both the histogram and the frequency plot to display the data are almost same: dividing the data range into units/bins, counting the number of data points within the units/bins, and displaying the data as the height or area within a bar graph (Walpole and Myers, 1985). Besides similarity, both the histogram and the frequency plot slightly differ from each other. The frequency plot represents the relative density of the data points by the relative height of the bars, while in a histogram, the area within the bar represents the relative density of the data points. A more distinct difference between the two plots can be seen by using unequal box sizes (USEPA, 1996). Structure/patterns of the histogram and frequency plot reveal about the symmetry and variability of the data. If the data are symmetric, then the structure of these plots will be symmetric around a central point such as mean or median. The histogram and frequency plots will generally indicate if the data are skewed and the direction of the skewness (Dixon and Massey Jr., 1983). Step-by-step procedures for generating a frequency plot and histogram are given below:

Step 1: Select suitable data intervals that cover the range of entire observations of time series. The data intervals should be of equal widths, if possible. A rule of thumb is to have between 7 to 11 intervals (USEPA, 1996).

If necessary, specify an endpoint convention, i.e., what to do with cases that fall on interval endpoints.

Step 2: Count the number of data points within each data interval. For a frequency plot with equal interval sizes, the number of data points represents the height of the boxes on the frequency plot.

Step 3: Decide the horizontal axis based on the data range of time series. The vertical axis for a frequency plot is the number of data points, while the vertical axis of the histogram is based on percentages.

Step 4: For a histogram, compute the percentage of data points within each data interval by dividing the number of data points within each data interval (Step 3) by the total number of time series data.

Step 5: When the data intervals are not of equal widths for a histogram, select a common unit that corresponds to the X-axis. Compute the number of common units in each interval and divide the percentage of data points within each data interval (Step 4) by this number.

Step 6: Plot the data intervals using boxes, against the results of Step 5 for a histogram or the intervals against the number of data points in a data interval (Step 2) for a frequency plot.

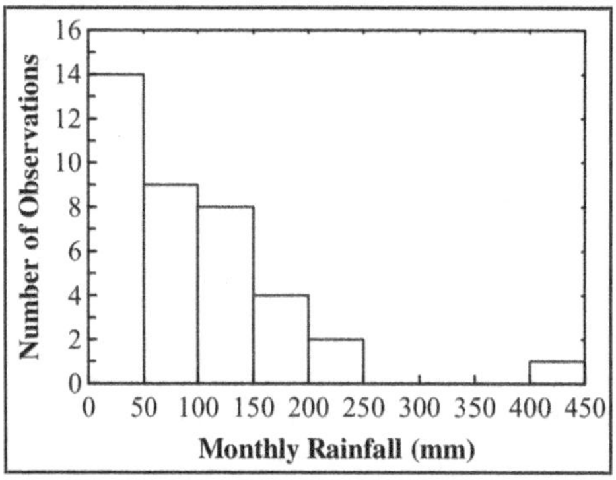

Fig. 3.1. Example of a frequency plot.

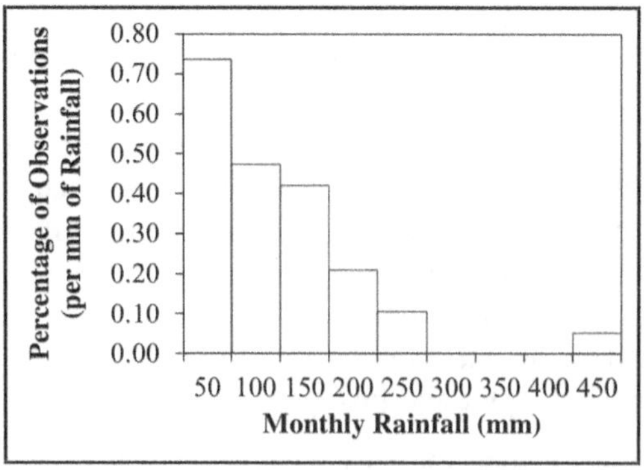

Fig. 3.2. Example of a histogram.

3.1.2 Stem-and-Leaf Plot

The 'stem-and-leaf plot' is useful to show both the data values themselves and information about the distribution of the time series data. It is a useful method for storing large datasets in a compact form while, at the same time, sorting the data in ascending order (Walpole and Myers, 1985). A stem-and-leaf plot (Fig. 3.3) is more useful in analyzing time series data than a histogram, because it not only allows a visualization of the data distribution but also enables the data to be reconstructed and lists the data points in the ascending order (USEPA, 1996). However, the stem-and-leaf plot is one of the more subjective visualization techniques as it requires the analyst to make some random choices while deciding data interval on the stem. Therefore, some practice or trial and error is necessary before a useful plot can be created by this technique. As a result, the stem-and-leaf plot should only be used to display the data distribution and their characteristics (USEPA, 1996).

0	09	12	17	22	22	24	26	27	30	32	33	43	45	47
0	61	65	72	82	84	86	93	98	99					
1	03	06	14	21	28	36	47	48						
1	53	54	65	83										
2	36	37												
2														
3														
3														
4	38													

Fig. 3.3. Example of a Stem-and-Leaf plot.

In the stem-and-leaf plot, each data point is partitioned in two parts: the stem of the data point and the leaf. The leading digit of the numerical value of the data point becomes the stem while the trailing digits become the leaf in the order that corresponds to the order of magnitude from left to right. The stem is displayed on the vertical axis and the data points make up the leaves on the horizontal axis. The stem can be changed by increasing or decreasing the digits that are used, dividing the groupings of one stem (i.e., all numbers which start with the numeral 4 can be divided into smaller groupings), or multiplying the data by a constant factor (i.e., multiply the data by 10 or 100). Figure 3.3 presents a stem-and-leaf plot for a 38-year monthly rainfall time series with 9 and 438 mm as minimum and maximum rainfall amounts of monthly rainfall. Digits on the left side of vertical bar indicate stem with interval of 50 mm while digits on the right side of the bar represent leaves. Thus, the top-most leaves of Fig. 3.3 contain rainfall data below 50 mm while the second row of leaves contains rainfall data between 50 and 100 mm. In similar manner, the last row of leaves contain rainfall data less than 450 mm.

A stem-and-leaf plot roughly displays the data distribution and, hence, helps identifying the underlying probability distribution. For example, the stem-and-leaf plot of normally distributed data should be close to bell shaped. Also, the data distribution displayed by the stem-and-leaf plot may be used to assess skewness or symmetry of the data. The top half of the stem-and-leaf plot should be a mirror image of the bottom half of the stem-and-leaf plot for symmetric data distribution. The data that are skewed to the left will have the higher data density in the top half of the plot and less data density over the bottom half of the plot. Consider a hydrologic time series $X_1, X_2, ..., X_n$ with n data points. To develop a stem-and-leaf plot, following steps are used:

Step 1: Arrange n data points of time series in order of increasing magnitude from the smallest to the highest. The ordered data are generally labelled (from smallest to largest) as $X_{(1)}, X_{(2)}, ..., X_{(n)}$.

Step 2: Decide either one or more of the leading digits to be the stem values. As an example, for the value 236, 2 could be used as the stem value as it is the leading digit.

Step 3: List the stem values from smallest to largest at the left (along a vertical axis). Enter the leaf (the remaining digits) values in order from lowest to highest to the right of the stem. Using the value 236 as an example, if 2 is the stem then 36 will be the leaf.

3.1.3 Box and Whisker Plot

A 'box and whisker plot' or simply 'box plot' is a schematic diagram (Fig. 3.4) used for visualizing important statistical quantities (such as quartiles) of the time series data. Box plots are useful in situations where it is not necessary or feasible to depict all the details of a time series distribution. The box and whisker plot consists of a central box with a square or a line in the box and two lines extending out from each end of the box called *whiskers*. The square or line within the box represents the median (USEPA, 1996; USEPA, 2006).

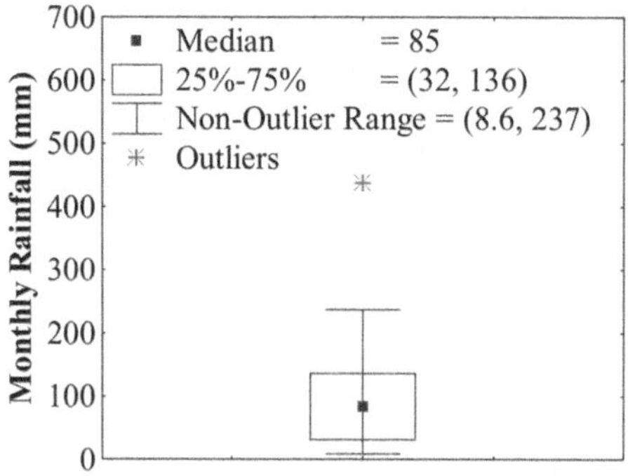

Fig. 3.4. Example of a Box and Whisker plot.

The bottom and top horizontal lines in the box in a 'box and whisker plot' indicate the 25th and 75th percentile, respectively, of the statistics computed from the observed data. The length of the central box indicates the spread of the bulk of the data (central 50%), while the length of the whiskers shows the extent of the rest of the data (USEPA, 2006). The box plot divides the entire data into four sections, each section containing 25% of the data. The whisker extends to the most extreme data value within 1.5 times the interquartile range of the data and indicates how tails of the distribution are stretched. The width of the box has no specific meaning; the plot can be made quite narrow without affecting its visual impact. The values beyond the ends of the whiskers are unusually small or large data points, which are called *outliers* and are displayed by a '*' on the plot. A 'box and whisker plot' can be used to evaluate the symmetry of the data (USEPA, 2006). If the data distribution is symmetrical, the box is divided in two equal halves by the median, length of both the upper and the lower whiskers will be the same and the number of extreme data points will be distributed equally on either end of the plot. Since the 'box and whisker plot' cannot be made so easily manually, STATISTICA software may be used for creating this plot. The following steps are used for generating a box and whisker plot:

Step 1: Choose the vertical scale of the plot based on the maximum and minimum values of the time series data. Select a width for the box plot keeping in mind that the width has no particular meaning and is only a visualization tool. If the width is labelled as W, the horizontal scale of the plot ranges from $-\frac{1}{2}W$ to $+\frac{1}{2}W$.

Step 2: Compute the upper quartile [$Q_{(0.75)}$ or the 75th percentile] and the lower quartile [$Q_{(0.25)}$ or the 25th percentile] based on time series data. Compute the sample mean and median (X_m) for the time series data. Then, compute the interquartile range (IQR) where IQR $= Q_{(0.75)} - Q_{(0.25)}$.

Step 3: Draw a box through four points $[-\frac{1}{2}W, Q_{(0.75)}]$, $[-\frac{1}{2}W, Q_{(0.25)}]$, $[\frac{1}{2}W, Q_{(0.25)}]$ and $[\frac{1}{2}W, Q_{(0.75)}]$. Draw a line from $[\frac{1}{2}W, Q_{(0.5)}]$ to $[-\frac{1}{2}W, Q_{(0.5)}]$ and mark point $(0, X_m)$ with (+). The line or point $(0, X_m)$ indicates median of the data.

Step 4: Compute the upper end of the top whisker by finding the largest data value X less than $Q_{(0.75)} + 1.5 \times [Q_{(0.75)} - Q_{(0.25)}]$. Draw a vertical line from $[0, Q_{(0.75)}]$ to $(0, X_m)$. Compute the lower end of the bottom whisker by finding the smallest data value Y greater than $Q_{(0.25)} - 1.5 \times [Q_{(0.75)} - Q_{(0.25)}]$. Draw a vertical line from $[0, Q_{(0.25)}]$ to $(0, Y)$.

Step 5: For all points $X^* > X$ (outliers and extremes), place an asterisk (*) at the point $(0, X^*)$. For all points $Y^* < Y$ (outliers and extremes), place an asterisk (*) at the point $(0, Y^*)$.

3.1.4 Ranked Data Plot

A 'ranked data plot' is a useful graphical method that is easy to construct and interpret, and does not depend upon any assumptions about a model for the time series data. It is not subjective as the user does not have to make any choice regarding the data to construct a ranked data plot (such as unit/bins for a histogram or frequency plot). Additionally, a ranked data plot displays every data point of a time series instead of summary of the data (USEPA, 2006).

A 'ranked data plot' is a plot of the data arranged in ascending order at evenly spaced intervals (Fig. 3.5). This plot is very similar to the 'quantile plot' dealt in Section 3.1.5. Both the 'ranked data plot' and the 'quantile plot' can be used to determine the density of the data points and to check whether the data are skewed (USEPA, 2006). However, a ranked data plot is quite easier to generate than a quantile plot. It should be noted that a quantile plot contains information on the quartiles of the data, while a 'ranked data plot' contains information on the data themselves.

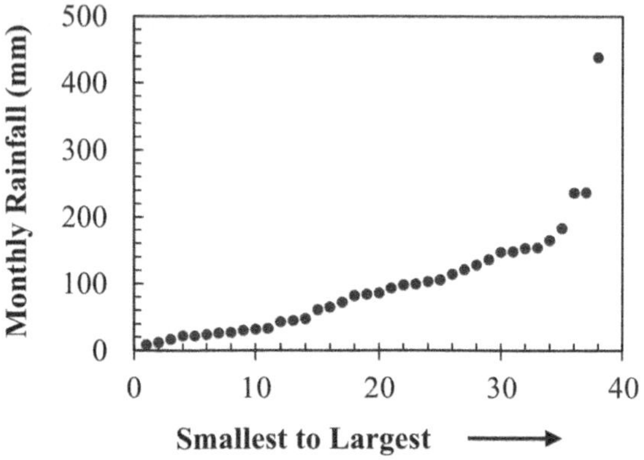

Fig. 3.5. Example of a ranked data plot.

Ranked data plots can be used to determine the density of the data values, i.e., if all the data values are close to the centre of the data with relatively few values in the tails or if there is a large amount of values in one tail with the rest evenly distributed (USEPA, 2006). The density of the data points is displayed by the slope of the line. Generally, a large amount of data values has a flat slope, i.e., the graph rises slowly and a small amount of data values has a large slope, i.e., the graph rises quickly. Thus, the user can decide whether the data points are either evenly distributed or lie in large clusters of points. It can be seen from Fig. 3.5 that the graph rises slowly up to a point where the slope increases and the graph rises relatively quickly. This means that there is a large amount of small values of the data points and relatively few large values of the data points.

Moreover, a 'ranked data plot' can be used to determine whether the time series data are skewed or symmetric (USEPA, 2006). A 'ranked data plot' of the right-skewed data extends more sharply at the top resulting in a convex shape of the graph, whereas a 'ranked data plot' of the left-skewed data increases sharply near the bottom resulting in a concave shape of the graph. However, if the data are symmetric, the top portion of the graph will stretch to upper right corner and similarly, the bottom portion of the graph stretches to lower left, creating an S-shape. Figure 3.5, as an example, shows a 'ranked data plot' of the right-skewed time series data.

Ranked data plots can be created as follows: consider a hydrologic time series $X_1, X_2, ..., X_n$ with n data points. Arrange the time series data in ascending order and let $X_{(i)}$, for $i = 1$ to n, be the data listed in order from smallest to largest such that $X_{(1)}$ is the smallest, $X_{(2)}$ is the second smallest, and $X_{(n)}$ is the largest data point. Then, plot the ordered $X_{(i)}$ values at equally spaced intervals along the horizontal axis to generate a ranked data plot. The entire procedure can be executed easily using MS-Excel software.

3.1.5 Quantile Plot

A 'quantile plot' is a graph of the quantiles of the data (Fig. 3.6). It is very similar to the 'ranked data plot' and makes no assumptions about a model for the data. It is not subjective and displays every data point of a time series instead of a summary of the data. The basic quantile plot is visually identical to a ranked data plot except for its horizontal axis, which varies from 0.0 to 1.0, with each point plotted according to the fraction of the points it exceeds (Walpole and Myers, 1985; USEPA, 1996). This allows the addition of vertical lines indicating the quartiles or, many other quantiles of interest. Quantile plots can be generated as follows: consider a hydrologic time series $X_1, X_2, ..., X_n$ with n data points. Arrange the time series data in ascending order and let $X_{(i)}$, for $i = 1$ to n, be the data listed in order from smallest to largest such that $X_{(1)}$ is the smallest, $X_{(2)}$ is the second smallest, and $X_{(n)}$ is the largest data point. For each i, compute the fraction $F_i = (i - 0.5)/n$. The 'quantile plot' is a plot of the pairs $[F_i, X_{(i)}]$, with straight lines connecting consecutive points.

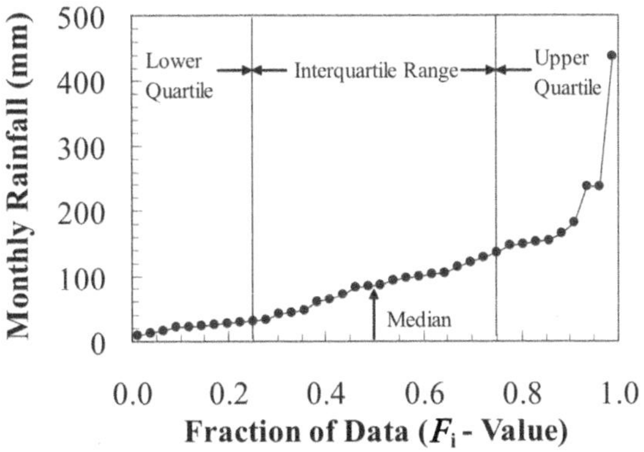

Fig. 3.6. Example of a quantile plot.

A 'quantile plot' can be used to evaluate the quantile information such as the median, quartiles, and interquartile range of the data points. Also, it can be used to know the density of the data points, i.e., are all the data points close to the centre with relatively few values in the tails or are there a large amount of data points in one tail with the rest evenly distributed? The density of the data points is displayed by the slope of the graph or line. Similar to the 'ranked data plot', a large amount of data points result in a flat slope, i.e., the graph rises slowly, while a small amount of data points result in a large slope, i.e., the graph rises quickly. A 'quantile plot' can also be used to check skewness or symmetry of the data points. A 'quantile plot' of the right-skewed data is steeper at the top right than at the bottom left, as shown in Fig. 3.6. A quantile plot of the left-skewed data increases sharply near the bottom left of the graph. If the data are symmetric about the centre point (mean or median), the top portion of the graph will stretch to the upper right corner in the same way the bottom portion of the graph stretches to the lower left, creating an S-shape similar to the ranked data plot.

3.1.6 Normal Probability Plot

There are two types of quantile-quantile plots (*q-q* plots). One is an empirical quantile-quantile plot, which involves plotting the quantiles of two hydrologic time series against each other. The other type of a quantile-quantile plot involves plotting the quantiles of a time series against the quantiles of a particular probability distribution. This is a technique to determine if the time series data were generated by the theoretical distribution (USEPA, 2006). The most common of these plots for hydrologic time series data is the 'normal probability plot', which is also known as a 'normal *q-q* plot'. However, the discussion about the 'normal probability plot' holds good for other *q-q* plots as well. Being a graphical method, the normal probability plot is a visual technique to roughly determine how well the time series data is modelled by a normal distribution (Dixon and Massey Jr., 1983).

A 'normal probability plot' is the plot of the quantiles of a hydrologic dataset against the quantiles of the standard normal distribution using normal probability graph paper (Fig. 3.7). This can be accomplished by plotting the sample quantiles against standard normal quantiles, or by plotting the sample

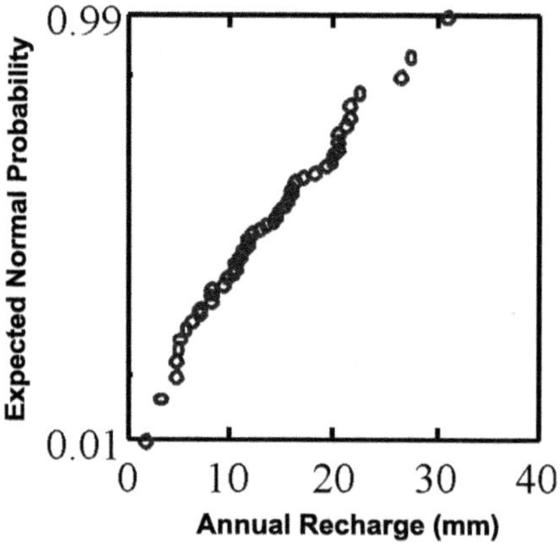

Fig. 3.7. Normal probability plot.

quantiles on a normal probability paper. If the graph is approximately linear (i.e., correlation coefficient is reasonably high excluding outliers), it indicates that the data are normally distributed and a formal statistical method of normality test should be used to confirm the result. On the other hand, if the graph of a 'normal probability plot' is not linear, the departures from linearity provide important information about how the data distribution deviates from a standard normal distribution.

Furthermore, if the graph of a 'normal probability plot' is non-linear, it may be used to evaluate the degree of symmetry (or asymmetry) displayed by the data (Dixon and Massey Jr., 1983; USEPA, 2006). Shape of the graph is convex for the right-skewed data, whereas shape of the graph is concave for the left-skewed data. If the data points in the upper tail of the graph fall above and the data points in the lower tail of the graph fall below the quartile line, the data are too slender to be well modelled by a normal distribution, i.e., there are fewer data points in the tails of the dataset than what is expected from a standard normal distribution. In contrast, if the data points in the upper tail of the graph fall below and the data points in the lower tail of the graph fall above the quartile line, the tails of the data points are too heavy to be well modelled using a standard normal distribution, i.e., there are more data points in the tails of the data than what is expected from a standard normal distribution. A normal probability plot can also be used to identify expected outliers/ extremes in the datasets. A data point or a few data points much larger or much smaller than the remaining data points will cause the other data points to be compressed into the middle of the graph, thereby destroying the resolution.

Considering a hydrologic time series $X_1, X_2, ..., X_n$ with n data points, the following is the step-by-step procedure to develop a normal probability plot:

Step 1: Compute the absolute frequency, AF_i for each data point. The absolute frequency is the number of times each data point occurs in the time series. For distinct data points, the absolute frequency is 1. For non-distinct data points, count the number of times a data point occurs. For example, consider the data 2, 3, 4, 4. The absolute frequency of data point 2 is 1 and that of data point 3 is 1. Similarly, the absolute frequency of data point 4 is 2 as 4 appears 2 times in the dataset.

Step 2: Compute the cumulative frequencies, CF_i. The cumulative frequency is the number of data points that are less than or equal to X_i, i.e., $CF_i = \sum_{j=1}^{i} AF_j$. For the data given in Step 2, the cumulative frequency for data point 2 is 1, the cumulative frequency for data point 3 is 2 (1+1), and the cumulative frequency for data point 4 is 4 (1+1+2).

Step 3: Finally, compute $Y_i = 100 \times \frac{CF_i}{(n+1)}$ and plot the pairs (Y_i, X_i) on a normal probability paper as illustrated in Fig. 3.7. As mentioned above, if the graph of these pairs approximately forms a straight line, the data are probably normally distributed. Otherwise, the data distribution may not be normal.

3.2 Statistical Methods

Decision making about the normality of time series data based on graphical methods alone is subjective. For extremely non-normal data, it is easy to make such decision. However, such a decision is not straightforward in many cases. Therefore, statistical methods are usually necessary to test the assumption of normality. The statistical methods commonly used for checking normality of the time series are described in subsequent sections. Of the total eleven statistical tests discussed ahead, the Kolmogorov-Smirnov, Anderson-Darling and Cramér-von Mises tests for normality are based on the empirical distribution function (EDF) and are often referred to as EDF tests (Stephens, 1986).

3.2.1 Chi-Square Test

The 'chi-square test' is used to test if a sample of data came from a population with a specific distribution (Snedecor and Cochran, 1980). An attractive feature of the chi-square goodness-of-fit test is that it can be applied to any univariate distribution for which you can calculate the cumulative distribution function. The chi-square goodness-of-fit test is applied to the binned data (i.e., data put into classes). This is actually not a restriction because for the non-binned data, a histogram or frequency table can be calculated before using the chi-square test. However, the value of the chi-square test statistic is dependent on how the data is binned (Snedecor and Cochran, 1980). Another disadvantage of this test is that it requires a sufficient sample size so that the chi-square approximation is valid.

For using the 'chi-square test', the time series data are divided into k bins and the test-statistic is defined as follows (Snedecor and Cochran, 1980):

$$\chi^2 = \sum_{i=1}^{k} (O_i - E_i)^2 / E_i \qquad (1)$$

where O_i = observed frequency for the bin i and E_i = expected frequency for the bin i. The expected frequency is calculated as

$$E_i = N\{F(Y_U) - F(Y_L)\} \qquad (2)$$

where F = cumulative distribution function for the distribution being tested, Y_U = upper limit for class i, Y_L = lower limit for class i and N = size of the sample.

The test-statistic approximately follows a chi-square distribution with $(k - c)$ degrees of freedom, where k is the number of non-empty cells and c is the number of estimated parameters (including location, scale and shape parameters) for the distribution plus 1. For example, for a 3-parameter Weibull distribution, $c = 4$. Therefore, the hypothesis that the data are from a population with the specified distribution is rejected, if $\chi^2 > \chi^2_{(\alpha,k-c)}$, where $\chi^2_{(\alpha,k-c)}$ is the critical test-statistic value with $k - c$ degrees of freedom and a significance level of α.

As mentioned above, the 'chi-square test' is sensitive to the choice of bins. There is no optimal choice for the bin width because the optimal bin width depends on the distribution (Snedecor and Cochran, 1980). It should be noted that the 'chi-square test' is an alternative to the Anderson-Darling and Kolmogorov-Smirnov tests. The 'chi-square test' can be applied to discrete distributions such as binomial and Poisson distributions, but the application of Kolmogorov-Smirnov and Anderson-Darling tests are restricted to continuous distributions only. For the chi-square approximation to be valid, the expected frequency should be at least 5 (Snedecor and Cochran, 1980). Generally, this test is not valid for small samples, and if some of the counts are less than 5, it may be necessary to combine some bins in the tails.

3.2.2 Kolmogorov-Smirnov Test

The Kolmogorov-Smirnov (K-S) test is an empirical distribution function (EDF) test in which the theoretical cumulative distribution function of the test distribution is compared with the EDF of the time series data (Conover, 1980; Armitage and Colton, 1998a). The K-S test was first proposed by Kolmogorov and then modified by Smirnov. This test finds difference between cumulative distribution of the time series data and the expected cumulative normal distribution, and computes its *P*-value for the largest discrepancy. The test-statistic is defined as (Massey Jr., 1967):

$$D = \text{supx}|Fn(x) - F(x, \mu, s)| \tag{3}$$

where $F(x, \mu, s)$ is theoretical cumulative distribution function of the normal distribution function and $Fn(x)$ is the empirical distribution function of the data.

Large values of D indicate presence of non-normality in the time series. The table of critical values $D_\alpha(n)$ of the distribution of D for various sample sizes (n) and significance levels (α) is given in Massey Jr. (1967). If the population parameters (i.e., μ and s) are known, the original K-S test can be used. However, if they are not known, they can be replaced by sample estimates (Massey Jr., 1967; Conover, 1980).

It is worth to mention that the K-S test is strongly criticized by the researchers due to ambiguous results (Steinskog et al., 2007). Particularly, conclusions based on the results of not rejecting normality could be very misleading. D'Agostino (1986) emphasized that the K-S test should not be applied if population parameters have to be estimated (a usual case).

3.2.3 Lilliefors Test

The 'Lilliefors test' was named after Hubert Lilliefors, a professor of statistics at the George Washington University and is an adaptation of the 'Kolmogorov-Smirnov test' (Lilliefors, 1967, 1969). The 'Lilliefors test' compares the cumulative distribution of data to the expected cumulative normal distribution. The 'Lilliefors test' is different from the 'K-S test' as unknown population parameters are estimated, while the test-statistic is the same. The critical values of the two tests are different, which results in different decisions (Mendes and Pala, 2003). The 'Lilliefors test' is more powerful than the 'chi-square test' for large sample sizes and is recommended by the US Environmental Protection Agency (USEPA, 1996).

The 'Lilliefors test' is used to test the null hypothesis that the data come from a normally distributed population. The mean and variance of the normally distributed population are estimated. The procedures for applying the test are as follows:

Step 1: Estimate the population mean and variance of the time series data.

Step 2: Find the maximum discrepancy between the empirical distribution function and the cumulative distribution function (CDF) of the normal distribution with the estimated mean and estimated variance. The test-statistics for the 'Lilliefors test' is similar to that for the 'K-S test' shown in Eqn. (3).

Step 3: Finally, the null hypothesis should be rejected if the maximum discrepancy is large enough to be statistically significant, which is the criteria for testing the null hypothesis by 'K-S test'. In 'Lilliefors test', since the population parameters and CDF are estimated based on sample data, the hypothesized CDF moves closer to the data themselves. As a result, the computed maximum discrepancy becomes smaller than it would have been if the null hypothesis had singled out just one normal distribution. Thus, probability distribution of the 'Lilliefors test', assuming the null hypothesis is true, is stochastically smaller than that for the Kolmogorov-Smirnov test. This is the Lilliefors correction to 'K-S test'. To date, tables for this distribution have been prepared by Monte Carlo methods only (Lilliefors, 1967, 1969).

3.2.4 Anderson-Darling Test

The Anderson-Darling test is used to test if a sample of data came from a population with a normal distribution. It is a modification of the Kolmogorov-Smirnov (K-S), which gives more weight to tails compared to the K-S test (Stephens, 1974). The K-S test is distribution-free in the sense that the critical values do not depend on the normal distribution. However, in the Anderson-Darling test, the critical values are dependent on a given distribution, which makes it a more sensitive test; though the disadvantage is that critical values need to be calculated for each distribution. The Anderson-Darling test is an alternative to the chi-square and Kolmogorov-Smirnov tests. It makes use of the fact that in case of a hypothesized underlying distribution, and assuming the data does arise from this distribution, the data can be transformed to a uniform distribution. Thereafter, the transformed sample data can be tested for uniformity (Shapiro, 1980).

The test-statistic (A) for the Anderson-Darling test to evaluate, if the data ($Y_1<Y_2<\ldots<Y_n$) comes from a distribution with cumulative distribution function (CDF) F, is given as (Stephens, 1986):

$$A^2 = -n - S \tag{4}$$

where $$S = \sum_{k=1}^{n} \frac{2k-1}{n} [\ln F(Y_k) + \ln(1 - F(Y_{n+1-k}))] \tag{5}$$

Note that the time series data need to be arranged in decreasing order (i.e., $Y_1<Y_2<\ldots<Y_n$) before computing the test-statistic, A. The value of A thus computed is compared with the corresponding critical value of the theoretical distribution. This test is a one-sided test and the hypothesis that the distribution is normal is rejected if the value of A is greater than the critical value.

Despite excellent theoretical properties of the Anderson-Darling test, it has a serious flaw when applied to real-world time series data. The Anderson-Darling test is severely affected by ties in the data because of poor precision (Stephens, 1986). When a significant number of ties exist in a dataset, the Anderson-Darling test will often reject the data as non-normal, irrespective of how well the data fit the normal distribution.

3.2.5 Cramér-von-Mises Test

The Cramér-von-Mises test is an alternative to the Kolmogorov-Smirnov test and the Anderson-Darling test. Let $x_1, x_2, \ldots, x_n$ be the observed values of a hydrologic time series in increasing order. The Cramér-von-Mises statistic is computed as (Stephens, 1986):

$$W^2 = \sum_{i=1}^{n} \left(F(x_i) - \frac{2i-1}{2n} \right)^2 + \frac{1}{12n} \tag{6}$$

where $F(x)$ = distribution function of x and n = sample size of the time series. If the value of the test-statistic is larger than the corresponding critical value, the hypothesis that the data come from the distribution F is rejected.

3.2.6 Shapiro-Wilk Test

The Shapiro-Wilk (S-W) test is one of the most powerful and omnibus normality test (Shapiro, 1980; Gilbert, 1987; USEPA, 2006). This test is similar to computing a correlation between the quantiles of the standard normal distribution and the ordered data points of a hydrologic time series. If the normal probability plot is approximately linear (i.e., the data follow a normal probability distribution), the test-statistic value will be relatively high. In contrast, if the normal probability plot is nonlinear, the test-statistic value will be relatively low. The S-W test has been recommended by the US Environmental Protection Agency (USEPA, 2006) as well as in many statistical texts (Gilbert, 1987) for testing normality in the environmental time series. In recent years, the S-W test has become the preferred test of normality due to its good power properties as compared to a wide range of alternative tests (Mendes and Pala, 2003).

The test-statistic (*W*) of the S-W test examines whether a random sample, $x_1, x_2, \ldots, x_n$ of a hydrologic variable comes from (specifically) a normal distribution. This test-statistic is given as follows (Shapiro and Wilk, 1965):

$$W = \frac{\left(\sum_{i=1}^{n} a_i \, x_{(i)}\right)^2}{\sum_{i=1}^{n} (x_i - \bar{x})^2} \qquad (7)$$

where $x_{(i)}$ = ordered (increasing ordered) sample values and a_i = constants generated from the means, variances and covariances of the order statistics of a sample of size *n* from a normal distribution (Pearson and Hartley, 1972).

Small values of the test-statistic, *W* indicate departure from normality. Percentage points for the *W*, obtained by Monte Carlo simulations, are given in Pearson and Hartley (1972). Since the computation of *W* is not much easier, available statistical packages such as GraphPad Prism, Dataplot, DataQUEST or STATISTICA can be used for analyzing the time series data by using Shapiro-Wilk test.

3.2.7 Probability Plot Correlation Coefficient

Probability plot correlation coefficient (PPCC) test is considered as an extension of the Shapiro-Wilk test. It is also known as 'Filliben's test-statistic' (Filliben, 1975). It measures the linearity of the data on a normal probability paper. Like the S-W test, if the normal probability plot is approximately linear (i.e., the hydrologic data follow a normal distribution curve), the correlation coefficient value will be relatively high (USEPA, 1992). On the other hand, if the normal probability plot contains several data points deviating from linearity (i.e., the data do not follow a normal distribution curve), the correlation coefficient will be relatively low. Although the Filliben's test-statistic is easier to compute than the test-statistic of the S-W test, it is still difficult to compute by hand. Therefore, statistical software like DataQUEST can be used to calculate the Filliben's test-statistic.

3.2.8 Coefficient of Variation

The coefficient of variation (CV), a well-known term in statistics, can be used to quickly decide whether or not the time series data follow a normal distribution curve by comparing the value of sample CV with 1. However, checking normality based on the CV is somewhat a weak approach. This method is valid only for some hydrologic and environmental applications if the data represent a non-negative characteristic such as rainfall amount or pollutant concentration (USEPA, 1992). If the value of CV is greater than 1, the data should not be modelled with a normal distribution curve. However, the opposite statement is not correct, i.e., we cannot conclude that the data can be modelled with a normal distribution curve if the CV is less than 1 (USEPA, 1992). The CV test is generally recommended to be used along with other statistical tests or when the graphical representation of data indicates extreme departures from normality.

3.2.9 Range Tests

The range tests are based on the fact that almost entire area of a normal distribution curve lies within ±5 standard deviations from the mean. There are two types of range tests: Studentized range test, and Geary's test. Both of these tests use a ratio of an estimate of the sample range to the sample standard deviation (Madansky, 1988; USEPA, 1996). A brief description about these range tests is provided below.

3.2.9.1 Studentized Range Test

The Studentized range test uses the ratio of range of a sample to the sample standard deviation. Tables of critical values for sample sizes up to 1000 are available for checking whether the absolute value of this ratio is significantly large (Madansky, 1988). The Studentized range test does not perform well if the data points are not symmetric or if the tails of the data points are heavier than that for the normal distribution. Also, this test may be sensitive to outlier or extreme data points. Unfortunately, lognormally distributed data, which are quite common in hydrological and environmental applications, have these characteristics. If the data appear to be lognormally distributed, this test should not be used (USEPA, 2006). In most cases, the Studentized range test performs the same as the Shapiro-Wilk test but is much easier to apply.

3.2.9.2 Geary's Test

Test-statistic of the Geary's test is defined as the ratio of mean deviation of a sample to the sample standard deviation. This ratio indicates whether time series data follows a standard normal distribution or deviates from the normal distribution (Madansky, 1988). This test is not as strong as the Shapiro-Wilk test or the Studentized range test. However, since the Geary's test-statistic is based on the normal probability distribution, critical values for all possible sample sizes are available. For a random sample $x_1, x_2, \ldots x_n$, assuming that it follows normal distribution, the Geary's test-statistic (U) is defined as follows (Walpole and Myers, 1985):

$$U = \frac{\sqrt{\pi/2}\ \sum_{i=1}^{n} |x_i - \bar{x}|/n}{\sqrt{\sum_{i=1}^{n}(x_i - \bar{x})^2 \Big/ n}} \tag{8}$$

It should be noted that the denominator of Eqn. (8) is a reasonable estimator of standard deviation whether the distribution is normal or non-normal. The numerator is a good estimator of standard deviation if the distribution is normal, but may overestimate or underestimate standard deviation when there are departures from the normality. Thus, the values of U differing considerably from 1.0 indicate that the hypothesis of normality should be rejected (Walpole and Myers, 1985).

3.2.10 Jarque Bera Test

Jarque and Bera (1980) proposed a normality test using classical skewness and kurtosis coefficients. The Jarque-Bera (JB) test is a goodness-of-fit measure of departure from normality, based on the sample kurtosis and skewness. The test-statistic JB is defined as (Jarque and Bera, 1987):

$$\text{JB} = \frac{n}{6}\left(s^2 + \frac{(k-3)^2}{4}\right) \tag{9}$$

where n = number of observations, s = sample skewness and k = sample kurtosis.

The major disadvantage of the Jarque-Bera test is that asymptotic convergence of the test-statistic is very slow. Therefore, decisions for testing normality based on the quantile function of the chi-square distribution can lead to serious errors (Bowman and Shenton, 1975; Jarque and Bera, 1987; Lehmann, 1999).

3.2.11 D'Agostino Pearson Omnibus Test

The D'Agostino Pearson (DAP) Omnibus test first analyzes time series data to determine skewness (to quantify the asymmetry of the data distribution) and kurtosis (to quantify the shape of the data distribution). Thereafter, it calculates how far each of the two values differs from the value expected with a normal distribution, and computes a single P-value from the sum of the squares of these discrepancies (D'Agostino, 1986). This test is a combination of the D'Agostino skewness test and Anscombe-Glynn kurtosis test. The test-statistic (K^2) of the DAP Omnibus test is expressed as (D'Agostino et al., 1990):

$$K^2 = Z^2\left(\sqrt{b_1}\right) + Z^2(b_2) \tag{10}$$

where $Z^2(\sqrt{b_1})$ and $Z^2(b_2)$ are the standard normal deviates equivalent to transformations of $\sqrt{b_1}$ (skewness) and b_2 (kurtosis) (Armitage and Colton, 1998b). The test-statistic (K^2) has approximately a chi-square distribution with 2 degrees of freedom under the assumption that two summands [i.e., $Z^2(\sqrt{b_1})$ and $Z^2(b_2)$] are independent and the population is normally distributed. The assumption of independence cannot be held up for small and moderate sample sizes. Thus, the fact that K^2 is chi-distributed under the null hypothesis does not hold true for the most common sample sizes. Test-statistics of the DAP test is not easy to compute manually, and therefore, statistical software may be used.

References

Adeloye, A.J. and Montaseri, M. (2002). Preliminary streamflow data analyses prior to water resources planning study. *Hydrological Sciences Journal*, **47(5):** 679-692.

Armitage, P. and Colton, T. (1998a). Encyclopedia of Biostatistics, Volume 2. Wiley, New York, pp: 1759-1760.

Armitage, P. and Colton, T. (1998b). Encyclopedia of Biostatistics, Volume 4. Wiley, New York, pp. 3075-3079.

Bowman, K.O. and Shenton, L.R. (1975). Omnibus test contours for departures from normality based on $\sqrt{b_1}$ and b_2. *Biometrika*, **62:** 243-250.

Conover, W.J. (1980). Practical Nonparametric Statistics (Second Edition). John Wiley, New York, NY.

D'Agostino, R.B. (1986). Tests for the normal distribution. *In:* R.B. D'Agostino and M.A. Stephens (editors), Goodness-of-Fit Techniques. Marcel Dekker, New York, USA.

D'Agostino, R.B., Belanger, A. and D'Agostino Jr., R.B. (1990). Suggestion for using powerful and informative tests of normality. *The American Statistician*, **44:** 316-321.

Dixon, W.J. and Massey Jr., F.J. (1983). Introduction to Statistical Analysis. 4th Edition, McGraw-Hill, New York, USA.

Filliben, J.J. (1975). The probability plot correlation coefficient test for normality. *Technometrics*, **17:** 111-117.

Gilbert, R.O. (1987). Statistical Methods for Environmental Pollution Monitoring. Van Nostrand Reinhold, New York, USA.

GraphPad. (2007). Normality Tests: Use with Caution. http://www.graphpad.com/library/BiostatsSpecial/article_197.htm (accessed on 14 January 2011).

Jarque, C.M. and Bera, A.K. (1980). Efficient tests for normality, homoscedasticity and serial independence of regression residuals. *Economics Letters*, **6(3):** 255-259.

Jarque, C.M. and Bera, A.K. (1987). A test for normality of observations and regression residuals. *International Statistical Review*, **55(2):** 163-172.

Lehmann, E.L. (1999). Elements of Large Sample Theory. Springer, New York.

Lilliefors, H.W. (1967). On the Kolmogorov-Smirnov test for normality with mean and variance unknown. *Journal of the American Statistical Association*, **62:** 399-402.

Lilliefors, H.W. (1969). Correction to the paper "On the Kolmogorov-Smirnov test for normality with mean and variance unknown". *Journal of the American Statistical Association*, **64:** 1702.

Madansky, A. (1988). Prescriptions for Working Statisticians. Springer-Verlag, New York, NY.

Massey Jr., F.J. (1967). The Kolmogorov-Smirnov test for goodness of fit. *Journal of the American Statistical Association*, **46:** 68-78.

Mendes, M. and Pala, A. (2003). Type I Error Rate and Power of Three Normality Tests. *Pakistan Journal of Information and Technology*, **2:** 135-139.

Pearson, E.S. and Hartley, H.O. (editors) (1972). Biometrika Tables for Statisticians, Volume 2. Cambridge University Press, U.K.

Shapiro, S.S. (1980). How to Test Normality and Other Distributional Assumptions. Volume 3, The ASQC Basic References in Quality Control: Statistical Techniques, American Society for Quality Control, Milwaukee, WI.

Shapiro, S.S. and Wilk, M.B. (1965). An analysis of variance test for normality (complete samples). *Biometrika*, **52(3&4):** 591-611.

Snedecor, G.W. and Cochran, W.G. (1980). Statistical Methods. The Iowa State University Press, Ames, Iowa.

Steinskog, D.J., Tjøstheim, D.B. and Kvamstø, N.G. (2007). A cautionary note on the use of the Kolmogorov-Smirnov test for normality. *Monthly Weather Review*, American Meteorological Society, **135:** 1151-1157.

Stephens, M.A. (1974). EDF statistics for goodness of fit and some comparisons. *Journal of the American Statistical Association*, **69:** 730-737.

Stephens, M.A. (1986). Tests based on EDF statistics. *In:* R.B. D'Agostino and M.A. Stephens (editors), Goodness-of-Fit Techniques. Marcel Dekker, New York, USA.

Thode, H.C. (2002). Testing for Normality. Marcel Dekker, New York, USA, 368 pp.

USEPA (1992). Guidance Document on the Statistical Analysis of Ground-Water Monitoring Data at RCRA Facilities. EPA/530/R-93/003, United States Environmental Protection Agency (USEPA), Office of Solid Waste, Washington D.C.

USEPA (1996). Guidance for Data Quality Assessment: Practical Methods for Data Analysis. EPA QA/G-9, United States Environmental Protection Agency (USEPA), Office of Research and Development, Washington D.C.

USEPA (2006). Data Quality Assessment: Statistical Methods for Practitioners. Guidance Document EPA QA/G-9S, United States Environmental Protection Agency (USEPA), Office of Environmental Information, Washington D.C.

Walpole, R. and Myers, R. (1985). Probability and Statistics for Engineers and Scientists. 3rd Edition, MacMillan, New York, USA.

4

Methods for Time Series Analysis

Natural time series, including hydrologic, climatic and environmental time series, which satisfy the assumptions of homogeneity, randomness, non-periodic, non-persistence and stationarity, seem to be the exception rather than the rule (Rao et al., 2003). In fact, for all water resources studies involving the use of hydrologic time series data, preliminary statistical analyses must always be carried out to confirm whether the hydrologic time series possess all the required assumptions/characteristics (Adeloye and Montaseri, 2002). Nevertheless, most time series analysis is performed using standard methods after relaxing the required conditions one way or another in the hope that the departure from these assumptions is not large enough to affect the analysis results (Rao et al., 2003). A comprehensive survey of the past studies on the hydrologic time series analysis (Machiwal and Jha, 2006) revealed that no studies considered all the aspects of time series analysis. Major work is reported dealing with only linear trend analysis, and the homogeneity, stationarity, periodicity, and persistence, which are equally important characteristics of the hydrologic time series, have been ignored. In most past studies on time series analysis, only regression and/or Kendall's rank correlation tests are applied for trend detection. Esterby (1996) and Hess et al. (2001) presented an overview of selected trend tests. Thus, very limited studies are reported to date concerning a detailed analysis of homogeneity, stationarity, periodicity and persistence in the hydrologic time series.

In the literature, several statistical tests/methods are available to determine a particular characteristic of the time series. It has been seen that choosing a specific statistical test for a particular characteristic of the time series is dependent on the knowledge of data analyst or researcher rather than on the assumptions/requirements of the test. Use of one or two statistical tests for time series analysis is quite common for easy decision making. However, Machiwal and Jha (2008) recommended that an adequate number of statistical tests must be applied for detecting a particular time series characteristic and the results should be analyzed critically to arrive at a reliable decision. Based on the extensive literature search, it was found that a single reference/source does not exist where detailed and easy-to-understand scientific information about all the available statistical methods can be found. This chapter fulfils this gap by presenting in-depth information about various statistical tests available for the time series analysis. A total of 28 statistical tests/methods are presented in this chapter, of which eight tests are for checking homogeneity, three tests for checking stationarity, fifteen tests for detecting presence or absence of trend, one test for checking periodicity, and one test for checking persistence in the hydrologic time series.

4.1 Methods for Checking Homogeneity

Homogeneity or consistency implies that all the collected hydrologic time series data belong to the same statistical population having a time invariant mean. Therefore, the tests to check the homogeneity or consistency of data series are based on evaluating the significance of changes in the mean value. The features of three homogeneity tests namely, the von Neumann test, Cumulative Deviations, and the Bayesian test are discussed in Buishand (1982) and Jayawardena and Lau (1990).

Buishand (1982, 1984) presents a detailed methodology for the above-mentioned three homogeneity tests, which can serve as major guidelines about these tests. Kanji (2001) in his excellent collection of 100 statistical tests, has reported various homogeneity tests for multiple comparisons (e.g., Tukey, Link-Wallace, Dunnett, Bartlett, and Hartley tests). However, it has drawbacks that the objectives of the tests are not clear and that original references are lacking. On the other hand, some researchers (e.g., Radziejewski et al., 2002) have considered a few homogeneity tests for trend detection. Such studies may create confusion about the general perceptions of homogeneity and trend for the researchers with no access to good literature in this line. It should be noted here that the homogeneity tests for multiple comparisons (e.g., Bartlett, Dunnett, Link-Wallace, Hartley, and Tukey tests) have not gained a wide popularity in hydrology and climatology. In the hydrologic time series analysis, multiple comparison tests are still *contemporary*, while these tests are considered as *classical* in geotechnical studies (e.g., Phoon et al., 2003). Detailed procedures for applying the homogeneity tests are described ahead.

4.1.1 The von Neumann Test

The von Neumann ratio (N) is the most widely used test for testing a time series for the absence or presence of homogeneity. It is closely related to the first-order serial correlation coefficient (WMO, 1966) and can be defined as follows:

$$N = \sum_{t=1}^{n-1} (x_t - x_{t+1})^2 \Big/ \sum_{t=1}^{n} (x_t - \bar{x})^2 \qquad (1)$$

where x_t =hydrologic variable constituting the sequence in time, n = total number of hydrologic records, and $\bar{x}$ = average of x_t.

Under the null hypothesis of constant mean, i.e., homogenous time series, the expected value of the von Neumann ratio is 2. However, it tends to be < 2 for the non-homogenous time series. The values of von Neumann ratio for normally distributed samples can be found in Owen (1962).

4.1.2 Cumulative Deviations Test

Tests for homogeneity are based on the adjusted partial sums or cumulative deviations from the mean, which are expressed as (Buishand, 1982):

$$S_k^* = \sum_{t=1}^{k}(x_t - \bar{x}), \quad k = 1, 2, \ldots\ldots, n \tag{2}$$

Rescaled adjusted partial sums (S_k^{**}) are obtained by dividing S_k^*'s by the sample standard deviation (D_x).

$$S_k^{**} = S_k^* / D_x, \quad k = 1, 2, \ldots\ldots, n \tag{3}$$

with

$$D_x^2 = \frac{1}{n}\sum_{t=1}^{n}(x_t - \bar{x})^2 \tag{4}$$

The values of S_k^{**}'s are not dependent on the unit of the hydrologic variable, and hence homogeneity tests are based on the rescaled adjusted partial sums.

Sensitivity to the departures from homogeneity is defined by the following statistic:

$$Q = \underset{0 \le k \le n}{\text{Max}} \left| S_k^{**} \right| \tag{5}$$

High values of Q are an indication for non-homogeneity in the time series. Another statistic which can be used for testing homogeneity is the range (R). It is defined as:

$$R = \underset{0 \le k \le n}{\text{Max}} \left| S_k^{**} \right| - \underset{0 \le k \le n}{\text{Min}} \left| S_k^{**} \right| \tag{6}$$

Critical values of Q for some specified values of n are given by Buishand (1982), which are based on the 19,999 synthetic sequences of Gaussian random numbers. For $n \to \infty$, the critical values of Q can be obtained from the Kolmogorov-Smirnov goodness-of-fit statistic table (Doob, 1949). The critical values of the distribution of R under the null hypothesis are given by Wallis and O'Connell (1973) in a graphical form, while Buishand (1982) presents salient critical values in a tabular form.

4.1.3 Bayesian Test

The Bayesian test was developed by Chernoff and Zacks (1964), which was modified later by Gardner (1969). The Gardner's test-statistic ($\tilde{G}$) for a two-sided test on a shift in the mean at an unknown point can be written as (Gardner, 1969):

$$\tilde{G} = \sum_{k=1}^{n-1} p_k \left(S_k^* / \sigma_Y\right)^2 \tag{7}$$

where p_k = prior probability that the shift occurs just after k^{th} observation. Here, it is assumed that the population variance (σ_Y^2) is known. If σ_Y^2 is not known, it can be replaced with the sample variance. For p_k independent of k, the test-statistic, U can be expressed as:

$$U = \frac{1}{n(n+1)} \sum_{k=1}^{n-1} \left(S_k^{**}\right)^2 \tag{8}$$

However, for p_k proportional to $[k(n-k)]^{-1}$, the test-statistic can be written as:

$$A = \sum_{k=1}^{n-1} \left(Z_k^{**}\right)^2, \quad k = 1, 2, \ldots\ldots, n \tag{9}$$

where Z_k^{**} = weighted rescaled partial sums, which can be computed using the following formula:

$$Z_k^{**} = \left[\{k(n-k)\}^{-1/2} S_k^*\right] / D_x \tag{10}$$

Large values of U and A test-statistics indicate departures from the homogeneity, which is judged based on their critical values (Buishand, 1982).

Buishand (1982) reported that the tests based on the cumulative deviations are superior to the von Neumann test for a model with only one change in the mean. The tests were applied to the annual rainfall data of 264 rainfall stations in the Netherlands, and departures from homogeneity were found. The von Neumann test provided almost the same results as the tests based on the cumulative deviations.

4.1.4 Tukey Test for Multiple Comparisons

This test is used to examine the significance of all possible differences among different population means. The size of the different samples may be unequal but all populations should be normally distributed with equal variances. Hence, it is a parametric test, which depends upon the distribution parameters. To apply this test on a hydrologic time series x_t ($t = 1, 2, \ldots, n$), the entire series is firstly divided into 'K' subseries of equal or unequal sample size. The total variance of the samples is then calculated by the following expression (Kanji, 2001):

$$s^2 = \frac{\sum_{i=1}^{K} (n_i - 1) s_i^2}{n - K} \tag{11}$$

where s_i = variance of the ith sample, and n = total sample size. Now the test-static or limit is computed as (Kanji, 2001):

$$W = \frac{q\,s}{n_t^{1/2}} \tag{12}$$

where q is the Studentized range. The critical values for q can be obtained at degrees of freedom (ν) from the standard table available in textbooks on statistics (e.g., Sachs, 1972; Kanji, 2001). The degree of freedom (ν) can be computed as:

$$\nu = \left(\sum_{i=1}^{K} n_i \right) - K \tag{13}$$

Here, n_t is expressed as:

$$n_t = \frac{K}{\left(\frac{1}{n_1} + \frac{1}{n_2} + \cdots + \frac{1}{n_K} \right)} \tag{14}$$

If the limit (W) exceeds by the absolute difference between any two sample means, it suggests that the corresponding population means differ significantly.

4.1.5 Link-Wallace Test

The Link-Wallace test is employed for the purpose similar to the Tukey test; however it has the limitation that the sample size of all populations must be equal. It is a parametric test based on the assumption that the 'K' populations are normally distributed with equal variances. This test can be used to examine the homogeneity of any hydrologic time series, x_t (t = 1, 2, ..., n) after dividing the entire series into 'K' subseries of equal sample size n_k. The test-statistic (K_L) is defined as (Kanji, 2001):

$$K_L = \frac{n_k\, w(\overline{x})}{\sum_{i=1}^{K} w_i(x)} \tag{15}$$

where $w_i(x)$ is the range of the x values for the ith sample and $w(\bar{x})$ is the range of the sample means. Critical values of the test-statistic can be found in Sachs (1972) or Kanji (2001). If computed values of the test-statistic are greater than the critical values, the null hypothesis of equal variances is rejected at 5% significance level. Furthermore, the critical value of the sample mean differences (D) is expressed as (Kanji, 2001):

$$D = \frac{K_{\text{critical}} \sum_{i=1}^{K} w_i(x)}{n_k} \tag{16}$$

where K_{critical} is the critical value of the test-statistic. If the mean differences for a sample are greater than the value of 'D', the null hypothesis of equal means is rejected.

4.1.6 Dunnett Test

This test is used to investigate the significance of the differences in means, when several samples are compared with a control one. This is a parametric test with a limitation that the samples of equal sizes (n_s) are drawn independently from normally distributed populations with equal variances. To apply this test to a hydrologic record x_t ($t = 1, 2, \ldots, n$), the entire series is divided into '$K + 1$' subseries. The variance within the $K + 1$ groups is calculated as (Kanji, 2001):

$$s^2 = \frac{s_0^2 + \sum_{i=1}^{K} s_i^2}{(K-1)(n_s-1)} \tag{17}$$

where s_0^2 is the variance of the control subseries and s_i^2 is variance of the ith subseries. The standard deviation of the differences between control and remaining subseries is then calculated as:

$$s(\bar{d}) = \sqrt{2s^2/n_s} \tag{18}$$

The test-statistic (D_j) is given by (Kanji, 2001):

$$D_j = \frac{|\bar{x}_i - \bar{x}_0|}{s(\bar{d})} \quad (i = 1, 2, \ldots, K) \tag{19}$$

Critical values of the test-statistic can be obtained from Kanji (2001). If an observed value is larger than the tabulated value, it can be concluded that the corresponding difference in mean between the subseries 'i' and control is significant. Similarly, each subseries may be considered as control and significance of differences between all the sample means can be examined.

4.1.7 Bartlett Test

Techniques for comparing means of normally distributed populations generally assume that the populations have the same variance. Before using ANOVA, it should be confirmed whether this assumption of homogeneity of variance is reasonable. The Bartlett test is widely used for equal variances. The step-by-step procedure for applying this test to a hydrologic time series x_t ($t = 1, 2, \ldots, n$) is given below.

Step 1: Fragment the entire series into 'K' subseries with n_i size of each ith series ($i = 1, 2, \ldots, K$).

Step 2: Setup the null hypothesis that the variances of all subseries are equal and alternative hypothesis of unequal variances.

Step 3: Compute the sample variance (s_i^2) of each subseries as:

$$s_i^2 = \sum_{j=1}^{n_i} \frac{\left(x_{ij} - \bar{x}_i\right)^2}{n_i - 1} \tag{20}$$

Step 4: Compute the overall variance as (Kanji, 2001):

$$s^2 = \frac{\sum_{i=1}^{K} (n_i - 1) s_i^2}{\sum_{i=1}^{K} (n_i - 1)} \tag{21}$$

Step 5: Calculate the Bartlett test-statistic, which is defined as (Kanji, 2001):

$$B = \frac{2.30259}{C} \left[\sum_{i=1}^{K} (n_i - 1) \log s^2 - \sum_{i=1}^{K} (n_i - 1) \log s_i^2 \right] \tag{22}$$

where 'C' is a bias correction factor and is mathematically expressed as follows:

$$C = 1 + \frac{1}{3(K+1)} \left\{ \sum_{i=1}^{K} \frac{1}{(n_i - 1)} - \frac{1}{\sum_{i=1}^{K} (n_i - 1)} \right\} \tag{23}$$

Step 6:

Case A: If $n_i > 6$, 'B' will approximate to a χ^2-distribution with 'K–1' degrees of freedom. Critical values of the test-statistic can be obtained from the standard tables of χ^2-distribution.

Case B: If $n_i \leq 6$, the test-statistic becomes $BC = M$, for which the critical values can be obtained from Kanji (2001).

In both the cases, the null hypothesis of equal variances is rejected if the computed value of the test-statistic is greater than its critical values.

4.1.8 Hartley Test

This is another parametric test to examine the significance of the differences between the variances of '*K*' normally distributed populations. The size of the '*K*' samples should be approximately equal. The entire hydrologic time series x_t ($t = 1, 2, \ldots, n$) is divided into '*K*' subseries with approximately equal size. The sample variances of all the subseries are then calculated from Eqn. (20). The test-statistic (F) is defined as (Kanji, 2001):

$$F_{max} = \frac{s_{max}^2}{s_{min}^2} \tag{24}$$

where s_{max}^2 is the largest of the '*K*' subseries variances, and s_{min}^2 is the smallest of the '*K*' subseries variances. The critical values of this test-statistic are given in Kanji (2001). If the computed value of the test-statistic exceeds its critical value, the null hypothesis of equal variances is rejected.

4.2 Methods for Checking Stationarity

A time series is said to be *strictly stationary*, if its statistical properties do not vary with changes of time origin. That is, if two non-overlapping time intervals are selected from a given time series, the two subseries will look almost the same. In fact, both the subseries will differ from one another, but will be scattered around the same mean value. Therefore, a stationary time series cannot have any trend or periodic component. This is the reason that sometimes trend and periodicity tests are used to check the stationarity of hydrologic time series. There are two general approaches for checking stationarity: *parametric* and *nonparametric*. The review of the literature reveals that the parametric approach is usually used by the researchers working in the time domain, such as economists, who make certain assumptions about the nature of their data. On the other hand, the nonparametric approach is more commonly used by the researchers working in the frequency domain, such as electrical engineers, who often treat the system as a black box and cannot make any basic assumptions about the nature of the system. However, in hydrology, both parametric and nonparametric approaches are used. It should be noted that the nonparametric tests are not based on the assumption that the population is normally distributed (Bethea and Rhinehart, 1991). Hence, the nonparametric tests are more widely applicable than the parametric tests which often require normality in the data. Nevertheless, the nonparametric tests are reported to be less powerful than the parametric tests. To arrive at the same conclusion with the same confidence level, the nonparametric tests require 5 to 35% more data than the parametric tests (Bethea and Rhinehart, 1991).

Only a couple of studies are reported wherein *t*-test has been used to examine the stationarity of hydrologic time series (e.g., Jayawardena and Lai,

1989). The Mann-Whitney test for detecting a shift in the mean or median of hydrological time series has been applied by McCuen and James (1972), Lazaro (1976), Lettenmaier (1976), Helsel and Hirsch (1988), Kiely et al. (1998), Kiely (1999) and Yue and Wang (2002). Also, stationary stochastic models such as AR (Autoregressive), MA (Moving Average), or ARMA (Autoregressive Moving Average) models are frequently used to characterize the standardized time series (Hipel and McLeod, 1994). However, the standardization procedure does not ensure stationarity in the transformed series (Salas, 1993). Moreover, some researchers (Appel and Brandt, 1983; Lovell and Boashash, 1987; Imberger and Ivey, 1991; Chen and Rao, 2002) have developed segmentation algorithms to determine stationary segments and to estimate the parameters characterizing each segment in order to establish piecewise stationary time series models. Two parametric tests and one nonparametric test for checking stationarity in a time series are described in the following sections.

4.2.1 Student's *t*-test

For applying the *t*-test, the series is divided into a number of subseries, and *t*-test is performed to check whether the statistical character of each subseries is significantly different from that of the original series. The null hypothesis that the means of each subseries do not significantly differ from the population mean is examined using the test-statistic, '*ts*', which is defined as follows:

$$ts = \frac{(\bar{x}_t - \mu)\sqrt{N-1}}{\sigma} \tag{25}$$

where $\bar{x}_t$ = mean of the subseries, μ = mean of the series, σ = standard deviation of the series, and N = number of the subseries.

Critical values for the *ts*-statistic can be obtained from standard texts on statistics (e.g., Shahin et al., 1993; Haan, 2002). If the calculated value of '*ts*' is found less than its critical value, the null hypothesis cannot be rejected.

4.2.2 Simple *t*-test

This is a parametric test, which assumes that the annual hydrologic series x_t (t = 1, 2, ..., n) is uncorrelated and normally distributed with mean μ and standard deviation σ. The series is divided into two subseries of sizes n_1 and n_2 such that $n = n_1 + n_2$. The first subseries x_t (t = 1, 2, ..., n_1) has a mean μ_1, and standard deviation σ and the second subseries x_t ($t = n_1$+1, n_1+2, ..., n), is assumed to have mean μ_2 and standard deviation σ. The simple *t*-test can be used to examine the null hypothesis $\mu_1 = \mu_2$ when the two subseries have the same standard deviation. Rejection of the null hypothesis is considered as a detection of a shift. The test-statistic is defined as (Snedecor and Cochran, 1980):

$$ts = \frac{|\bar{x}_2 - \bar{x}_1|}{S\sqrt{\frac{1}{n_1} + \frac{1}{n_2}}} \tag{26}$$

and

$$S = \sqrt{\frac{(n_1 - 1)s_1^2 + (n_2 - 1)s_2^2}{n-2}} \tag{27}$$

where $\bar{x}_1, \bar{x}_2, s_1^2$ and s_2^2 are the estimated means and variances of the first and the second subseries, respectively. Critical values of this test-statistic can be taken from the Student's *t*-distribution standard tables for '*n*-2' degrees of freedom and 5% level of significance. If the computed value of the test-statistic is greater than its critical value, the null hypothesis that both the subseries belong to the same population is rejected.

4.2.3 Mann-Whitney Test

For applying this nonparametric test, the annual hydrologic series x_t (t =1, 2, ..., n) is divided into two subseries $x_1, x_2, \ldots, x_{n1}$ and $x_{n1+1}, x_{n1+2}, \ldots, x_n$ of sizes n_1 and n_2, respectively such that $n_1+n_2 = n$. A new series, z_t (t = 1, 2, ..., n) is defined by arranging the original data (x_t) in increasing order of magnitude. The test-statistic to test the null hypothesis that the mean of the first subseries is equal to the mean of the second subseries is defined as (Snedecor and Cochran, 1980):

$$u_c = \frac{\sum_{t=1}^{n_1} R(x_t) - n_1(n_1 + n_2 + 1)/2}{[n_1 n_2 (n_1 + n_2 + 1)/12]^{1/2}} \tag{28}$$

where $R(x_t)$ is the rank of the observation x_t in ordered series z_t. The null hypothesis of equal means is rejected, if the computed value of this test-statistic is greater than its critical value obtained from the tables of standard normal distribution at 5% significance level.

4.3 Methods for Detecting Trend

A common deterministic component in a time series is a trend. A trend is a tendency for successive values to be increasing or decreasing over time (Haan, 2002). Changes in hydrologic conditions by natural and/or artificial factors can introduce linear or nonlinear trends into a hydrologic time series. The trend in a time series can be expressed by a suitable linear or nonlinear model; the linear model is widely used in hydrology (Shahin et al., 1993). The simplest of linear trend detection models is Student's *t*-test (Hameed et al., 1997), which requires that the series under testing should be normally distributed. Thus, whether or not the sample data follow a normal distribution has to be examined prior to applying the Student's *t*-test to assess the statistical significance of these two types of trends (Hoel, 1954). Unfortunately, some researchers (e.g., Fanta et al., 2001) ignore this important check. If normality is violated, the nonparametric test such as the Mann-Kendall test (Mann, 1945; Kendall, 1975) is commonly applied to assess the statistical significance of trends. This test detects a monotonic trend in the mean or median of a time series. As mentioned earlier, the nonparametric tests are more suitable for non-normal data and censored data compared to the parametric *t*-test (Helsel and Hirsch, 1988; Hirsch and Slack, 1984). The application of the nonparametric Mann-Kendall test for detecting monotonic trends in hydrological time series is reported by Hirsch et al. (1982), Hirsch and Slack (1984), Burn (1994), Burn and Elnur (2002), Lettenmaier et al. (1994), Gan (1992, 1998), Lins and Slack (1999), Douglas et al. (2000), Zhang et al. (2001), Yue et al. (2003), and others. Another important trend test is the Spearman Rank Order Correlation test, which has been applied by Khan (2001) and Adeloye and Montaseri (2002). However, in some hydrologic studies, the Kendall's Rank Correlation test has been preferred (Jayawardena and Lai, 1989; Zipper et al., 1998; Kumar, 2003).

Various parametric and nonparametric statistical tests have been reported in the literature for detecting trend in a hydrologic time series. The parametric statistical tests are: turning point, Kendall's phase, Kendall's rank, regression, Wald-Wolfowitz total number of runs, sum of squared lengths, and inversion tests (Shahin et al., 1993). The nonparametric tests are: Mann-Kendall test for a linear and/or nonlinear trend (Salas, 1993), Hotelling-Pabst test (Conover, 1971), and Sen test (Gilbert, 1987). Some more rank correlation tests have been suggested by Kanji (2001). Dahmen and Hall (1990) present salient established methods to detect the presence of a significant trend in the hydrologic time series.

Most of the tests (i.e., Turning Point, Kendall's Phase, Wald-Wolfowitz Total Number of Runs, Sum of Square Lengths, Adjacency, Difference Sign, Run Test on Successive Differences, Wilcoxon-Mann-Whitney, and Inversions tests) have not attracted the attention of hydrologists, which may be due to the availability of some sound trend detection tests. Esterby (1996) and Hess et al. (2001) present an excellent overview of the statistical methods for trend detection and estimation in environmental time series (e.g., water quality and atmospheric deposition monitoring data). Hess et al. (2001) evaluated six methods of trend detection using real-world data and provided recommendations based on a simulation study. It should be noted that the *t*-test adjusted for seasonality and the Seasonal Kendall tests are more powerful than the remaining four tests viz., the Spearman Partial Rank Correlation test, Ordinary Least Square Regression, Generalized Least Square Regression, and the Kolmogorov-Zurbenko test. However, all the trend detection tests, which are currently available, sound and widely employed in the hydrologic time series analysis, have not been reported by Hess et al. (2001). Some additional statistical tests for trend detection can be found in Mahé et al. (2001). A brief description of 15 tests for trend detection is presented in the succeeding sections.

4.3.1 Regression Test

The most commonly used approach for trend detection is to formulate a linear model between the data and time in the following form (Hameed et al., 1997):

$$x_t = \alpha + \beta \cdot t + \varepsilon_t \tag{29}$$

where x_t (t = 1, 2, ..., n) = observed value at time t, α and β = regression coefficients, and ε_t = a random error (white noise) with a mean of zero and variance of S_v^2.

The data of n years are substituted in the normal equations obtained by the least squares technique, and the parameters $\hat{\alpha}$ and $\hat{\beta}$ are estimated. The sum of squares of the residuals is given by:

$$SS_{res} = \sum_{t=1}^{n}(x_t - \bar{x})^2 - \hat{\beta}^2 \sum_{t=1}^{n}(t - \bar{t})^2 \tag{30}$$

The standard error of regression is calculated as:

$$s = [SS_{res}/(n-2)]^{1/2} \tag{31}$$

The ts-statistic is then computed as:

$$ts = \hat{\beta}/s_{\hat{\beta}} \tag{32}$$

where,

$$s_{\hat{\beta}} = \frac{s}{\left[\sum_{t=1}^{n}(t-\bar{t})^2\right]^{1/2}} \tag{33}$$

If the calculated value of the ts-statistic is less than its critical value at 5% level of significance with n–2 degrees of freedom, the null hypothesis of trend-free series cannot be rejected.

The main problem with the above approach is that it does not distinguish between the trend and the persistence (Hameed et al., 1997). This test can be misleading if seasonal cycles are present, the data are not normally distributed, and the data are serially correlated (Gilbert, 1987).

4.3.2 Spearman Rank Order Correlation Test

To overcome the problem associated with the linear model for trend detection, the Spearman rank order correlation (SROC) nonparametric test (McGhee, 1985) is used to check the existence of long-term trend. Among the available nonparametric trend tests, the World Meteorological Organization (WMO,

1988) recommends the SROC test for detecting trend in flow volumes. The procedures for applying the SROC test are given below (McGhee, 1985).

Let the data series x_t ($t = 1, 2, \ldots, n$) be observed in time t.

Step 1: Assign ranks R_{xt} to x_t such that the largest x_t has $R_{xt} = 1$ and the least x_t has a rank $= n$. If there are ties in the x_t, then assign to each of the ties a rank equal to the mean of the ranks that would have been used had there been no ties.

Step 2: Compute the difference,

$$d_t = R_{xt} - t \tag{34}$$

Step 3: Compute the coefficient of trend (r_s) using the following expression:

$$r_s = 1 - \frac{6 \sum_{t=1}^{n} d_t^2}{n\left(n^2 - 1\right)} \tag{35}$$

Under the null hypothesis (H_0) that the time series has no trend, it can be shown that the statistic, ts has a Student's t-distribution with n–2 degrees of freedom. Here, ts is defined as:

$$ts = r_s \sqrt{\frac{n-2}{1-r_s^2}} \tag{36}$$

Step 4: Calculate the value of ts from Eqn. (36) and get the critical value of the t-distribution for the chosen significance level, α and n–2 degrees of freedom. For a two-tailed test, denote the critical value by $\pm ts_{\alpha/2,n-2}$

Step 5: Finally, compare the computed value of 'ts' with its critical value. Reject null hypothesis (H_0) if $ts > ts_{\alpha/2,\, n-2}$ or $ts < -ts_{\alpha/2,\, n-2}$.

4.3.3 Turning Point Test

Let's assume that a turning point occurs in the series x_t ($t = 1, 2, \ldots, n$) at any time t ($t = 2, 3, \ldots, n-1$) if x_t is larger than each of x_{t-1} and x_{t+1} or x_t is smaller than x_{t-1} and x_{t+1}. This situation has four chances of occurrence in six different possibilities of the occurrence of x_{t-1}, x_t and x_{t+1}, assuming that all three elements have different values. Accordingly, the chance of having a turning point in a sequence of three values is 4/6 or 2/3, for all the values of 't' except for $t = 1$ and $t = n$. In other words, the expected number of turning points ($\bar{p}$) in the given random series can be expressed as:

$$\bar{p} = 2(n-2)/3 \tag{37}$$

For the same random series, variance is given by (Kendall, 1973):

$$\text{var}\,(\bar{p}) = (16n - 29)/90 \tag{38}$$

The test-statistic is represented by the standard normal variate (z), and is given as:

$$z = \frac{|p - \bar{p}|}{\sqrt{\mathrm{var}(\bar{p})}} \tag{39}$$

where p is observed number of turning points.

The computed standard normal variate is then compared with the standard normal variate obtained from the standard table at a given level of significance. If the calculated value of 'z' is within the region of acceptance, the hypothesis of no trend is accepted. If a trend is detected, it can be removed by regression technique (i.e., fitting a suitable equation).

The turning point test is easy to apply, especially when the time series is plotted graphically. It is an effective test for checking randomness against systematic oscillation. But if the turning points tend to bunch together, the Kendall's phase test is more suitable. However, the difficulty with this test is that a comparison of observed and theoretical numbers of phases based on the chi-square criterion is invalidated by the fact that the lengths of phases are not independent. The distribution of the phase lengths does not tend to normality for large lengths of a series, but the number of phases does so (Kendall, 1973).

4.3.4 Kendall's Phase Test

The phase is defined as the interval between any two successive turning points. Let the length of the phase be denoted by 'd'. The expected number of phases (n_p) of length 'd' in a random series of length 'n' is given as (Kendall, 1973):

$$n_p = \frac{2(n-d-2)\left(d^2+3d+1\right)}{(d+3)!} \tag{40}$$

Once n_p is calculated, the observed number of phases and the expected number of phases for a given length is compared. If this difference is large, the series is not considered to be random.

Among the above-mentioned trend tests, the superiority of one over other is mainly associated with the extent of adaptability of a given test to the structure of the time series to be examined. The turning points and number of phases tests are practically out-dated due to the availability of more powerful tests (Shahin et al., 1993), which are described below.

4.3.5 Wald-Wolfowitz Total Number of Runs Test

Let the objective be to test whether the data sample x_t (t = 1, 2, ..., n) is random based on the runs of the data with respect to the median of the observation. The step-by-step procedure for using the Wald-Wolfowitz test is as follows (Shahin et al., 1993):

Step 1: Determine the median of the data. For this, sort the data sample in ascending order such that $x_1 \leq x_2 \leq \ldots \leq x_n$. Now, for an integer k, the sample median ($\hat{x}_{0.5}$) is computed as:

$$\hat{x}_{0.5} = \begin{cases} x_{k+1} & \text{for } n = 2k+1 \\ 0.5\ (x_k + x_{k+1}) & \text{for } n = 2k \end{cases} \tag{41}$$

Step 2: Examine all the data of the series to check whether or not it exceeds the median. A plus sign (+) or minus sign (–) is assigned to every data of the series according to whether its value is greater than or less than the median, respectively. If the median coincides with an observed value (n is odd), neither a plus nor a minus sign is assigned to such a value, implying that the total number of observations is reduced by 1.

Step 3: Count the number of runs ($\underline{U}^+$) of plus signs. A run is defined as a sequence of the entries of same sign until it is interrupted by opposite sign.

Step 4: The mean (μ_{U+}) and the variance (σ^2_{U+}) of the statistic $\underline{U}^+$ are calculated by the following formulae:

$$\mu_{U+} = \frac{1}{2} n + 1 \tag{42}$$

$$\sigma^2_{U+} = \frac{n(n-2)}{4(n-1)} \tag{43}$$

Step 5: Compute the test-statistic (z) using the following formula:

$$z = \frac{\left| \underline{U}^+ - \mu_{U+} \right| - 0.5}{\sigma_{U+}} \tag{44}$$

Step 6: Under the null hypothesis (H_0) that the sequence of (+) and (–) signs is random, z follows a standard normal distribution. Hence, obtain the critical value of the standard normal distribution for a given significance level α and denote it by $\pm z_{\alpha/2}$.

Step 7: If value of z calculated in Step 5 is greater than its critical value, the null hypothesis is rejected.

It should be noted that the Wald-Wolfowitz test does not take into account the length of the runs, and considerable information is ignored. Hence, this test is not very powerful and not efficient, but can be used to determine whether the observations of a random variable are independent. If the observations of a random variable are independent, the time series is said to have no trend (i.e., trend-free).

4.3.6 Sum of Squared Lengths Test

This test considers the length of runs while testing a series for trend. The runs of different lengths are counted. A run consists of a sequence of like signs as defined in the Wald-Wolfowitz test. The test-statistic ($\underline{N}$), the sum of the squares of the run-lengths, is given by

$$\underline{N} = \sum_{j} j^2 n_j \tag{45}$$

where j = length of the run, and n_j = number of runs of length j.

Critical values of the test-statistic at 5% level of significance and $n/2$ degrees of freedom can be obtained from Himmelblau (1969). If the calculated test-statistic values are greater than its critical values, the null hypothesis of trend-free series is rejected. Note that the Sum of Squared Lengths test is more powerful than the Wald-Wolfowitz test (Himmelblau, 1969).

4.3.7 Adjacency Test

The adjacency test is applied to test the null hypothesis that the fluctuations in a series are random in nature. The limitation of this test is the assumption that the observations are obtained independently of each other and under similar conditions (Kanji, 2001). For a time series x_t (t = 1, 2, ..., n), the test-statistic 'z' for $n > 25$ is computed as follows:

$$z = \frac{|L|}{\sigma} \tag{46}$$

wherein L for $n > 25$ is given as:

$$L = 1 - \frac{\sum_{t=1}^{n-1} (x_{t+1} - x_t)^2}{2 \sum_{t=1}^{n} (x_t - \bar{x})^2} \tag{47}$$

It should be noted that L for $n > 25$ follows a normal distribution with zero mean and variance (σ^2) as:

$$\sigma^2 = \sqrt{\frac{(n-2)}{(n-1)\ (n+1)}} \tag{48}$$

Critical values of 'z' can be obtained from the tables for standard normal distribution available in the textbooks on statistics.

For $n \leq 25$, the test-statistic 'z' is computed with L as:

$$L = \frac{\sum_{t=1}^{n-1}(x_{t+1} - x_t)^2}{\sum_{t=1}^{n}(x_t - \bar{x})^2} \tag{49}$$

Critical values of the test-statistic 'z' for $n \leq 25$ can be obtained from Hart (1942). If the computed values of 'z' are less than its critical values, there is no reason to reject the null hypothesis.

4.3.8 Difference Sign Test

The difference sign test is used to examine whether fluctuations in a time series observations are independent of the order in the sequence. This test is applied with the assumption that the number of observations is large and that the observations are obtained under similar conditions (Kanji, 2001). In order to apply this test, first of all, a sequence of successive differences ($x_{t+1} - x_t$) is formed from the sequence of 'n' observations ($x_1, x_2, \ldots, x_n$). Thereafter, the number of '+' signs (n_+) in this derived sequence is counted. For large values of 'n', n_+ may be assumed to follow a normal distribution with the mean (μ_{n+}) and variance (σ^2_{n+}) given as:

$$\mu_{n+} = \frac{(n-1)}{2} \tag{50}$$

$$\sigma^2_{n+} = \frac{(n+1)}{12} \tag{51}$$

The test-statistic or standard normal variate (z) is defined as:

$$z = \frac{|\, n_+ - \mu_{n+} - 0.5 \,|}{\sigma_{n+}} \tag{52}$$

The test-statistic is computed using Eqn. (52) and is compared with 1.64 (i.e., critical value of 'z' at 5% level of significance). If the computed value of 'z' is greater than its critical value, the null hypothesis is rejected, which suggests that the time series has a trend.

4.3.9 Run Test on Successive Differences

The necessary condition for applying this test is that the observations in the sample are obtained under similar conditions. Null hypothesis (H_0) is made that the observations in a time series are independent of the order in the sequence, which is tested by the run test on successive differences. From the sequence of observations x_t ($t = 1, 2, \ldots, n$), a sequence of successive differences ($x_{t+1} - x_t$) is formed (i.e., each observation has the preceding one subtracted from it). The test-statistic (K) is defined as the number of runs of '+' and '–' signs in the sequence of differences.

For $5 \leq n \leq 40$, critical values of the test-statistic can be obtained from Kanji (2001). For $n > 40$, 'K' may be assumed to follow a normal distribution with the mean (μ_k) and variance (σ_k^2) given as:

$$\mu_k = \frac{2n-1}{3} \tag{53}$$

$$\sigma_k^2 = \frac{16n-29}{90} \tag{54}$$

Now, the test-statistic (z) for the case when $n > 40$ is computed by the following expression:

$$z = \frac{|K - \mu_k - 0.5|}{\sigma_k} \tag{55}$$

Critical values of 'z' can be obtained from the standard normal distribution tables available in statistics books. For both the cases of time series sizes, if the computed test-statistic values are less than its critical values, the null hypothesis cannot be rejected. That is, the time series is considered to be random.

4.3.10 Wilcoxon-Mann-Whitney Rank Sum Test

This is a nonparametric test (i.e., distribution-free), which is applicable if the observations are random and independent. It is used to examine whether the occurrence of increasing or decreasing successive values of a time series is random. Consider a time series x_t ($t = 1, 2, \ldots, n$). First, successive observations in the sequence are coded with a '+' or '–' sign by comparing two successive values, and the ranks (1, 2, 3, …, n) are assigned to all the observations of the series. Thereafter, the number of '+' and '–' signs are counted and the larger of two numbers is noted down (say n_1). If n_2 be the number of opposite signs, then $n = n_1 + n_2$. Now, from the integers describing the natural order of signs, the rank sum 'R_1' of 'n_2' signs is determined. Finally, the value of 'R_2' statistic is calculated by the following expression:

$$R_2 = n_2(n+1) - R_1 \tag{56}$$

The smaller of 'R_1' and 'R_2' is used as the test-statistic. The critical values of the test-statistic can be found in Natrella (1963). If the computed value of the test-statistic is greater than its critical value, the null hypothesis of no trend is rejected.

4.3.11 Inversions Test

This test is used to examine the linear trend in a time series. It is almost similar to the Kendall's rank correlation test (described ahead). The number of x_j-values ($j > i$) each smaller than a chosen x_i-value is counted for all i's

($i = 1, 2, \ldots, n-1$) and summed up. The total number of inversions denoted by $\underline{I}^*$ has a mean (μ_{I*}) and variance (σ^2_{I*}) given as follows (Shahin et al., 1993):

$$\mu_{I*} = \frac{n(n-1)}{4} \tag{57}$$

and

$$\sigma^2_{I*} = \frac{2n^3 + 3n^2 - 5n}{72} \tag{58}$$

As $n \to \infty$, the approximate standard normal variate (z) can be computed as:

$$z = \frac{\underline{I}^* - \mu_{I*}}{\sigma_{I*}} \tag{59}$$

If the computed z-value is within the acceptable range, i.e., ±1.96, which are the critical values for the two-sided test at 5% significance level, the null hypothesis that the time series is trend-free cannot be rejected.

4.3.12 Kendall's Rank Correlation Test

The Kendall's rank correlation test is mostly preferred for trend detection in hydrologic time series (Jayawardena and Lai, 1989; Zipper et al., 1998; Kumar, 2003). In this test, a null hypothesis of no trend is initially assumed, and then the test is carried out to reject or accept the hypothesis. If a series x_t ($t = 1, 2, \ldots, n$) is to be examined, the number of times ($\underline{p}$) that $x_j > x_i$ is counted in all pairs of observations (x_i, x_j). The Kendall's test-statistic (τ) is defined as follows:

$$\tau = \frac{4\underline{p}}{n(n-1)} - 1 \tag{60}$$

The test-statistic is then expressed as a standard normal variate in the following form (Kendall, 1973):

$$z = \frac{\tau}{\sqrt{\mathrm{Var}(\tau)}} \tag{61}$$

where

$$\mathrm{Var}(\tau) = \frac{2(2n+5)}{9n(n-1)} \tag{62}$$

If the value of 'z' lies within the limits ±1.96 at the 5% significance level, the null hypothesis of no trend cannot be rejected.

4.3.13 Mann-Kendall Test

This is a nonparametric test for exploring a trend in a time series without specifying the type of trend (i.e., linear or nonlinear). Mann (1945) originally used this test and Kendall (1975) subsequently derived the test-statistic distribution. This test has been found to be an excellent tool for trend detection (e.g., Hirsch et al., 1982; Gan, 1992). Considering the time series x_t ($t = 1, 2, \ldots, n$), each value of the series (x_t) is compared with all subsequent values (x_{t+1}) and a new series z_k is generated as follows (Salas, 1993):

$$\begin{aligned} z_k &= 1 \quad \text{for } x_t > x_{t'} \\ z_k &= 0 \quad \text{for } x_t = x_{t'} \\ z_k &= -1 \quad \text{for } x_t < x_{t'} \end{aligned} \tag{63}$$

where k is given by:

$$k = (t'-1)(2n-t')/2+(t-t') \tag{64}$$

The Mann-Kendall statistic (S) is defined as follows (Hirsch et al., 1982):

$$S = \sum_{t'=1}^{n-1} \sum_{t=t'+1}^{n} z_k \tag{65}$$

Thus, this statistic represents the number of positive differences minus the number of negative differences for all the differences under consideration.

Moreover, the above test-statistic for $n > 40$ may be written as (Hirsch et al., 1982):

$$u_c = \frac{S+m}{\sqrt{V(S)}} \tag{66}$$

where $$V(S) = \frac{1}{18}\left[n(n-1)(2n+5) - \sum_{i=1}^{g} e_i (e_i - 1)(2e_i + 5) \right] \tag{67}$$

In Eqns (66) and (67), $m = 1$ for $S < 0$ and $m = -1$ for $S > 0$, g is the number of tied groups, and e_i is the number of data in the ith tied group. The value of the test-statistic u_c is taken as zero for $S = 0$. Now, if the computed absolute value of u_c is greater than the critical value of the standard normal distribution, the hypothesis of an upward or downward trend cannot be rejected at the α significance level. It should be noted that Kendall (1975) suggested for using the Mann-Kendall test even for n values as low as 10 provided there are not too many tied values. Hirsch et al. (1982) reported the application of this test to seasonal time series.

4.3.14 Sen's Slope Estimation Test

Kendall slope (β), initially given by Sen (1968) and later extended by Hirsch et al. (1982), is a useful index to quantify monotone trend in the hydrologic time series (Hirsch et al., 1982; Gan, 1998). Sen's test for the estimation of slope requires a time series of equally spaced data. The slope is estimated by:

$$\beta_{gk} = \text{Median}\left(\frac{x_{igk} - x_{jgk}}{i - j}\right) \text{ for all } i < j \tag{68}$$

where β_{gk} = slope between data points x_{igk} and x_{jgk}, x_{igk} = data measurement at time i, x_{jgk} = data measurement at time j; and j = time after time i; g = season; and k = site. It is defined as the estimator β is the median overall combination of record pairs for the whole dataset, and is resistant/robust to the extreme observations or outliers. The positive value of the β connotes the slope of the upward trend and negative value for the downward trend.

4.3.15 Trend-Homogeneity Test

The trend results of the seasonal data are assumed to be homogeneous to find the overall trend by summing the trends over the seasons for a given station (Hirsch and Slack, 1984). However, presence of noticeable upward and downward seasonal trends may even result in an overall no trend in the series due to summing (Van Belle and Hughes, 1984). Hence, the test of homogeneity is conducted to interpret the seasonal and spatial variability of trend results and also their interactions. The χ^2-based homogeneity test partitions the total sum of square (χ^2_{Total}) into $\chi^2_{\text{Homogeneity}}$ and χ^2_{Trend} (e.g., Van Belle and Hughes, 1984; Gan, 1998; Kahya and Kalayci, 2004). The $\chi^2_{\text{Homogeneity}}$ is further partitioned into χ^2_{Season}, χ^2_{Station} and $\chi^2_{\text{Season-Station}}$, and are used to test the significance of heterogeneity of season, station and their interactions, respectively. The χ^2-based test statistics to carry out homogeneity test are given below:

$$\chi^2_{\text{Total},\, pq} = \sum_{s=1}^{p}\sum_{t=1}^{q} Z_{\text{st}}^2 \tag{69}$$

$$\chi^2_{\text{Homogeneity},\, pq-1} = \sum_{s=1}^{p}\sum_{t=1}^{q} (Z_{\text{st}} - \overline{Z})^2 \tag{70}$$

$$\chi^2_{\text{Season},\, p-1} = q\sum_{s=1}^{p} (Z_{\text{s}} - \overline{Z})^2 \tag{71}$$

$$\chi^2_{\text{Station},\, q-1} = p\sum_{t=1}^{q} (Z_{\text{t}} - \overline{Z})^2 \tag{72}$$

$$\chi^2_{\text{Season-Station},\, (p-1)(q-1)} = \sum_{s=1}^{p}\sum_{t=1}^{q} (Z_{\text{st}} - Z_{\text{s}} - Z_{\text{t}} + \overline{Z})^2 \tag{73}$$

$$\chi^2_{\text{Trend},\, 1} = pq\overline{Z}^2 \tag{74}$$

where the standardized Mann-Kendall test-statistics for *s* seasons and *t* stations has been defined as Z_{st}. It is to note that the Z_{st} is computed by using either Eqn. (65) for sample size (n) < 40 or Eqn. (66) otherwise. Here Z_s is the average of Z_{st} for season *s* over the stations; Z_t is the average of Z_{st} for station *t* over the seasons; and $\bar{Z}$ is the average of Z_{st} over the season-stations. The computed test-statistic values are compared with the critical values of the chi-square distribution at 5% significance level and specific degree of freedom (*df*). The *df* for all the test-statistics are denoted as subscript in left side expression [Eqns 69 to 74], for example, the *df* for χ^2_{Total} is pq and for $\chi^2_{\text{Homogeneity}}$ is $pq-1$. Finally, if the computed absolute value of a test-statistic is greater than the critical value of the chi-square distribution, the hypothesis of a trend cannot be rejected at the 5% significance level.

4.4 Methods for Checking Periodicity

Periodicity in the hydrologic time series can be detected if the time series are defined at time intervals less than a year; in most cases, six and 12 months periodicity is very common. The Fourier series has been mainly used for the detection of periodic components in the hydrologic time series (e.g., Maidment and Parzen, 1984; Kite, 1989; Jayawardena and Lai, 1989; Fernando and Jayawardena, 1994; Pugacheva et al., 2003). However, some researchers have suggested 'coherence plot' and 'periodic autocorrelation function' methods for testing the periodically correlated time series (e.g., Hurd and Gerr, 1991; Vecchia and Ballerini, 1991).

Periodicity is detected through harmonic analysis using the well-known Fourier series. If a periodicity exists in a trend-free time series, it can be represented by a Fourier series, which is expressed as follows (Stein and Weiss, 1971; Shahin et al., 1993; Howell, 2001):

$$x(t) = A_0 + \sum_{k=1}^{h} [A_k \sin(2\pi kt/P) + B_k \cos(2\pi kt/P)] \tag{75}$$

where $x(t)$ = harmonically fitted means at period $t(t = 1, 2, ..., P)$, A_0 = population mean, h = total number of harmonics [$h = P/2$ for even P and $(P+1)/2$ for odd P]. P = base period or period of the function and A_k and B_k are sine and cosine Fourier coefficients, respectively.

Here, A_o, A_k and B_k are computed as (Shahin et al., 1993):

$$A_0 = (1/P) \sum_{t=1}^{P} \bar{x}_t \tag{76}$$

$$A_k = (2/P) \sum_{t=1}^{P} \bar{x}_t \sin(2\pi kt/P), \quad k = 1, 2, ..., P/2-1 \tag{77}$$

$$B_k = (2/P)\sum_{t=1}^{P}\bar{x}_t \cos(2\pi kt/P), \quad k = 1, 2, ..., P/2-1 \tag{78}$$

For the monthly hydrologic records, $P = 12$, and hence $h = 6$. As it may not always be necessary to expand the Fourier series up to the maximum number of harmonics, the maximum number of significant harmonics (h^*) can be obtained by examining the cumulative periodogram, which determines the relative significance of each harmonic. To accomplish this task, firstly the variations caused by a periodic component (I_k), say k^{th} harmonic, is computed as (Shahin et al., 1993; Haan, 2002):

$$I_k = (1/2)(A_k^2 + B_k^2) \tag{79}$$

Secondly, the periodogram is obtained by plotting I_k against $2\pi k/P$ for $k = 1, 2, ..., N/2$. Thirdly, the cumulative periodogram (P_j) shows a rapidly rising part up to h^* and then gradually increases up to its maximum value of unity and is calculated as (Jayawardena and Lai, 1989):

$$P_j = \sum_{k=1}^{j}(A_k^2 + B_k^2)/2 \bigg/ \sum_{t=1}^{P}(\bar{x}_t - \mu)^2/P \tag{80}$$

where μ is the mean of $\bar{x}_t$. It should be noted that in the numerator of Eqn. (80), the terms under summation should be arranged in decreasing order of their magnitudes. Finally, the significance of different harmonics is tested using the Fisher's g-statistic, which is given as (Yevjevich, 1972; Shahin et al., 1993):

$$g_k = I_k \bigg/ \sum_{k=1}^{P/2}(A_k^2 + B_k^2)/2 \tag{81}$$

When the g-statistic (i.e., Eqn. 81) is multiplied by 100, it gives the percent contribution of the kth harmonic. It is worth mentioning that the quantity $\sum_{k=1}^{P/2}(A_k^2 + B_k^2)$ of Eqn. (81), when multiplied by 'P' is used as a measure of the variation caused by a given periodic component (Shahin et al., 1993). On the other hand, the total variations in a series are expressed as $\sum_{t=1}^{P}(\bar{x}_t - \mu)^2$, which is the denominator of Eqn. (80) multiplied by 'P'. Thus, if these two terms are equal, the total variations in the series can be explained by the periodic components only considering its all harmonics, and under such circumstances the cumulative periodogram is given as follows:

$$P_j = 100 \times \sum_{k=1}^{j} g_k \tag{82}$$

4.5 Methods for Persistence Testing

Persistence can sometimes be treated as periodicity. In many hydrologic time series studies, no distinction is made between persistence and randomness (McMohan and Mein, 1986; Aksoy, 2007). Therefore, the tests to examine the randomness of a hydrologic time series are used for detecting both trend and persistence (Machiwal and Jha, 2006). Generally, *randomness* or *non-persistence* is defined as the independence among data in a time series. On the contrary, the series is called *persistent* if the data in the series are dependent on each other. Practically, *persistence is a tendency of the successive values of a time series to 'remember' their antecedent values and to be influenced by them* (Giles and Flocas, 1984). Mathematically, *persistence is defined as the correlational dependency of order k between each ith element and the $(i\text{-}k)^{th}$ element of the series* (Kendall, 1973), and is measured by autocorrelation (i.e., correlation between two terms of the same time series). Here, '*k*' is usually called *time lag*. The detection of persistence can be made by autocorrelation technique (time domain) and/or spectral technique (frequency domain). However, the autocorrelation technique has been applied in several studies such as Mirza et al. (1998), Maidment and Parzen (1984), Schwankl et al. (2000), etc. Here, it is worth mentioning that some researchers (e.g., Jayawardena and Lai, 1989) have used the autocorrelation technique for testing periodicity in hydrologic time series. Such a misconception is quite common in the analysis of hydrologic time series (Machiwal and Jha, 2006).

The persistence test of a time series can be performed in two ways: (i) time domain (autocorrelation technique), and (ii) frequency domain (spectral technique). However, some investigators (e.g., Quimpo, 1968) suggest the application of autocorrelation technique only because the spectral technique alone cannot be used without knowing the autocorrelation in a series. This is due to the fact that the spectral density is a Fourier transform of the autocorrelation function. The autocorrelation and spectral techniques for examining the persistence in a time series are described below.

4.5.1 Autocorrelation Technique

The autocorrelation function in essence expresses the degree of temporal dependency among observations. It is actually a process of self-comparison, expressing the linear correlation between an equally-spaced series and the same series at a specified time lag or separation (Jenkins and Watts, 1968). If $x_0, x_1, x_2, ..., x_{n-1}$ is a realization of a stationary stochastic process, the covariance between x_t and its value x_{t+k}, separated by a time interval k, is known as the *population autocovariance* (γ_k) and is mathematically expressed as follows (Box and Jenkins, 1976):

$$\gamma_k = E[(x_t - \mu)(x_{t+k} - \mu)] \tag{83}$$

where x_t = value of variable at the tth location in time; k = time lag, μ = population mean and E = expectation operator. Furthermore, the *population autocorrelation function* (ρ_k) is defined as a ratio of the population autocovariance (ρ_k) to the population variance [Var(x_t)]. That is,

$$\rho_k = \frac{\gamma_k}{\mathrm{Var}(x_t)} \tag{84}$$

It should be noted that the *population autocorrelation* (ρ_k) can also be estimated by the *serial autocorrelation function* (r_k) from sample data using the following expression (Shahin et al., 1993; Haan, 2002):

$$r_k = \frac{\sum_{t=0}^{n-k}(x_t \cdot x_{t+k}) - 1/(n-k)\sum_{t=0}^{n-k} x_t \cdot \sum_{t=0}^{n-k} x_{t+k}}{\left[\sum_{t=0}^{n-k} x_t^2 - 1/(n-k)\left(\sum_{t=0}^{n-k} x_t\right)^2\right]^{1/2}\left[\sum_{t=0}^{n-k} x_{t+k}^2 - 1/(n-k)\left(\sum_{t=0}^{n-k} x_{t+k}\right)^2\right]^{1/2}} \tag{85}$$

For $k = 0$, Eqns (84 and 85) result in $\rho_0 = r_0 = 1$. As the lag (k) increases, the number of the pairs of elements used in calculating r_k decreases. It is a common practice to set the upper limit of lag between $0.1n$ to $0.25n$, depending on the size (n) of the series (Matalas, 1967). The detailed information about the internal structure of the time series can be obtained by examining autocorrelogram, which is drawn with an array of autocorrelation coefficients (i.e., ρ_0, ρ_1, ...) as ordinates and k as abscissa.

The upper and lower critical values of autocorrelation function can be obtained from the Anderson's test as follows (Anderson, 1942):

$$(r_k)_{\text{upper}} = \{1/(n-k)\}(-1 + z_{1-\alpha/2}\sqrt{n-k-1}) \tag{86}$$

$$(r_k)_{\text{lower}} = \{1/(n-k)\}(-1 - z_{1-\alpha/2}\sqrt{n-k-1}) \tag{87}$$

where $z_{1\text{-}\alpha/2}$ = standard normal variate at α significance level. If the value of r_k obtained from Eqn. (85) falls within the critical value given by either Eqn. (86) or Eqn. (87), the null hypothesis that (ρ_k) is zero is rejected. This indicates that the series is not purely random and some persistence exists.

4.5.2 Spectral Technique

The spectral analysis technique can be considered as an alternative to the autocorrelation technique where the spectral density function replaces the Fourier transformation of the autocorrelation function (Shahin et al., 1993). The spectrum of a time series can be defined by harmonic analysis. The basic function of the spectrum is to decompose a time series on a frequency basis, and then frequencies and amplitudes of the series can be estimated, if they are present (Kottegoda, 1980). Therefore, if one or more persistence are present in the time series, spectral technique is often used. A non-stationary periodic time series $x(t)$ may be expanded into a Fourier series using following expression:

$$x(t) = A_0 + \sum_{k=1}^{\infty} [A_k \sin(2\pi k f t) + B_k \cos(2\pi k f t)] \tag{88}$$

where f = frequency and rest of the parameters are the same as expressed in Eqn. (75). The frequency can be given as:

$$f = \frac{1}{P} = \frac{\omega}{2\pi} \tag{89}$$

where P = period of the function (base period) and ω = angular frequency. With reference to Eqn. (89), two spectral density functions $S(\omega)$ and $S(f)$ can be related as follows:

$$S(\omega) = 2\pi S(f) \tag{90}$$

The density function $S(\omega)$ is related to Fourier transformation $F(\omega)$ as follows:

$$S(\omega) = \frac{\partial}{\partial \omega} F(\omega) \tag{91}$$

Similarly, $$S(f) = \frac{\partial}{\partial f} F(f) \tag{92}$$

Considering an infinitesimal portion of frequency in the range of $(f, f + df)$, the spectrum $S(f)\, df$ represents the contribution of components with frequencies in the range $(f, f + df)$ to the total variance.

The spectral density function $S(f)$ for a discrete process can be written as (Box and Jenkins, 1976):

$$S(f) = \gamma_0 + 2\sum_{k=1}^{\infty} \gamma_k \cos(2\pi f k) \quad -\infty < f < \infty \tag{93}$$

However, the spectral density function for a continuous process can be written as:

$$S(f) = \int_{-\infty}^{\infty} \rho(k) \cos(2\pi f k)\, dk \quad -\infty < f < \infty \tag{94}$$

Negative frequencies obtained with above expressions can be avoided and integral of the normalized spectral density function can be maintained over the entire range at one, a one-sided spectral density function $G(f) = 2S(f)$ can be written, where f varies only over $(0, \infty)$ and zero elsewhere (Shahin et al., 1993). The function $G(f)$ may be expressed as:

$$G(f) = 2\int_0^{\infty} \rho(k)\cos(2\pi f k)dk \quad 0 < f < \infty \tag{95}$$

The above function may be discretized as follows:

$$G(f) = 2\left(\gamma_0 + 2\sum_{k=1}^{\infty} \gamma_k \cos(2\pi f k)\right) \quad 0 < f < \infty \tag{96}$$

The above analysis considers entire population of the variable, and for a finite time series of size n, the γ_k's are replaced with their sample estimates C_k's and γ_0 is replaced with C_0. Equation (96) for a finite time series may be reduced to the following:

$$\hat{G}(f) = 2\left(C_0 + 2\sum_{k=1}^{n-1} C_k \cos(2\pi f k)\right) \tag{97}$$

In Equation (97), an increase in value of k results in reduced precision in estimating the value of C_k. Hence, there is a need to give more weight to C_k values for small k's and less weight to C_k values for larger k's. This is normally done by introducing a set of weights λ_k, known as *lag window*, and by truncating the upper limit of the summation in Equation (97) to k values less than $(n-1)$. Generally, k is chosen equal to or less than $n/4$. Equation (97) can be more conveniently expressed in terms of *autocorrelation function* (ρ_k) rather in terms of *autocovariances* (C_k). The *serial correlation coefficient* (r_k) based on sample data is used as an estimate of the *autocorrelation coefficient* for the same lag k. Considering all these statements, Equation (97) can be rewritten as:

$$\bar{G}(f) = 2\left(r_0 + 2\sum_{k=1}^{m} \lambda_k r_k \cos(2\pi f k)\right) \tag{98}$$

For autocorrelation, $r_0 = 1$, Equation (98) can be written as:

$$\bar{G}(f) = 2\left(1 + 2\sum_{k=1}^{m} \lambda_k r_k \cos(2\pi f k)\right) \tag{99}$$

where $\bar{G}(f)$ = smoothed spectral ordinate.

Of the various available *lag window expressions*, the mostly used are 'Parzen window' (Parzen, 1963), 'Tukey window' (Blackman and Tukey, 1959), and 'hamming and hanning procedures' (Shahin et al., 1993).

4.6 Merits and Demerits of Time Series Methods

Based on the experiences of world-wide researchers and scientists in the area of time series analysis, the following merits and demerits of time series analysis methods can be identified. Firstly, the merits and demerits of time series analysis as a whole are described, and thereafter the merits and demerits of individual tests used for time series analysis are highlighted:

(i) The assumptions of the classical parametric tests viz., normality, linearity, and independence are usually not met by the hydrological time series data, especially in case of surface water quality data. Therefore, recently some nonparametric tests have been proposed to determine the trend in surface water quality time series (Kalayci and Kahya, 1998). At the same time, the statistical tests for trend detection in water quality are normally confounded by one or more of the following problems: missing values, censored data, flow relatedness, and seasonality.

(ii) In general, the parametric methods to assess significance of trend employ pre-specified models and associated tests, whereas the nonparametric methods generally apply rank tests to the data. Neither approach is suitable for exploratory analysis (Ramesh and Davison, 2002).

(iii) Cumulative Deviations test is superior to the classical von Neumann test for a model with only one change in the mean (Buishand, 1982).

(iv) The major limitation with all the multiple comparison tests of homogeneity (i.e., Tukey, Link-Wallace, Dunnett, Bartlett and Hartley tests) is the requirement that populations should be normally distributed with equal variances, which makes the tests parametric in nature. Although the Link-Wallace test, the Dunnett's test and the Hartley's test can be employed for the same purpose as the Tukey's test, the former three tests can be applied only when the sample size of all populations is equal, though methodology of the Hartley's test can still be followed in case sample size of all the populations are more or less similar.

(v) Of the three stationarity tests, both the *t*-tests are parametric in nature, which require normality assumption of the time series to be tested. However, the Mann-Whitney test is nonparametric in nature and is more robust as it can be applied to normal as well as normal non-normal time series.

(vi) Although the linear model (i.e., Regression test) is most commonly used for trend detection, it has a demerit that it does not distinguish between trend and persistence. The linear model can also be misleading if seasonal cycles are present, the data are not normally distributed, and/or the data are serially correlated (Gilbert, 1987). The Spearman Rank Order Correlation (SROC) test overcomes these demerits of the linear model. The merit of this test is its nearly uniform power for detecting linear as well as nonlinear trends (WMO, 1966; Dahmen and Hall, 1990). Among the trend tests, the superiority of one over other is mainly associated with the extent of adaptability of a chosen test to the structure of the time series to be tested.

(vii) The Turning Point test is easy to apply, especially when the time series is plotted graphically. It is an effective test for randomness against systematic oscillation. However, if the turning points tend to bunch together, the Kendall's Phase test is more relevant (Shahin et al., 1993). The demerit of the Kendall's phase test is that a comparison of observed and theoretical numbers of phases by the usual chi-square test is invalidated due to the fact that the lengths of phases are not independent. Also, the distribution of phase lengths does not tend to be normal for large lengths of a series, but the number of phases follows a normal distribution (Kendall, 1973). The Turning Point test and Kendall's phase test are practically out-dated due to the availability of much more powerful tests (Shahin et al., 1993).

(viii) The Wald-Wolfowitz test has demerit that it does not consider length of runs and significant information about the time series is ignored. Hence, this test is not very powerful nor efficient. The Sum of Squared Lengths test is more powerful than the Wald-Wolfowitz test (Himmelblau, 1969).

(ix) The Adjacency test for checking trends has demerit that it inherently considers that the time series data points are independent and the data are collected under uniform conditions (Kanji, 2001), which may not be true for the real-world time series.

(x) The Difference Sign test for trend detection has a demerit almost similar to the Adjacency test that the data points of the time series are collected under uniform conditions and the number of data points is large, which may not be true for short-term time series where data were collected under dissimilar conditions.

(xi) The Run test on Successive Differences has a demerit very similar to the Adjacency test that the time series data are collected under uniform conditions, which may not be true for real-world time series.

(xii) The Wilcoxon-Mann-Whitney test has merit that it is a nonparametric test (i.e., distribution-free) and hence, it may be applicable to normal and non-normal time series. However, this test has demerit that it considers that the time series data points are random and independent of time, which may not be true for an autocorrelated or persistent time series.

(xiii) The Kendall's Rank Correlation test is one of the most powerful tests for trend checking in the hydrologic time series.

(xiv) The Mann-Kendall test is a nonparametric test for trend detection in a time series without specifying whether the trend is linear or nonlinear. Hence, this test has an advantage of being applicable for non-normal as well as normal time series. The nonparametric nature of the test avoids testing of normality in the time series. However, the existence of serial correlation in a time series may affect the ability of the Mann-Kendall test to assess the site significance of a trend, and the presence of cross correlation among sites in a network may influence the ability of the test to evaluate the field significance of trends over the network (Yue et al., 2003). The effect of serial correlation on the Mann-Kendall test can be eliminated by using trend-free pre-whitening procedure as suggested by Yue et al. (2003).

References

Adeloye, A.J. and Montaseri, M. (2002). Preliminary streamflow data analyses prior to water resources planning study. *Hydrological Sciences Journal*, **47(5):** 679-692.

Aksoy, H. (2007). Hydrological variability of the European part of Turkey. *Iranian Journal of Science & Technology*, **31(B2):** 225-236.

Anderson, R.L. (1942). Distribution of the serial correlation coefficient. *Annals of Mathematical Statistics*, **13:** 1-13.

Appel, U. and Brandt, A.V. (1983). Adaptive sequential segmentation of piecewise stationary time series. *Information Sciences*, **29:** 27-56.

Bethea, R.M. and Rhinehart, R.R. (1991). Applied Engineering Statistics. Marcel Dekker, Inc., New York.

Blackman, R.T. and Tukey, J.W. (1959). The Measurement of Power Spectra from the Point of View of Communication Engineering. Dover Publication, U.K.

Box, G.E.P. and Jenkins, G.M. (1976). Time Series Analysis: Forecasting and Control. Holden-Day, San Francisco, 575 pp.

Buishand, T.A. (1982). Some methods for testing the homogeneity of rainfall records. *Journal of Hydrology*, **58:** 11-27.

Buishand, T.A. (1984). Tests for detecting a shift in the mean of hydrological time series. *Journal of Hydrology*, **73:** 51-69.

Burn, D.H. (1994). Hydrologic effects of climatic change in west-central Canada. *Journal of Hydrology*, **160(1-4):** 53-70.

Burn, D.H. and Elnur, M.A.H. (2002). Detection of hydrologic trends and variability. *Journal of Hydrology*, **255(1-4):** 107-122.

Chen, H.-L. and Rao, A.R. (2002). Testing hydrologic time series for stationarity. *Journal of Hydrologic Engineering, ASCE*, **7(2):** 129-136.

Chernoff, H. and Zacks, S. (1964). Estimating the current means of a normal distribution which is subjected to change in time. *Annals of Mathematical Statistics*, **35:** 999-1018.

Conover, W.J. (1971). Practical Non-Parametric Statistics. Wiley, New York.

Dahmen, E.R. and Hall, M.J. (1990). Screening of Hydrologic Data: Tests for Stationarity and Relative Consistency. ILRI Publication No. 49, Wageningen, the Netherlands, 60 p.

Doob, J.L. (1949). Heuristic approach to the Kolmogorov-Smirnov theorems. *Annals of Mathematical Statistics*, **20:** 393-403.

Douglas, E.M., Vogel, R.M. and Kroll, C.N. (2000). Trends in floods and low flows in the United States: Impact of spatial correlation. *Journal of Hydrology*, **240(1-2):** 90-105.

Esterby, S.R. (1996). Review of methods for the detection and estimation of trends with emphasis on water quality applications. *Hydrological Processes*, **10(2):** 127-149.

Fanta, B., Zaake, B.T. and Kachroo, R.K. (2001). A study of variability of annual river flow of the southern African region. *Hydrological Sciences Journal*, **46(4):** 513-524.

Fernando, D.A.K. and Jayawardena, A.W. (1994). Generation and forecasting of monsoon rainfall data. Proceedings of the 20th WEDC Conference on Affordable Water Supply and Sanitation, Colombo, Sri Lanka, pp. 310-313.

Gan, T.Y. (1992). Finding trends in air temperature and precipitation for Canada and north-eastern United States. *In*: G.W. Kite and K.D. Harvey (editors), Using Hydrometric Data to Detect and Monitor Climatic Change. Proceedings of the NHRI Workshop No. 8, National Hydrology Research Institute, Saskatoon, pp. 57-78.

Gan, T.Y. (1998). Hydroclimatic trends and possible climatic warming in the Canadian Prairies. *Water Resources Research*, **34(11):** 3009-3015.

Gardner Jr., L.A. (1969). On detecting changes in the mean of normal variates. *Annals of Mathematical Statistics*, **40:** 116-126.

Gilbert, R.O. (1987). Statistical Methods for Environmental Pollution Monitoring. Van Nostrand Reinhold. New York.

Giles, B.D. and Flocas, A.A. (1984). Air temperature variation in Greece, Part-I: Persistence, trend and fluctuations. *International Journal of Climatology*, **4:** 531-539.

Haan, C.T. (2002). Statistical Methods in Hydrology. Second edition, Iowa State University Press, Ames, Iowa, USA, 496 pp.

Hameed, T., Marino, M.A., DeVries, J.J. and Tracy, J.C. (1997). Method for trend detection in climatological variables. *Journal of Hydrologic Engineering, ASCE*, **2(4):** 157-160.

Hart, B.I. (1942). Significance levels for the ratio of the mean square successive difference to the variance. *Annals of Mathematical Statistics*, **13:** 445-447.

Helsel, R.D. and Hirsch, R.M. (1988). Discussion of "Applicability of the t-test for detecting trends in water quality variables" by Robert H. Montgomery and Jim C. Loftis. *Water Resources Bulletin*, **24(1):** 201-204.

Hess, A., Iyer, H. and Malm, W. (2001). Linear trend analysis: A comparison of methods. *Atmospheric Environment*, **35(30):** 5211-5222.

Himmelblau, D.M. (1969). Process Analysis by Statistical Methods. John Wiley and Sons, New York.

Hipel, K.W. and McLeod, A.I. (1994). Time Series Modeling of Water Resources and Environmental Systems. Elsevier, Amsterdam, the Netherlands, pp. 463-465.

Hirsch, R.M. and Slack, J.R. (1984). Nonparametric trend test for seasonal data with serial dependence. *Water Resources Research*, **20(6):** 727-732.

Hirsch, R.M., Slack, J.R. and Smith, R.A. (1982). Techniques of trend analysis for monthly water quality data. *Water Resources Research*, **18(1):** 107-121.

Hoel, P.G. (1954). Introduction to Mathematical Statistics. Second edition, John Wiley and Sons, Inc., New York.

Howell, K.B. (2001). Principles of Harmonic Analysis. Chapman and Hall/CRC Press, Washington, D.C., 769 pp.

Hurd, H.L. and Gerr, N.L. (1991). Graphical methods for determining the presence of periodic correlation. *Journal of Time Series Analysis*, **12:** 337-350.

Imberger, J. and Ivey, G.N. (1991). On the nature of turbulence in a stratified fluid, Part II: Applications to lakes. *Journal of Physical Oceanography*, **21(5):** 659-680.

Jayawardena, A.W. and Lai, F. (1989). Time series analysis of water quality data in Pearl river, China. *Journal of Environmental Engineering, ASCE*, **115(3):** 590-607.

Jayawardena, A.W. and Lau, W.H. (1990). Homogeneity tests for rainfall data. *Hong Kong Engineer*, The Hong Kong Institution of Engineers, **18:** 22-25.

Jenkins, G.M. and Watts, D.G. (1968). Spectral Analysis and Its Applications. Holden-Day, San Francisco, 525 pp.

Kahya, E. and Kalayci, S. (2004). Trend analysis of streamflow in Turkey. *Journal of Hydrology*, **289(1-4):** 128-144.

Kalayci, S. and Kahya, E. (1998). Detection of water quality trends in the rivers of the Susurluk Basin. *Turkish Journal of Engineering and Environmental Sciences*, **22(6):** 503-514.

Kanji, G.K. (2001). 100 Statistical Tests. Sage Publication, New Delhi, 111 pp.

Kendall, M.G. (1973). Time Series. Charles Griffin and Co. Ltd., London, U.K.

Kendall, M.G. (1975). Rank Correlation Methods. Charles Griffin and Co. Ltd., London, U.K.

Khan, A.R. (2001). Analysis of Hydro-Meteorological Time Series in the Upper Indus Basin: Searching Evidence for Climatic Change. International Water Management Institute, Working Paper 23, Colombo, Sri Lanka, 31 p.

Kiely, G. (1999). Climate change in Ireland from precipitation and streamflow observations. *Advances in Water Resources*, **23:** 141-151.

Kiely, G., Albertson, J.D. and Parlange, M.B. (1998). Recent trends in diurnal variation of precipitation at Valentia on the west coast Ireland. *Journal of Hydrology*, **207:** 270-279.

Kite, G. (1989). Use of time series analyses to detect climatic change. *Journal of Hydrology*, **111:** 259-279.

Kottegoda, N.T. (1980). Stochastic Water Resources Technology. McMillan & Co. Ltd., London, U.K.

Kumar, V. (2003). Rainfall characteristics of Shimla district (H.P.). *Journal of Indian Water Resources Society*, **23(1):** 1-10.

Lazaro, T.R. (1976). Nonparametric statistical analysis of annual peak flow data from a recently urbanized watershed. *Water Resources Bulletin*, **12(1):** 101-107.

Lettenmaier, D.P. (1976). Detection of trends in water quality data from records with dependent observations. *Water Resources Research*, **12(5):** 1037-1046.

Lettenmaier, D.P., Wood, E.F. and Wallis, J.R. (1994). Hydro-climatological trends in the continental United States, 1948-88. *Journal of Climate*, **7:** 586-607.

Lins, H.F. and Slack, J.R. (1999). Streamflow trends in the United States. *Geophysical Research Letters*, **26(2):** 227-230.

Lovell, B. and Boashash, B. (1987). Evaluation of criteria for detection of changes in nonstationary signals. Proceedings of the 1st IASTED International Symposium on Signal Processing and its Applications (ISSPA), Brisbane, August 24-28, 1987, pp. 291-296.

Machiwal, D. and Jha, M.K. (2006). Time series analysis of hydrologic data for water resources planning and management: A review. *Journal of Hydrology and Hydromechanics*, **54(3):** 237-257.

Machiwal, D. and Jha, M.K. (2008). Comparative evaluation of statistical tests for time series analysis: Application to hydrological time series. *Hydrological Sciences Journal*, **53(2):** 353-366.

Mahé, G., L'Hôte, Y., Olivry, J.-C. and Wotling, G. (2001) Trends and discontinuities in regional rainfall of West and Central Africa: 1951-1989. *Hydrological Sciences Journal*, **46(2):** 211-226.

Maidment, D.R. and Parzen, E. (1984). Time patterns of water use in six Texas cities. *Journal of Water Resources Planning and Management, ASCE*, **110(1):** 90-106.

Mann, H.B. (1945). Non-parametric tests against trend. *Econometrica*, **13:** 245-259.

Matalas, N.C. (1967). Time series analysis. *Water Resources Research*, **3(3):** 817-829.

McCuen, R.H. and James, L.D. (1972). Nonparametric statistical methods in urban hydrologic research. *Water Resources Bulletin*, **8(5):** 965-975.

McGhee, J.W. (1985). Introductory Statistics. West Publishing Co., New York, USA.

McMohan, T.A. and Mein, R.G. (1986). River and Reservoir Yield. Water Resources Publication, Littleton, Colorado, USA.

Mirza, M.Q., Warrick, R.A., Ericksen, N.J. and Kenny, G.J. (1998). Trends and persistence in precipitation in the Ganges, Brahmaputra and Meghna river basins. *Hydrological Sciences Journal*, **43(6):** 845-858.

Natrella, M.G. (1963). Experimental Statistics. National Bureau of Standards Handbook No. 91, US Government Printing Office, Washington, D.C.

Owen, D.B. (1962). Handbook of Statistical Tables. Addison-Wesley Publishing Company, Reading, Mass., USA.

Parzen, E. (1963). Notes on Fourier Analysis and Spectral Windows. Technical Report 48, Stanford University, California, USA.

Phoon, K.-K., Quek, S.-T. and An, P. (2003). Identification of statistically homogeneous soil layers using modified Bartlett statistics. *Journal of Geotechnical and Geoenvironmental Engineering, ASCE*, **129(7):** 649-659.

Pugacheva, G., Gusev, A., Martin, I., Schuch, N. and Pankov, V. (2003). 22-year periodicity in rainfalls in littoral Brazil. Geophysical Research Abstracts, EGS-AGU-EUG Joint Assembly, Abstracts from the meeting held in Nice, France, April 6-11, 2003, p. 6797.

Quimpo, R.G. (1968). Autocorrelation and spectral analysis in hydrology. *Journal of the Hydraulics Division, ASCE*, **HY 2:** 363-373.

Radziejewski, M., Kundzewicz, Z.W. and Graczyk, D. (2002). Detectability of trends in long time series of river flow data – a runup effect. Proceedings of the 5th International Conference on HydroScience and Engineering, Warsaw, Poland, September 18-21, 2002. http://kfki.baw.de/conferences/ICHE/2002-Warsaw/ARTICLES/PDF/257C.pdf (accessed on 4 February 2004).

Ramesh, N.I. and Davison, A.C. (2002). Local models for exploratory analysis of hydrological extremes. *Journal of Hydrology*, **256(1-2):** 106-119.

Rao, A.R., Hamed, K.H. and Chen, H.-L. (2003). Nonstationarities in Hydrologic and Environmental Time Series. *Water Science and Technology Library*, **45**, 392 pp.

Sachs, L. (1972). Statistische Auswertungsmethoden. 3rd edition, Springer-Verlag, Berlin.

Salas, J.D. (1993). Analysis and modeling of hydrologic time series. *In*: D.R. Maidment (editor-in-chief), Handbook of Hydrology. McGraw-Hill, Inc., USA, pp. 19.1-19.72.

Schwankl, L.J., Raghuwanshi, N.S. and Walender, W.W. (2000). Time series modeling for predicting spatially variable infiltration. *Journal of Irrigation and Drainage Engineering, ASCE*, **126(5):** 283-287.

Sen, P.K. (1968). Estimates of the regression coefficient based on Kendall's tau. *Journal of the American Statistical Association*, **63(324):** 1379-1389.

Shahin, M., Van Oorschot, H.J.L. and De Lange, S.J. (1993). Statistical Analysis in Water Resources Engineering. A.A. Balkema, Rotterdam, the Netherlands, 394 pp.

Snedecor, G.W. and Cochran, W.G. (1980). Statistical Methods. Iowa State University Press, Ames, Iowa.

Stein, E.M. and Weiss, G. (1971). Introduction to Fourier Analysis on Euclidean Spaces. Princeton University Press, Princeton, N.J., 312 pp.

Van Belle, G. and Hughes, J.P. (1984). Nonparametric tests for trend in water quality. *Water Resources Research*, **20(1):** 127-136.

Vecchia, A.V. and Ballerini, R. (1991). Testing for periodic autocorrelations in seasonal time series data. *Biometrika*, **78:** 53-63.

Wallis, J.R. and O'Connell, P.E. (1973). Firm reservoir yield — How reliable are historic hydrological records? *Hydrological Sciences Bulletin*, **18:** 347-365.

WMO (1966). Climatic change. Technical Note 79, the World Meteorological Organization (WMO), Geneva, Switzerland.

WMO (1988). Analysing long time series of hydrological data with respect to climatic variability. WCAP-3, WMO/TD No. 224, the World Meteorological Organization (WMO), Geneva, Switzerland.

Yevjevich, V. (1972). Stochastic Processes in Hydrology. Water Resources Publication, Fort Collins, Colorado, 276 pp.

Yue, S. and Wang, C.Y. (2002). The influence of serial correlation on the Mann-Whitney test for detecting a shift in median. *Advances in Water Resources*, **25:** 325-333.

Yue, S., Pilon, P. and Phinney, B. (2003). Canadian streamflow trend detection: Impacts of serial and cross-correlation. *Hydrological Sciences Journal*, **48(1):** 51-64.

Zhang, X., Harvey, K.D., Hogg, W.D. and Yuzyk, T.R. (2001). Trends in Canadian streamflow. *Water Resources Research*, **37(4):** 987-998.

Zipper, C.E., Holtzman, G.I., Darken, P., Thomas, P., Gildea, J. and Shabman, L. (1998). An analysis of long-term water quality trends in Virginia. http://www.nwqmc.org/98proceedings/Papers/49-ZIPP.html (accessed on 24 January 2004).

5

Stochastic Modelling of Time Series

In practice, hydrologists often deal with a limited amount of recorded data (i.e., a *sample*) while analyzing a hydrologic time series. This *sample* consists of a limited number of realizations of the *population* of same hydrologic process. When a hydrologic time series is characterized with statistical and probabilistic parameters, it represents a probability of occurrence of one of its possible stages. This probabilistic occurrence of the hydrologic time series is considered as one realization. All possible realizations of the hydrologic process constitute a *population*. The concept of terms *sample* and *population* has already been explained in Chapter 2. The main intent of the most hydrologic studies is to understand and quantitatively describe the *population* as well as the process that generates it based on a limited number of *samples*. Also, future predictions and/or simulations about the hydrologic time series can be made by applying statistical tools and techniques using probabilistic or stochastic models based on the historical data. When a hydrologic time series is analyzed in this manner, the technique is known as '*stochastic modelling*' of time series and the parameters described with statistic and probabilistic terms are called '*stochastic parameters*'.

Stochastic models are used to model a time series without considering physical nature of the time series (Box and Jenkins, 1976; Shahin et al., 1993). In hydrology, common stochastic models are: pure random (or white noise) model, autoregressive (AR) model, moving average (MA) model, autoregressive moving average (ARMA) model, and autoregressive integrated moving average (ARIMA) model. In this chapter, common stochastic processes are discussed with a major emphasis on autoregressive integrated moving average process. Step-by-step procedure for stochastic modelling of the time series is explained.

5.1 Common Stochastic Processes

Different stochastic models mainly follow distinct stochastic processes. The stochastic processes associated with stochastic models are briefly described in this section.

5.1.1 Purely Random Process

A discrete hydrologic process is called a *purely random process* if the random data points of a variable x_t ($t = 1, 2, \ldots$) form a sequence of mutually independent, identically distributed data points of the same variable (Chatfield, 1980). It is also known as *white noise*. The definition of a purely random process reflects that it is strictly stationary. In practice, this type of stochastic process does not appear. The purely random process has the least practical importance; however, it is important as a building block for other processes.

5.1.2 Autoregressive (AR) Process

Most time series consist of data points that are serially dependent in the sense that one can estimate a coefficient or a set of coefficients that describes consecutive data points of the series from specific, time-lagged (previous) data points. This can be summarized by the following expression of autoregressive process (Box and Jenkins, 1976).

$$x_t = \xi + \phi_1\, x_{(t-1)} + \phi_2\, x_{(t-2)} + \phi_3\, x_{(t-3)} + \ldots + \varepsilon_t \qquad (1)$$

where x_t = data point of variable x at time t; $x_{(t-1)}$, $x_{(t-2)}$ and $x_{(t-3)}$ = data points of variable x at previous times $t-1$, $t-2$ and $t-3$, respectively; ξ = a constant (intercept or population mean); ϕ_1, ϕ_2 and ϕ_3 = autoregressive model parameters; and ε_t = random error component or random shock (white noise).

It is seen from Eqn. (1) that each data point of a time series is made up of a random error component and a linear combination of prior data points. For the population of a hydrologic variable, expression given in Eqn. (1) represents an infinite autoregressive process. However, in practice, population mean in Eqn. (1) is replaced with sample mean and the order of autoregressive process is reduced to p. Thus, Eqn. (1) can be rewritten as (Box and Jenkins, 1976):

$$x_t = \bar{x} + \phi_1\, x_{(t-1)} + \phi_2\, x_{(t-2)} + \ldots + \phi_p\, x_{(t-p)} + \varepsilon_t \qquad (2)$$

The order of the autoregressive process is defined by the highest value of p, for which $\phi_p \neq 0$. Thus, for $p = 1$, the autoregressive (AR) process is of the first order and for $p = 2$, the process is of the second order. The first and second order autoregressive processes can be simply denoted as AR(1) and AR(2), respectively. Similarly, the AR process of order p can be denoted as AR(p).

Stationarity Requirement: The autoregressive process will be stable only when the autoregressive model parameters lie within a certain range. Otherwise, past effects (influence of previous data points) would accumulate and the successive values of the variable x_t would move towards infinity, and therefore, the time series would not be stationary. For example, if there is only one autoregressive model parameter (ϕ_1) as the case of AR(1), then ϕ_1 must lie within ± 1 or $-1 < \phi_1 < +1$. If there is more than one autoregressive model parameter, similar kind of general restrictions on the parameter values can be defined (Box and Jenkins, 1976; Montgomery et al., 1990).

5.1.3 Moving Average (MA) Process

Apart from the serial dependence of the data points as in case of autoregressive process, each data point in the time series can also be affected by the past random error (or random shock) that cannot be taken care of by the autoregressive model. This can be expressed by moving average process as given below (Box and Jenkins, 1976).

$$x_t = \mu + \varepsilon_t - \theta_1\, \varepsilon_{(t-1)} - \theta_2\, \varepsilon_{(t-2)} - \theta_3\, \varepsilon_{(t-3)} - \ldots \qquad (3)$$

where x_t = data point of variable x at time t; μ = a constant or population mean; θ_1, θ_2, θ_3 = moving average model parameters; and ε_t, $\varepsilon_{(t-1)}$, $\varepsilon_{(t-2)}$, $\varepsilon_{(t-3)}$ = random error components of the data points at previous times t, $t-1$, $t-2$ and $t-3$, respectively.

According to Eqn. (3), each data point of the time series is made up of a random error component or random shock and a linear combination of random shocks involved in prior data points. For the population of a hydrologic variable, expression given by Eqn. (3) represents an infinite moving average process. However, Eqn. (3) can be rewritten for a sample by replacing population mean with sample mean and reducing the order of the moving average process from infinite to q, as shown (Box and Jenkins, 1976):

$$x_t = \mu + \varepsilon_t - \theta_1\, \varepsilon_{(t-1)} - \theta_2\, \varepsilon_{(t-2)} - \ldots - \theta_q\, \varepsilon_{(t-q)} \qquad (4)$$

The order of the moving average process is defined by the highest value of q, for which $\theta_q \neq 0$. Thus, for $q = 1$, the moving average (MA) process is of the first order and for $q = 2$, the process is of the second order. The first and second order moving average processes can be simply denoted as MA(1) and MA(2), respectively. Similarly, the MA process of order q can be denoted as MA(q).

Invertibility Requirement: There is a duality between the moving average process and the autoregressive process (Box and Jenkins, 1976; Montgomery et al., 1990) such that a MA(q) process is not uniquely determined by its autocorrelation function. However, a unique relationship between moving average process and their autocorrelation function is required, since the coefficients of the MA(q) process can only be estimated by empirical autocorrelation function. Box and Jenkins (1976) resolved the problem by introducing a concept of 'invertibility' condition. The 'invertibility' condition states that the expression for the MA(q) [Eqn. (4)] can be rewritten (inverted) into an autoregressive form (of infinite order) if the parameters of the moving average model make the model 'invertible'. Note that the invertibility condition of a moving average process is analogous to the stationarity condition of an autoregressive process.

5.1.4 Autoregressive Moving Average (ARMA) Process

The autoregressive AR(p) and moving average MA(q) processes are special cases of an autoregressive moving average process. An autoregressive moving average (ARMA) process of order (p,q) denoted by ARMA(p,q) represents a real stochastic process x_t with the following expression:

$$x_t = \mu + \phi_1 x_{(t-1)} + \phi_2 x_{(t-2)} + \ldots + \phi_p x_{(t-p)} + \varepsilon_t - \theta_1 \varepsilon_{(t-1)} - \theta_2 \varepsilon_{(t-2)} - \ldots - \theta_q \varepsilon_{(t-q)} \tag{5}$$

An ARMA(p,0) process with $p \geq 1$ is obviously an AR(p) process, whereas an ARMA(0,q) process with $q \geq 1$ is an MA(q) process.

5.1.5 Autoregressive Integrated Moving Average (ARIMA) Process

Stochastic modelling and forecasting of a time series requires adequate knowledge about mathematical techniques for identifying patterns in time series data and for expressing the physical process in terms of the mathematical model. However, the physical processes are very complex in nature, the patterns of time series data are unclear, and individual data points involve considerable error. Hence, it is highly challenging in practice to explore the hidden patterns in the data and also to generate forecasts. Box and Jenkins (1976) developed an autoregressive integrated moving average (ARIMA) model and successfully demonstrated their applications in forecasting of physical processes. The ARIMA modelling is inherently a very powerful technique and contains great flexibility. However, the ARIMA modelling requires a great deal of experience because it is complex to understand and it is not easy to use. The ARIMA modelling may often produce satisfactory results but the results entirely depend on the analyst's/scientist's level of expertise (Bails and Peppers, 1982).

The general ARIMA model includes autoregressive as well as moving average parameters, and explicitly includes differencing in the formulation of the model (Box and Jenkins, 1976). Three specific parameters of a general ARIMA model are: the autoregressive parameters (p), the number of differencing passes (d), and moving average parameters (q). Box and Jenkins (1976) denoted the autoregressive integrated moving average process as ARIMA (p,d,q), which means the ARIMA model contains 'p' autoregressive parameters and 'q' moving average parameters which were computed for the series after it was differenced 'd' times.

5.2 Methodology for ARIMA Model Application

Methodology for applying any of the four stochastic models (AR, MA, ARMA and ARIMA) is almost the same. This chapter describes methodology for applying stochastic models to time series with reference to the ARIMA model, since this model contains fundamental parameters of other stochastic models (AR, MA and ARMA). In addition, the ARIMA(p,d,q) model can be easily transformed to AR, MA and ARMA models by adjusting the model parameters. The methodology for stochastic modelling of the time series involves four basic steps (Box and Jenkins, 1976): (i) identification or selection of model, (ii) estimation of model parameters/coefficients, (iii) evaluation or diagnostic check of the model, and (iv) forecasting. It is necessary for a time series to be stationary in nature, free from any kind of trends, and adjusted for seasonality before proceeding to stochastic modelling. The four steps of the methodology are elucidated in subsequent sections. An application-oriented methodology of ARIMA models without mathematical descriptions can be found in McDowall et al. (1980).

5.2.1 ARIMA: Identification of the Model

Number of Differencing Passes: The input time series for an ARIMA model needs to be stationary, i.e., the time series should have a constant mean, variance, and autocorrelation through time. A non-stationary time series is first required to be made stationary. The most common way of making time series stationary is simply differencing the series repetitively till it becomes stationary. Sometimes, the time series is transformed for stabilizing the variance of the series by applying suitable transformations; mostly logarithmic transformation is applied. The number of times the series is differenced to attain stationarity is known as number of differencing passes and is indicated by the parameter 'd' of ARIMA(p,d,q) model. To get an idea of the expected number of differencing passes to make the time series stationary, time plot and autocorrelogram of the series can be critically examined. Significant changes in level (strong upward or downward) suggests that the time series require first-order non-seasonal (lag = 1) differencing. However, strong changes of slope suggest that the time series require second-order non-seasonal differencing. If there are seasonal patterns, the time series usually require respective seasonal differencing. In an autocorrelogram, when the estimated autocorrelation coefficients decline slowly at longer lags, first-order differencing of the time series is usually needed. It is suggested for the newly practising analysts to avoid unnecessary differencing of the time series as sometimes the time series may require little or no differencing. An over-differenced time series may produce less stable coefficient/parameter estimates.

Order of the ARIMA Model: Number of autoregressive (p) and moving average (q) parameters (i.e., order of the model) are also decided in identification step. The order of the model is selected in such a way that the model should be effective and parsimonious. A parsimonious model will have the fewest parameters and the greatest number of degrees of freedom among all the stochastic models that fit to the time series. It is observed that the number of AR and MA parameters hardly exceed two in most of the studies.

In addition to help deciding required number of differencing passes, the time plots of the data series, correlograms of autocorrelation function (ACF), and partial autocorrelation function (PACF) can also assist analysts in selecting order of a stochastic model. Though the decision cannot be straightforward and requires not only vast experience but also a good deal of testing with alternative stochastic models and their parameters. Autocorrelation function and partial autocorrelation functions are discussed below.

Autocorrelation refers to the correlation of a time series with its own past and future data points. Autocorrelation is also sometimes called as 'lagged correlation' or 'serial correlation', which refers to the correlation between members of a series of numbers arranged in time. Positive autocorrelation might be considered as a specific form of *persistence*, a tendency for a system to remain in the same state from one data point to the next. Autocorrelation analysis has been discussed in Chapter 4. The autocorrelation function (ACF) is expressed as Eqns (84) and (85) in Chapter 4 for 'population' and 'sample' of the time series, respectively.

Partial autocorrelation function (PACF) is the partial correlation coefficients between the time series and lags of the time series over time. The partial autocorrelation at lag k is the autocorrelation between x_t and x_{t-k} that is not accounted for by lags 1 through $k - 1$. The partial autocorrelation of an AR(p) process is zero at lag more than or equal to ($p + 1$). Detailed algorithm and mathematical expressions for computing the PACF can be found in Box and Jenkins (1976) and Brockwell and Davis (1991).

Pankratz (1983) formulated general guidelines for identifying one of the five basic stochastic models based on the shape/characteristics of autocorrelogram (ACF) and partial autocorrelogram (PACF) (Table 5.1). A majority of time series patterns can be satisfactorily approximated using one of the five basic models mentioned in Table 5.1. Further details and suggestions for selecting order of the stochastic model can be found in Box and Jenkins (1976), Hoff (1983), McCleary and Hay (1980), McDowall et al. (1980), and Vandaele (1983).

The selection of the correct orders p and q of an ARIMA model is fairly challenging. In this situation, few criteria have been proposed in the literature to select such a pair (p,q) of the parameters minimizing some function, which is based on the variance estimate ($\hat{\sigma}^2_{p,q}$) of the estimated model parameters.

Table 5.1. Guidelines for identifying model parameters based on the characteristics of ACF and PACF

Sl. No.	*Model parameter*	*Characteristics of ACF*	*Characteristics of PACF*
1	One autoregressive (p)	Exponential decay	Spike at lag 1, no correlations for other lags
2	Two autoregressive (p)	A sine-wave shape pattern or a set of exponential decays	Spikes at lags 1 and 2, no correlation for other lags
3	One moving average (q)	Spike at lag 1, no correlation for other lags	Damps out exponentially
4	Two moving average (q)	Spikes at lags 1 and 2, no correlation for other lags	A sine-wave shape pattern or a set of exponential decays
5	One autoregressive (p) and one moving average (q)	Exponential decay starting at lag 1	Exponential decay starting at lag 1

One of the common functions is Akaike Information Criterion (AIC) (Brockwell and Davis, 1991), which is expressed as:

$$\text{AIC}\ (p,q) = \log(\hat{\sigma}^2_{\text{p,q}}) + 2\,\frac{p+q+1}{n+1} \tag{6}$$

One minimization function is Bayesian Information Criterion (Brockwell and Davis, 1991) expressed as:

$$\text{BIC}\ (p,q) = \log(\hat{\sigma}^2_{\text{p,q}}) + \frac{(p+q)\log(n+1)}{n+1} \tag{7}$$

Another criterion is proposed by Hannan and Quinn (1979), which is given as:

$$\text{HQ}\ (p,q) = \log(\hat{\sigma}^2_{\text{p,q}}) + \frac{2\,(p+q)\,c\log\{\log(n+1)\}}{n+1} \text{ with } c>1 \tag{8}$$

It is worth mentioning that the variance estimate ($\hat{\sigma}^2_{\text{p,q}}$) becomes small as ($p+q$) increases. Hence, the additive terms in the above criteria serve as penalties for large values of p and q, and help to prevent over-fitting of the data by selecting p and q too large. There is no specific reason to use a certain criterion for a specific condition. However, it is to be noted that AIC has the tendency not to underestimate the model order and the BIC is generally to be preferred for larger samples (Schlittgen and Streitberg, 2001).

Order of Seasonal ARIMA Model: A seasonal ARIMA is a generalization and extension of the regular ARIMA process discussed earlier in this chapter. The seasonal ARIMA or SARIMA model is used for a time series where a pattern repeats seasonally over time. In addition to the non-seasonal or regular parameters of the ARIMA model, seasonal model parameters for a specified lag (selected in the identification step) need to be estimated. Analogous to the simple or regular ARIMA model parameters, there are three seasonal model parameters: seasonal autoregressive (p_S), seasonal differencing (d_S), and seasonal moving average parameters (q_S). The SARIMA model is usually denoted as ARIMA$(p,d,q)(p_S,d_S,q_S)$, which describes a model that includes 'p' regular AR parameters and 'p_S' seasonal AR parameters, and, 'q' regular MA parameters and 'q_S' seasonal MA parameters, and these parameters for the time series were computed when the series was differenced 'd' times and 'd_S' time seasonally differenced. The seasonal lag used for the seasonal parameters is usually determined during the identification phase and must be explicitly specified.

The general guidelines for the selection of regular model parameters to be estimated (based on ACF and PACF) also apply to seasonal model parameters. The main difference is that in seasonal series, ACF and PACF will show sizable coefficients at multiples of the seasonal lag (in addition to their overall patterns reflecting the non-seasonal components of the series).

5.2.2 ARIMA: Estimation of Model Parameters

Once ARIMA model has been identified and selection of model order is over, the next step is estimation of model parameters. The model parameters are estimated by using function minimization procedures, in order to minimize the sum of squared residuals. There are different methods for estimating the ARIMA model parameters. It is supposed that all the estimation methods should produce very similar values of the model parameters, but a particular estimation method may be more or less efficient for any given ARIMA model. Generally, the model parameter estimation make use of a function minimization algorithm (e.g., quasi-Newton method for nonlinear estimation) to maximize the likelihood/probability of the observed time series for the given values of the model parameters. In practice, sum of squares of the residuals for the given respective parameters are computed for the function minimization. The sum of squares of residuals can be computed by any of three methods: (i) the approximate maximum likelihood method (McLeod and Sales, 1983), (ii) the approximate maximum likelihood method with backcasting, and (iii) the exact maximum likelihood method (Melard, 1984).

All the methods for computing the sum of squares of residuals are equally efficient in the most real-world time series applications. However, the method of approximate maximum likelihood with no backcasts is the fastest among three methods, and should particularly be used for estimating the model parameters of very long time series with more than 30,000 data points. The exact maximum likelihood method proposed by Melard (1984) may become inefficient when used to estimate the model parameters with long seasonal lags (e.g., annual lags). A general recommendation is to first use the approximate maximum likelihood method in order to establish initial parameter estimates that may be close to the real final values and then the exact maximum likelihood method may be employed to get final estimates of the model parameters with certainly a few iterations.

The ARIMA models may also include a constant in addition to the standard parameters of AR and MA models. However, interpretation of the statistically significant constant depends on the type of the model that is to be fit. When the AR parameters are not present in the ARIMA model, the expected value of the constant simply represents mean of the series. Whereas, if the autoregressive parameters are present in the series, the constant represents the intercept. If the series is differenced, the constant represents the mean or intercept of the differenced series. Thus, if the series is differenced once, and there are no autoregressive parameters in the model, the constant represents the mean of the differenced series.

5.2.3 ARIMA: Evaluation of the Model

When the first two steps of ARIMA modelling are complete then orders p and q of the model and respective AR and MA parameters are known in order to model an ARIMA(p,d,q) process underlying the data. Before proceeding to make forecasting by using the ARIMA model, it is essential to apply diagnostic check of the model. The approximate values of the t test-statistics are computed from the parameter standard errors. If the test-statistics are not found significant, the respective parameter can in most cases be dropped from the model without affecting substantially the overall fit of the model.

Another straightforward way for evaluating the reliability of the selected ARIMA model is to check the accuracy of generated forecasts. A comparison of the forecasts with the observed (measured) data points can reveal how efficient the model is in making forecasts. A good model should not only provide sufficiently accurate forecasts, but it should also be parsimonious and produce statistically independent residuals that contain only noise and no systematic components. The correlogram of the residuals should not reveal any serial dependencies. One more approach is to plot the residuals of the original (observed) series and inspect them for any systematic trends, and to examine the autocorrelogram of residuals. There should not be any serial dependency between residuals.

The *portmanteau lack-of-fit test* is generally applied to evaluate the model fitness. The *portmanteau lack-of-fit test-statistic*, Q is defined as follows (Box and Pierce, 1970):

$$Q = n\sum_{k=1}^{L} r_k^2(\eta) \tag{9}$$

where $r_k(\eta)$ = autocorrelation coefficient of the residual series at lag k, and L = maximum lag considered. If η_t is independent, then Q, which is approximately chi-squared distributed with $L - p - q$ degrees of freedom, should be less than $\chi^2 (L - p - q)$.

The portmanteau lack-of-fit test checks whether the estimated residuals $\hat{\varepsilon}_t$, $t = 1, 2, \ldots, n$, behave approximately like realizations from a white noise process.

The major concern here is that the residuals are systematically distributed across the series (e.g., they could be negative in the first part of the series and approach zero in the second part) or that they contain some serial dependency which may suggest that the ARIMA model is inadequate. The analysis of ARIMA residuals constitutes an important test of the model. The estimation procedure assumes that the residuals are not autocorrelated and that they are normally distributed.

5.2.4 ARIMA: Forecasting

When the selected ARIMA model successfully passes the evaluation step, the estimated model parameters are then used in the last stage of forecasting to compute new values of the time series and confidence intervals for the predicted values. Usually, the forecasts are made for future such that these computed new values are beyond the data points included in the input time series. It is worth mentioning that if the estimation process is performed on transformed or differenced time series, then the series needs to be integrated before the forecasts are generated. Integration is the inverse process of differencing, which is performed in order to express the forecasts in values compatible with the input time series data. This integration feature is represented by the letter 'I' in the name of the model (ARIMA = Autoregressive Integrated Moving Average).

REFERENCES

Bails, D.G. and Peppers, L.C. (1982). Business Fluctuations: Forecasting Techniques and Applications. Englewood Cliffs, Prentice-Hall, NJ.

Box, G.E.P. and Jenkins, G.M. (1976). Time Series Analysis: Forecasting and Control. Holden-Day, San Francisco, 575 pp.

Box, G.E.P. and Pierce, D.A. (1970). Distribution of the residual autocorrelations in autoregressive integrated moving average time series models. *Journal of the American Statistical Association*, **65:** 1509-1526.

Brockwell, P.J. and Davis, R.A. (1991). Time Series: Theory and Methods. 2nd Edition, Springer Series in Statistics, Springer, New York.

Chatfield, C. (1980). The Analysis of Time Series: An Introduction. 2nd Edition, Chapman and Hall, London, U.K.

Hannan, E.J. and Quinn, B.G. (1979). The determination of the order of an autoregression. *J.R. Statist. Society*, **B 41:** 190-195.

Hoff, J.C. (1983). A Practical Guide to Box-Jenkins Forecasting. Lifetime Learning Publications, London, U.K.

McCleary, R. and Hay, R.A. (1980). Applied Time Series Analysis for the Social Sciences. Beverly Hills, Sage Publications, CA.

McDowall, D., McCleary, R., Meidinger, E.E. and Hay, R.A. (1980). Interrupted Time Series Analysis. Beverly Hills, Sage Publications, CA.

McLeod, A.I. and Sales, P.R.H. (1983). An algorithm for approximate likelihood calculation of ARMA and seasonal ARMA models. *Applied Statistics*, **32:** 211-223.

Melard, G. (1984). A fast algorithm for the exact likelihood of autoregressive-moving average models. *Applied Statistics*, **33:** 104-119.

Montgomery, D.C., Johnson, L.A. and Gardiner, J.S. (1990). Forecasting and Time Series Analysis, 2nd edition. McGraw-Hill, New York.

Pankratz, A. (1983). Forecasting with Univariate Box-Jenkins Models: Concepts and Cases. Wiley, New York.

Schlittgen, J. and Streitberg, B.H.J. (2001). Zeitreihenanalyse. Oldenbourg, Munich, Germany.

Shahin, M., Van Oorschot, H.J.L. and De Lange, S.J. (1993). Statistical Analysis in Water Resources Engineering. A.A. Balkema, Rotterdam, the Netherlands, 394 pp.

Vandaele, W. (1983). Applied Time Series and Box-Jenkins Models. Academic Press, New York.

6

Current Status of Time Series Analysis in Hydrological Sciences

Time series analysis has been successfully applied in the fields like geology, ocean engineering, seismology, hydrology, climatology, etc. The hydrological and climatological time series studies have been carried out for analyzing the historic rainfall data (e.g., Henderson, 1989; De Michele et al., 1998; Mirza et al., 1998; Pagliara et al., 1998; Abaurrea and Cebrian, 2003; Pugacheva et al., 2003; Astel et al., 2004), streamflow data (Avinash and Ghanshyam, 1988; Capodaglio and Moisello, 1990; Radziejewski et al., 2000; Fanta et al., 2001; Adeloye and Montaseri, 2002; Chen and Rao, 2002), flood data (Grew and Werrity, 1995; Changnon and Kunkel, 1995; Westmacott and Burn, 1997; Robson et al., 1998; Reed et al., 1999; Lins and Slack, 1999; Loukas and Quick, 1996, 1999; Cayan et al., 1999; Jain and Lall, 2001; Douglas et al., 2000; Adamowski and Bocci, 2001; Zhang et al., 2001; Cunderlik and Burn, 2002), infiltration data (Schwankl et al., 2000), and surface water quality data (Jayawardena and Lai, 1989; Higashino et al., 1999) as well as for generating synthetic rainfall data in semi-arid regions (Janos et al., 1988), determining water consumption patterns (Maidment and Parzen, 1984), detecting trends in evapotranspiration and wind speed (Hameed et al., 1997; Raghuwanshi and Wallender, 1997), and for detecting climate change or variability (Kite, 1989; Khan, 2001).

A comprehensive literature survey (Machiwal and Jha, 2006) revealed that both theoretical and applied research on time series analyses have been conducted in the hydrological and climatological (meteorological) sciences in the past. The theoretical research basically involves advancement of an existing time series test or development of a novel method for analyzing the hydrologic time series while the applied research mainly highlights application of the existing approaches of time series analysis to hydrologic time series for different purposes. In climatological studies, time series analysis has been applied to precipitation, maximum and minimum air temperature, water temperature, evapotranspiration and climate change. In hydrologic studies, streamflow, groundwater flow and water quality are the variables, which have employed application of the time series methods. This chapter discusses the current status of time series analysis in hydrological sciences; it has been largely drawn from Machiwal and Jha (2006) with updates. Although the reviewed literature is extensive, only major relevant reviews in the context of this book are included in this chapter.

6.1 Theoretical Research on Hydrologic Time Series

Sen (1968) studied a simple and robust estimator (point as well as interval) of BETA based on the Kendall's rank correlation tau. Various properties of these estimators were studied and compared with those of the least squares and some other nonparametric estimators. Statistical tests for monotonic trend in seasonal hydrologic time series are commonly confounded by some of the following problems: non-normal data, missing values, seasonality, censoring (detection limits), and serial dependence. Hirsch and Slack (1984) presented an extension of the Mann-Kendall trend test for such data. Because the suggested test is based entirely on ranks, it is robust against non-normality and censoring. Seasonality and missing values present no theoretical or computational obstacles to its application. Monte Carlo experiments indicated that, in terms of Type I error, it is robust against serial correlation except when the data have strong long-term persistence [e.g., ARMA (1,1) monthly processes with phi greater than 0.6] or short records (approximately five years). When there is no serial correlation, it is less powerful than a related simpler test not robust against serial correlation.

Anh et al. (1997) developed a new class of stochastic models to represent the properties of time series (i.e., long-range dependence and small-scale behaviour) from various fields, such as geophysics, meteorology, hydrology, and air pollution. An efficient estimation procedure is described, which was tested on two concentration time series collected in an environmental wind tunnel. These time series simulated two different types of odour sources and possessed quite different statistical properties that were well described by the new model.

Hamed and Rao (1998) studied the effects of autocorrelation on the variance of the Mann-Kendall trend test-statistic. A theoretical relationship was derived to calculate the variance of the Mann-Kendall test statistic for autocorrelated data. The special cases of AR(1) and MA(1) dependence were discussed as examples. Based on the modified value of the variance of the Mann-Kendall trend test statistic, a modified nonparametric trend test suitable for the autocorrelated data is proposed. The modified test was applied to rainfall and streamflow data to demonstrate its performance compared to the original Mann-Kendall trend test. The accuracy of the modified test was found to be superior to that of the original Mann-Kendall trend test without any loss of power.

Tsakalias and Koutsoyiannis (1999) developed a new approach for the computer-aided exploration and analysis of hydrologic time series with a focus on identification of multiple stage-discharge relationships in a river section, analyses for homogeneity and temporal consistency, detection of outliers, shifts and trends. To demonstrate the developed methodology, initially a mathematical representation was proposed based on the set theory. It was demonstrated that an exhaustive search of all candidate solutions is intractable. Therefore, a heuristic algorithm is proposed, which emulates the exploratory data analysis of the human expert. This algorithm encodes a number of search strategies in a pattern directed computer program, and results in an automatic determination of a satisfactory solution.

Anderson et al. (1999) used periodic ARMA, or PARMA time series to model periodically stationary time series. The innovations algorithm was developed for periodically stationary processes. Thereafter, the algorithm was used to obtain parameter estimates for the PARMA model. These estimates were proved to be weakly consistent for PARMA processes whose underlying noise sequence has either finite or infinite fourth moment. Since many time series from the fields of economics and hydrology exhibit heavy tails, the results regarding the infinite fourth moment case are of particular interest.

Haywood and Wilson (2000) proposed a method for investigating the evolution of trend and seasonality in observed time series. A general model was fitted to a residual spectrum, using components to represent the seasonality. The method was applied to model two time series and the resulting forecasts and seasonal adjustment for one series are presented.

Darken et al. (2000) developed a methodology for testing the equivalence of two modified Kendall's tau nonparametric correlation coefficients. Several estimators of the variance tau_{mod} (i.e., bootstrap estimate, the standard null-case variance estimate, and a delta method variance estimate) were evaluated using simulation. The variance estimators and their corresponding Wald-type tests were studied under different conditions, including the presence of varying degrees of serial correlation, different distributions, and different percentages of tied data. The power study revealed that in the presence of serial correlation, a new method for estimating variance, called the *effective sample size bootstrap*, allowed the hypothesis test to consistently hold its level while no other methods of variance estimation did so. Finally, it was demonstrated how this test can be used to detect changes in trend of water-quality variables over time.

Perreault et al. (2000) proposed a Bayesian method for the analysis of two types of sudden change at an unknown time-point in a sequence of energy inflows modelled by independent normal random variables. To our knowledge, this study is the first of its kind in hydrology from a Bayesian perspective. Even if this model is quite simple, no analytic solutions for parameter inference are available. It is shown that the Gibbs sampler is particularly suitable for change-point analysis, and Markovian updating scheme is used. Finally, a case study involving annual energy inflows of two large hydropower systems of Canada is presented.

Yue et al. (2002a) investigated the interaction between a linear trend and a lag-one autoregressive [AR(1)] model using Monte Carlo simulation. Simulation analysis indicated that the existence of serial correlation alters the variance of the Mann-Kendall (MK) statistic estimate, and the presence of a trend alters the magnitude of serial correlation. Furthermore, it was found that the commonly used pre-whitening procedure for eliminating the effect of serial correlation on the MK test leads to inaccurate assessments of the significance of a trend. Therefore, it was suggested that firstly trend should be removed prior to ascertaining the magnitude of serial correlation. Both the suggested approach and the existing approach were employed to assess the significance of a trend in the serially correlated annual mean and annual minimum streamflow data of some pristine river basins in Ontario, Canada. It was concluded that the researchers might have incorrectly identified the possibility of significant trends by using the already existing approach.

Yue et al. (2002b) studied the efficacy of the two nonparametric rank-based statistical tests (the Mann-Kendall test and Spearman's rho test) by Monte Carlo simulation. These two tests were used to assess the significance of trends in annual maximum streamflow data of 20 pristine basins in Ontario, Canada. The results indicated that their effectiveness depends on the pre-assigned significance level, magnitude of trend, sample size, and the amount of variation within a time series. Thus, the bigger the absolute magnitude of trend or larger the sample size, the more powerful are the tests; but as the amount of variation in a time series increases, the power of the tests decreases. When a trend is present, the power is also dependent on the distribution type and skewness of the time series. It was also found that these two tests have practically similar power in detecting a trend.

Clarke (2002) described a model in which the Gumbel distribution has a (possibly) time-variant mean. The time-trend in mean value was determined by a single parameter β estimated by Maximum Likelihood (ML). The large-sample variance of the ML estimate was compared with the variance of the trend calculated by linear regression; the latter was found to be 64% greater. The simulated samples from a standard Gumbel distribution were given superimposed linear trends of different magnitudes, and the efficacy of three trend-testing methods viz., Maximum Likelihood, Linear Regression, and the nonparametric Mann-Kendall test was compared. The ML test was found always more powerful than the Linear Regression or Mann-Kendall test regardless of the value (positive) of the trend β; the MK test was found least powerful for all the values of β.

Ducré-Robitaille et al. (2003) evaluated eight homogenization techniques for the detection of discontinuities in the temperature series using simulated datasets reproducing a vast range of possible situations. The simulated data represented homogeneous series and the series having one or more steps. Although the majority of the techniques considered in this study performed very well, two methods are reported to work slightly better than the others: the standard normal homogeneity test without trend, and the multiple linear regression technique.

Yue et al. (2003) proposed a trend-free pre-whitening (TFPW) procedure to remove serial correlation from the time series, and hence to eliminate the effect of serial correlation on the Mann-Kendall (MK) test. An additional bootstrap test with the preservation of the cross correlation structure of a network was proposed to assess the field significance of upward and downward trends over the network separately. At the significance level of 5%, the site significance of trends in the Canadian annual minimum, mean, and maximum streamflows with 30-, 40- and 50-year records was assessed by the MK test using the TFPW procedure (TFPW-MK). It was found that: (a) the 30-year annual minimum and mean daily flows significantly decreased in the regions of southern British Columbia (BC), around the centre of Prairie Provinces, and in Atlantic Provinces, but they significantly increased in the region of northern BC and Yukon Territory; and (b) the annual maximum flow significantly decreased across southern Canada. The field significance of trends over the whole country was evaluated by the bootstrap test at the significance level of 5% and none of the three flow regimes experienced field-significant changes.

Yue and Wang (2004) proposed effective sample size (ESS) to modify the MK statistic for eliminating the effect of serial correlation on the MK test. This study investigated the ability of ESS to eliminate the influence of serial correlation on the MK test by Monte Carlo simulation. Simulation demonstrated that when no trend exists within time series, ESS can effectively limit the effect of serial correlation on the MK test. When trend exists within time series, the existence of trend will contaminate the estimate of the magnitude of sample serial correlation, and ESS computed from the contaminated serial correlation cannot properly eliminate the effect of serial correlation on the MK test. However, if ESS is computed from the sample serial correlation that is estimated from the detrended series, ESS can still effectively reduce the influence of serial correlation on the MK test.

Zhang et al. (2010a) developed a novel approach to identify trend patterns of streamflows when trend is gradual or abrupt. The proposed approach uses repeated monotonic trend tests with varying beginning and ending times. The sensitivity of trends with respect to the period of time was then employed to characterize the trend pattern. The proposed approach was demonstrated by applying to watersheds within the Susquehanna River Basin, US. It was observed that the new approach is capable of characterizing trend patterns. A comparison with the results of single monotonic trend tests showed that the novel approach is also useful for the exploration of all available data in contrast to a single monotonic trend test that only shows trends for a specified time period.

6.2 Application of Time Series Analysis in Climatology

6.2.1 Precipitation/Precipitation with Other Climatic Data

Bhuiya (1971) tested the assumption of stationarity of standardized hydrologic series after subtraction of the periodic and trend components. Based on the Fourier series representation of the periodic process, a test for stationarity was developed. The first order periodicity was explained by the periodicity of the first moment of the hydrologic variable, whereas the periodicity in the covariance was explained by the harmonization of the stochastic component. Finally, observed monthly runoff and precipitation records were tested for stationarity using their raw and transformed series.

Buishand (1979) used monthly rainfall data from urbanized and rural areas in the western part of The Netherlands to investigate the effects of urbanization on the rainfall regime. A trend test in the sequence of differences between urban and rural rainfall amounts is described. For the urban areas of Amsterdam and Rotterdam, some evidence was found for an increase in precipitation.

Buishand (1982) discussed the features of five tests on the cumulative deviations from the mean, which are often used for the analysis of homogeneity. Some of the tests showed optimal properties in testing the null hypothesis of homogeneity against a shift in the mean at an unknown point. These tests together with the classical von Neumann ratio were applied to the 30-year annual rainfall records of The Netherlands. For a large number of records, strong indications for a change in the mean were found. There were only small differences between the various test-statistics with respect to the number of records for which the null hypothesis was rejected.

Buishand (1984) presented methodology for testing the shift in the mean of hydrological time series with great attention to the likelihood ratio and Bayesian tests, and critical values were also derived for the Bayesian tests. Furthermore, testing for a systematic change in a linear regression model was also presented. The proposed methodology was demonstrated using the runoff and precipitation data of the Colorado River Basin, USA, and the Thames River at Teddington, U.K., respectively.

Boroneant et al. (1995) examined the time series of seasonal and annual precipitation totals for selected representative stations in the southern part of Romania. The common period was 1951-1993. The homogeneity was tested by the Alexandersson's test. In order to detect the trend and the change points in the series, the linear test, Mann-Kendall's and Pettitt's tests were used. Common change points were identified for different stations.

Brázdil and Stepánek (1995) examined the continuous air temperature and precipitation measurements in Brno during 1848-1993 in different parts of the city and then homogenized on the present station, Brno-Turany by Maronna-Yohai and Alexandersson tests with the reference stations of Vienna,

Kremsmünster and Prague. Results of the statistical analysis of homogenized series are presented.

Fortuniak (1995) used the daily precipitation totals and mean daily temperature for the period 1956-1990 from 10 Polish meteorological stations (Gdansk, Szczecin, Suwalki, Poznan, Lódz, Warszawa, Wroclaw, Kraków, Przemysl and Zakopane) to test their periodicity. The annual course of temperature was removed by subtracting the 35th Fourier harmonic. The classical Blackman and Tukey test was used to detect the cyclic behaviour of the analysed series. The power spectrum of temperature for each station exhibited two significant peaks: around 7.4 years and 193 days. For the precipitation, the power spectra were found different for each station and it was impossible to find a characteristic cycle for the entire region.

Nieplová (1995) applied five statistical homogeneity tests (Student's, Bartlett's, Kruskal-Wallis's, Abbe criterion, and Spearman rank correlation method) and the Craddock's relative homogeneity test to the annual and monthly air temperature means, precipitation totals and relative air humidity means of 40 years and longer series. It was found that most of inhomogeneities were caused by changed observation terms and by the relocation of measuring stations. These results were used for selecting stations for long-term monitoring of climate change in Slovakia.

Kristev and Koleva (1995) studied the variation of some important characteristics of the snow cover pattern in Bulgaria during the period of 1935/6-1992/3. The basic data used were number of days with snow cover, dates of the first and last days with snow cover and maximum snow depths. The presence of some forms of trend in the data was examined by the Spearman test and the Mann-Kendall rank test statistics.

Walanus-Gliwice (1995) analyzed the periodicity by using the Fast Fourier Transform (FFT). The water stages of Vistula River at Szczucin, discharges of Warta (Poland) and Tisa (Hungary) rivers, Dnieper River (Ukraina), precipitation from Warsaw, Cracow, Wroclaw and other towns, air temperature, dendroclimatological curves and the thickness of yearly strata (Warws) from Gosciaz Lake (Central Poland) were analyzed. The 3.5-year (3.5±0.15 yr) periodicity of unknown origin in the river discharge and the precipitation was confirmed. For rivers, the 3.5-yr signal was found much less in Szczucin, but it was visible. For the precipitation, the signal was still less, especially in comparison to the more dominant seasonal periodicity. The periodicity in rivers' discharges was of higher amplitude than that in the precipitation. Finally, a detailed analysis revealed that the 3.5-yr peak in frequency domain should be treated as a random event.

Aulenbach et al. (1996) evaluated the trends in precipitation and surface water quality at a network of 15 small watersheds (<10 km^2) in USA using the seasonal Kendall test for monotonic trends and a graphical smoothing technique for the visual identification of trends. A relation between precipitation and surface water trends was not evident either for individual inorganic solutes or for solute combinations at most sites. The only exception was chloride, which had the same trend at 60% of the sites. The smoothing technique indicated that the short-term patterns in precipitation chemistry were not reflected in surface waters. The magnitude of short-term variations in surface water quality was generally larger than the overall long-term trend.

Kothyari et al. (1997) analyzed rainfall and temperature (i.e., long-term monsoon rainfall, number of rainy days during the monsoon season, and annual maximum temperature) from three stations at Agra, Dehradun and Delhi for evaluating the changes in regimes in the upper and middle parts of the Ganga basin in northern India. The nonparametric methods employed to identify trends showed that the total monsoon rainfall and the number of rainy days during the monsoon season have a declining trend, whereas the annual maximum temperature has a rising trend. These changes were found to have started around the second half of the 1960's. The results of this study suggested a possible change in the climatic regime of the Ganga basin, which has far-reaching implications for the Indian economy.

Giakoumakis and Baloutsos (1997) investigated trends in the historic hydrological time series of aerial annual precipitation and mean annual runoff from the Evinos River basin in western Greece. Using different tests for randomness, a statistically significant long-lasting decreasing trend was found in the precipitation records, whereas a significant fluctuating "local" trend was detected in the runoff records. Further, it was demonstrated that the effects of precipitation change on the mean annual runoff can be quantified through a magnification factor.

Angel and Huff (1997) examined the stationarity and trends of precipitation records for the period 1901 to 1994 at 304 sites in the midwestern United States. The results of this study indicated that the stations are more likely to experience their heaviest rainfall events in more recent years. An analysis of the geographic distribution of changes in the annual maximum rainfall time series showed an increase in areas across the Midwest. It was concluded that the rainfall frequency studies should be updated on a regular basis for maximum usefulness.

Mirza et al. (1998) highlighted the importance of analyzing the trends and persistence in precipitation time series. The annual precipitation time series for the Ganges, Brahmaputra and Meghna river basins were examined for trends by the Mann-Kendall rank statistic, Student's *t*-test and the regression analysis, and for persistence by the first order autocorrelation technique. The results indicated that the precipitation in the Ganges basin is almost stable, while in the Brahmaputra basin, decreasing and increasing trends were found in two subdivisions. One of the three subdivisions in the Meghna basin has a decreasing trend, while the two subdivisions have an increasing trend. The Markovian persistence was not found present in the Ganges basin, but it was present in the two common subdivisions of the Brahmaputra and Meghna basins.

Tarhule and Woo (1998) analyzed the rainfall records (i.e., annual total rainfall, number of rainy days, dates of onset, termination and duration of rainy seasons as well as monthly rainfall, monthly number of rainy days and various categories of rainfall above certain intensities) at 25 locations in northern Nigeria to examine the abrupt changes and trends using the Pettitt and the Mann-Kendall tests. It was found that an abrupt change occurred in the time series of annual rainfall and number of rainy days, which affected the areas north of latitude 11° N. However, the sub-periods before and after the change points was considered to be homogenous. The series of duration of rainy seasons exhibited no significant trends or jumps. It was concluded that recent changes in rainfall over the Sahel were driven by a reduction in the frequency of rainy days of high rainfall intensities during August and September. The continuation of agricultural activities in the Sahel despite massive reductions in annual rainfall is attributed to the fact that the high intensity rainfall does not contribute significantly to crop growth.

Johann et al. (1998) proposed a method for filling gaps in the precipitation time series obtained from the Emschergenossenschaft and Lippeverband (EG/LV) in northwest Germany. Several neighbouring stations of the EG/LV raingauge network were considered. Various time intervals based on deterministic and statistical approaches were investigated, but the intervals between 5 and 120 min are discussed in more detail. Based on representative examples, it was demonstrated how the time intervals influence the quality of the estimated (gap-filling) rainfall data.

Moraes et al. (1998) investigated change in the patterns of streamflow and precipitation and its possible relation to man-induced changes in the Piracicaba River basin of Brazil. With an area of 12,400 km^2, this basin is a typical example of new landscape resulting from development in tropical and sub-tropical regions: establishment of intensive industrial and agricultural processes were followed by significant population growth and water management. Statistical analyses were performed using the precipitation, evapotranspiration and streamflow data from 1947 to 1991. The precipitation and evapotranspiration data showed significant increasing trends for the entire basin. Out of eight streamflow gauging stations, four stations showed a significant decreasing trend. The cause of these trends was attributed to the export of water from the basin to the metropolitan region of São Paulo city.

De Luís et al. (2000) examined the spatial and temporal rainfall characteristics in the Valencia region, Western Mediterranean Basin (east Spain) using the WMO normal period 1961-1990. The dense and homogeneous daily precipitation database comprising 97 rainfall stations were collected and mean values, interannual variability and spatial diversification of total and monthly rainfalls were studied. Trends were analyzed using both parametric and nonparametric tests. The spatial distribution of rainfall patterns was established and the homogeneous areas with similar rainfall patterns were detected based on the Cramér-von Mises test-statistic. The kriging interpolation technique was used for characterizing the magnitude of detected changes.

Brunetti et al. (2000) analyzed the daily precipitation data of northern Italy for trend detection. The nonparametric Mann-Kendall test was applied to the mean anomaly series of some precipitation intensity statistics over five stations: Genoa (1833-1998), Milan (1858-1998), Mantova (1868-1997), Bologna (1879-1998) and Ferrara (1879-1996). It was found that in northern Italy, the number of rainy days has a more significant negative trend than the corresponding precipitation amount. Consequently, the precipitation intensity has a positive trend. The increase in precipitation intensity was found to cause a significant positive trend in the total precipitation contributed by heavy precipitation events (i.e., daily precipitations >25 mm and >50 mm). Furthermore, the trend was mainly caused by past 60-80 years, and was particularly evident during the periods of 1930-1945 and 1975-1995.

Molénat et al. (2000) analyzed the hydrological and hydrochemical behaviour of three agricultural catchments located in different regions of France (Kervidy, Melarchez and Mercube). The time series were considered as input or output data and the spectral analysis was performed. The input data for hydrology and chemistry were respectively rainfall and nitrate leaching, and the output data were streamflow and nitrate concentration in the stream. It appears that nitrate concentrations measured at the outlet of the three catchments exhibit a strong and unique one-year periodicity. This periodicity is due to the hydrological regime and the time distribution of the nitrate availability in the soil. Moreover, a cross-spectral analysis was performed between the input and output data for each catchment and the major processes that govern water and nitrate transfer and the characteristic time scale of these processes were identified. It was concluded that the spectral and cross-spectral methods are valuable techniques for identifying the main transfer processes operating in different catchments.

Sharma et al. (2000) studied basinwide land-use, climatic and hydrologic trends over the Kosi Basin (54,000 km^2) in the central Himalayan region. The analysis of anthropogenic inputs showed that the population of the basin has increased at a compound rate of about 1% per annum during past four decades. A comparison of the land-use data of the 1960s and 1978-1979 did not reveal noticeable trends in the land-use change. On the other hand, the analysis of the meteorological and hydrological time series from 1947 to 1993 indicated an increasing tendency of temperature and precipitation. However, the statistical tests of hydrologic trends indicated an overall decrease in the discharges of the Kosi River and its major tributaries. The decreasing trends of streamflow were found more significant during low-flow months. Further, the statistical analysis of homogeneity suggested that the climatic as well as the hydrologic trends are more localized in nature and do not have a distinct basinwide significance.

Brunetti et al. (2001) analyzed the seasonal and annual precipitations and the number of rainy days in northeastern Italy during 1920-1998. The precipitation intensity was analyzed by using both the mean precipitation amount per wet day and dividing the precipitation into heavy and non-heavy classes. In addition, the return period of extreme events was calculated for the 30-years and its variations were examined. The results indicated a negative trend in the number of wet days associated with an increase in the contribution of heavy rainfall events to the total precipitation. This finding is in agreement with the reality (i.e., a reduction in the return period of extreme events since 1920).

Kripalani and Kulkarni (2001) prepared regional rainfall anomaly time series using the 118-year (1881 to 1998) data of three regions, India, northern China and southern Japan. All the three series (India, China and Japan) were subjected to selected statistical tests. The analysis of the results revealed that though there are year-to-year fluctuations in rainfalls, the Mann-Kendall rank statistic suggested no significant long-term trends. However, the application of the Cramer's statistic to study the short-term climate variability depicts decadal variability with certain epochs of above and below normal rainfall over each region. The epochs tend to last for about three decades over India and China, and about five decades over Japan. The turning points for China follow those of India after about a decade.

Adamowski and Bougadis (2003) estimated trends for different durations of annual extreme rainfall by using the regional average Mann-Kendall S trend test. The method of L-moments was employed to delineate homogeneous regions. The trend test was modified to account for the observed autocorrelation, and the bootstrap method was used to account for the observed spatial correlation. Numerical analysis was performed for 44 rainfall stations in Ontario, Canada for a 20-year time frame using the data from homogeneous regions. Depending on the rainfall durations, four and five homogeneous regions were delineated. At the 5% significance level, approximately 23% of the regions tested had a significant trend, predominantly for short-duration storms. The serial dependency was observed in 2-3% of datasets and the spatial correlation was found in 18% of the regions. The presence of serial and spatial correlation was found to have significant impacts on trend determination.

Xu et al. (2003) detected long-term trends in the spatially averaged Japanese precipitation time series by applying the parametric t-test and the nonparametric Mann-Kendall and Mann-Whitney tests. The results indicated that despite several step changes in the Japanese precipitation, the time series did not exhibit significant evidence of monotonic trend during the past century. Further, it was found that if the magnitude of the step change reaches one or two times of its standard deviation, the previous 50-year records together with five years or more new data will be available for detecting the possible trend. This finding is useful for the detection of step changes in the regions where the precipitation has near-normal distributions.

Oguntunde et al. (2006) investigated hydrological variability and trends in the Volta River basin in West Africa over the period 1901-2002. Potential (Ep) and actual evaporation (E), rainfall variability index (d), Budyko's aridity index (*IA*), evaporation ratio (*C*E) and runoff ratio (CQ) were analyzed by using Mann-Kendall and Sen's slope estimation trend tests to detect monotonic trend direction and magnitude of change over time. Rainfall variability index showed that 1968 was the wettest year ($d = +1.75$), while 1983 was the driest ($d = -3.03$), with the last three decades being drier than any other comparable period in the hydrological history of the Volta basin. An increase of 0.2 mm/yr^2 ($P < 0.05$) was observed in E_p for the 1901-1969 sub-series while an increase of 1.8 mm/yr^2 ($P < 0.01$) was recorded since 1970. Rainfall increased at the rate of 0.7 mm/yr^2 or 49 mm/yr between 1901 and 1969, whereas a decrease of 0.2 mm/yr^2 (6 mm/yr) was estimated for 1970-2002 sub-series. Runoff increased significantly at the rate of 0.8 mm/yr^2 (23 mm/yr) since 1970. Runoff before dam construction was higher (87.5 mm/yr) and more varied ($CV = 41.5\%$) than the post-dam period with value of 73.5 mm/yr ($CV = 23.9\%$). A 10% relative decrease in P resulted in a 16% decrease in Q between 1936 and 1998. Since 1970, all the months showed increasing runoff trends with significant slopes ($P < 0.05$) in nine out of the 12 months.

Yu et al. (2006) evaluated long-term trends in seasonal and annual precipitations for 33 raingauge stations in Taiwan. Cumulative Deviations, Mann-Whitney-Pettitt and Kruskal-Wallis tests were employed to analyze the trends. Analytical results indicated that the annual rainfall has increased in northern Taiwan, declined in central and southern Taiwan, and exhibited no clear tendency in Eastern Taiwan. Almost all of these rainfall series changed significantly around 1960, which date divides historical rainfall series into two sample groups.

Kumar et al. (2010) studied monthly, seasonal and annual rainfall trends using monthly data series of 135 years (1871-2005) for 30 sub-divisions in India. Half of the sub-divisions showed an increasing trend in annual rainfall. However, the trend was statistically significant for only three (Haryana, Punjab and Coastal Karnataka) sub-regions. Similarly, only one sub-division (Chattisgarh) indicated a significant decreasing trend out of the 15 sub-divisions showing decreasing trend in annual rainfall. In India, the monsoon months of June to September account for more than 80% of the annual rainfall. During June and July, the number of sub-divisions showing increasing rainfall is almost equal to those showing decreasing rainfall. In August, the number of sub-divisions showing an increasing trend exceeds those showing a decreasing trend, whereas in September, the situation is the opposite. The majority of sub-divisions showed very little change in rainfall during non-monsoon months. The five main regions of India showed no significant trend in annual, seasonal and monthly rainfall in most of the months. For the whole of India, no significant trend was detected for annual, seasonal, or monthly rainfall. Annual and monsoon rainfall decreased, while pre-monsoon, post-monsoon and winter rainfall increased at the national scale. Rainfall in June, July and September decreased, whereas in August it increased, at the national scale.

Sahin and Cigizoglu (2010) performed homogeneity analysis of the maximum air temperature, the minimum air temperature, the mean air temperature, the total precipitation, the relative humidity and the local pressure of 232 stations of Turkey for the period 1974-2002. The homogeneity analysis was performed on the annual data using a relative test and four absolute homogeneity tests were used for the stations where non-testable series were found due to the low correlation coefficients between the test and the reference series. A comparison was accomplished by the graphics where relative and absolute tests provided different outcomes. Absolute tests failed to detect the inhomogeneities in the precipitation series at a significance level of 1%. Interestingly, most of the inhomogeneities detected on the temperature variables existed in the Aegean region of Turkey. It is considered that these inhomogeneities were mostly caused by non-natural effects such as relocation. Because of changes in topography at short distance in this region, non-random characteristics of the temperature series are intensified when relocation occurs even in small distances. The marine effect, which causes artificial cooling effect due to sea breezes, has important impact on temperature series and the orography allows this impact go through the inner parts in this region.

6.2.2 Air and Water Temperature

Esteban-Parra and Castro-Diez (1995) analyzed the longest annual and seasonal series of maximum, minimum and average temperatures of some localities in Spain. The homogeneity was checked by using the Thom and Bartlett tests. These methods are reported to yield different results in some cases. The analysis was performed to explore how the existence of actual trends and/or discontinuities in the series affects the sensitivity and have a repercussion on the results of the tests. This analysis suggested an adequate confidence level and the need of the use of relative homogeneity tests.

Webb (1996) analyzed the future trends in water temperatures from different parts of the world. The potential causes of trends in the thermal regimes of streams and rivers are many, but the existing database of water temperature was found inadequate to provide a global perspective on changes during the recent past. The data from Europe suggested that an increase of up to 1°C in the mean river water temperatures has occurred during the 20th century. However, this trend was not found continuous and correlated with simple hydrometeorological factors, rather it was found to be distorted by extreme hydrological events influenced by a variety of human activities. Predictive studies indicated that an accelerated rise in stream and river water temperatures will occur during the next century because of global warming. However, the forecasts are tentative because future climatic conditions are uncertain, and the interactions between climate and hydrological and vegetation changes are complex.

Tayanç and Toros (1997) studied the daily maximum temperature and temperature difference series (1951-1990) of four urban stations and their neighbouring rural stations in Turkey. The results indicated that there is a shift towards the warmer side in the frequency distributions of both the series, which is an indication of urban heat island. The seasonal analysis of individual 21.00 h temperature series suggested that the regional warming is the strongest in spring and the weakest in autumn and winter. Urban warming is detected to be more or less equally distributed over the year with a slight increase in autumn. Using the Mann-Kendall trend test for the temperature difference series, the urban heat island effect was found to be significant at all urban sites. On the other hand, no significant urban effects on the precipitation were found.

Kadioglu (1997) analyzed the mean annual temperature records of Turkey for the period 1939 to 1989. A warming trend was found from 1939 to 1989 but a cooling trend was detected from 1955 to 1989. These trends in the mean annual temperature series, however, were not found to be statistically significant. Comparatively greater warming effects were found in spring and minimum in winter. A regional increase in the mean minimum temperature around 1955 is attributed to the urban heat island effect. In general, the predictions of general circulation model (GCM) were consistent with a sign of trends only in the Turkish climatic records during the entire (1939 to 1989) period.

Keiser and Griffiths (1998) used a homogeneity test developed by Alexandersson (1986, 1995) and applied it to the mean monthly maximum, minimum, and mean temperature data from 22 stations in the northern Great Plains of USA. One of these stations, Valentine, is a first-order station and is used as the reference station. When Valentine station was adjusted for a possible inhomogeneity due to its move, it was found that the Valentine's adjustments have a distinct seasonal pattern. The testing of other stations against Valentine revealed that the position of a significant discontinuity in a station's monthly mean or annual temperature series is not always the same as in the corresponding monthly maximum and minimum series. In addition, a seasonal pattern similar to that of Valentine station was found in every station's adjustment values.

Tayanç et al. (1998) presented a combination of different methods (i.e., graphical analysis, nonparametric Kruskal-Wallis homogeneity test and Wald-Wolfowitz runs test) to test climatological time series for inhomogeneities. These methods were applied to the annual mean difference temperature series of 82 Turkish weather stations, and the inhomogeneity detection efficiencies of these tests were determined by a series of Monte Carlo simulation studies. It was concluded that the procedure is statistically rigorous, provides estimates of the time and magnitude of change in the mean, and is a valuable tool for testing time series.

Vincent and Gullett (1999) developed the Canadian historical temperature database (CHTD) to produce an improved historical climate change database based on the datasets of monthly mean maximum and minimum temperatures of 210 Canadian stations. The stations were selected based on the length of record, data completeness, and spatial distribution across the country. Relative homogeneity was assessed using a Canadian developed technique based on regression models. A bias in the minimum temperatures was identified and adjusted at principal stations located in eastern Canada. Although a bias was also detectable in western Canada, it was smaller and hence no bias adjustments were performed. The spatial presentation of linear trends before and after adjustments indicated overall improvement in the regional and national trends in terms of spatial consistency.

Serra et al. (2001) used the entropy concept and spectral power analysis to analyze the homogeneity, randomness, trends and their statistical significance, and time irregularities in the daily maximum and minimum temperature series (1917 to 1998) recorded at Fabra Observatory, Barcelona. The homogeneity, randomness and the statistical significance of trends in the time series were tested by using the adaptive Kolmogorov-Zurbenko filter, the von Neumann ratio test, and the Spearman and Mann-Kendall tests, respectively. The periodicities obtained from spectral power analyses were checked with the hypothesis of white-noise and Markov's red-noise stochastic processes. The most notable features, common to maximum and minimum temperatures, were the lack of randomness in the series and the different trends obtained for the periods 1917-1980 and 1917-1998, which were confirmed by the Spearman and sequential Mann-Kendall tests. Nevertheless, the maximum and minimum temperature series showed a very different behaviour based on the time irregularities in terms of entropy and periodicities.

Astatkie et al. (2003) used the daily average temperature data of 15 locations spatially distributed across Canada to test the presence of trend in variability (measured by the range, standard deviation and IQR) by using a bootstrap method. The length of the temperature series at these sites ranged from 30 to 151 years. The analysis was undertaken for the monthly, seasonal, and annual data. For calculating standard deviations, estimates of the annual mean temperatures were used to make the results invariant to the presence of trend in mean. The monthly and seasonal analysis revealed the presence of either increasing or decreasing variability for some months and some seasons. The results of the annual data analysis did not reveal appreciable variability, especially at sites where some months have an increasing trend while others have a decreasing trend. The results across sites did not exhibit a clear geographic pattern. However, consistently increasing trends in the variability were found in Toronto and St. John's during non-summer months, and mostly decreasing trend in Edmonton. The significance of trend in the variability measured by the range and standard deviation were consistent in less than 30% of the time across sites and across the monthly, seasonal and annual aggregations. There was not much agreement between the standard deviation and the IQR, which shows the importance of the choice for measuring variability.

Singh et al. (2008) estimated seasonal and annual trends of change in maximum temperature (T_{max}), minimum temperature (T_{min}), mean temperature (T_{mean}), temperature range (T_{range}), highest maximum temperature (H_{max}), and lowest minimum temperature (L_{min}) in northwest and central India by applying Mann-Kendall test. Monthly data of first four variables and annual data of last two variables were used for 90- to 100-year period for 43 stations over nine river basins. Of the nine river basins studied, seven showed a warming trend, whereas two showed a cooling trend. The Narmada and Sabarmati river basins experienced the maximum warming and cooling, respectively. The majority of basins showed increasing trend in T_{range}, H_{max} and L_{min}. Seasonal analysis of different variables shows that the greatest changes in T_{max} and T_{mean} were observed in the post-monsoon season, while T_{min} experienced the greatest change in the monsoon season. This analysis provides scenarios of temperature changes which may be used for sensitivity analysis of water availability for different basins, and accordingly in planning and implementation of adaptation strategies.

Feng et al. (2011) applied parametric regression test and nonparametric Mann-Kendall test to analyze trends in annual temperature of 16 hydrological stations in Nenjiang River basin, Northeastern China. The mean annual temperature data of 1956-2006 were used for the analysis. The results showed significant increasing trends of annual and seasonal mean temperature versus time. An overall increase of 2 °C in the temperature was observed over the past 50 years.

6.2.3 Evapotranspiration

Zaninovic and Gajic-Capka (2000) analyzed the variations and trends in some water balance components viz., soil water content, evaporation losses from the surface and subsurface soil layers, transpiration, groundwater recharge and runoff. These components were calculated by the Palmer method using the 1900-1995 data from Osijek, Croatia. Besides the meteorological inputs necessary for the water balance calculation (i.e., precipitation, temperature and relative humidity), the pedological characteristics of this area was also taken into account. Fluctuations have been considered by means of the 11-year binomial filtered series and the linear trends were tested by means of the Mann-Kendall rank test. A progressive analysis of the time series was also performed to obtain further insights into the trends of water balance components. The results suggested a significant increase in the potential evapotranspiration and evapotranspiration, but a decrease in the runoff and soil water content during the twentieth century.

Hobbins et al. (2001) analyzed the annual and seasonal trends in a monthly time series of actual evapotranspiration using the Mann-Kendall test within the context of the complementary relationship on a regional basis to establish that regional trends can be determined to originate in either the energy budget or the water budget, or both. The monthly time series of 27 years at a 5-km resolution over the conterminous United States was created by using a regional, seasonal Advection-Aridity model, which provided a tool for studies on climate change and variability based on comparison of intra-annual trend results with results from another study.

6.2.4 Climatic Change

Burn (1994) examined the impact of climatic change on the timing of spring runoff by using a nonparametric statistical trend test applied to the datasets of 84 natural rivers from the west-central region of Canada. The results indicated that a greater number of rivers exhibit earlier spring runoff than can be attributed to the chance occurrence. The impacts on the timing of spring runoff were found more prevalent in the recent portion of the records, which is consistent with what one would expect if the impacts are due to the greenhouse gas-induced climatic change.

Charvátová (1995) processed the time series of solar and volcanic activities as well as the time series of surface air temperature in connection with climatic change in recent centuries in central Europe. The subintervals used corresponded to the two types of solar motion, the ordered and chaotic. The exceptional and repeating behaviour of these phenomena in the intervals of the ordered motion of the Sun, in spacing of 180 years, and a quite different behaviour in the intervals of chaotic motion were demonstrated by means of statistical characteristics. The results enabled predictions because the solar motion can be computed well in advance.

Vassilev and Georgiev (1996) reported that climatic changes have already started in Bulgaria. Because water resources to a great extent depend upon climate dynamics, the linkage of climatic fluctuations and water resources were developed using the time-series analysis of climatic parameters and some river runoff characteristics and their correlation. The results of this study pointed to several problems of regional and local importance. It was concluded that the climatic change manifested during last 15 years has significantly influenced the river runoffs. A study of the recent prolonged drought and the influence of changes on water resources and human activity begins with characterizing climate-hydrology linkages.

Westmacott and Burn (1997) evaluated the possible effects of climate change on four hydrologic variables pertaining to the magnitude and timing of hydrologic events in the Churchill-Nelson River basin of west-central Canada. By using the Mann-Kendall trend test, and a regionalization procedure, the severity of climatic effects within the river basin was quantified, which was then used to create awareness about future consequences of water resource planning and management strategies. It was found that the magnitude of hydrologic events decreased during the study period, while the snowmelt runoff events occurred earlier. The only exceptions to this behaviour were the spring mean monthly streamflows, which exhibited increasing trends due to the potential for snow melting during the study period. The timing of a hydrologic event was greatly influenced by the changes in temperature. Further, the decreasing trends were found to be concentrated in the southern regions of the basin whereas the increasing trends were found primarily in northern regions.

Lubes-Niel et al. (1998) investigated the power and the robustness of some widely-used climatic variability tests with the help of simulation. In each case, 100 samples of fifty elements were generated based on the main characteristics of natural rainfall series. A shift in the mean was used to represent a possible climatic variation. The rank correlation test, Pettitt's test, Buishand's test, Lee and Heghinian's bayesian method, and the Hubert and Carbonnel's segmentation method were used for hydrometeorological series. Each simulation of 100 samples were used to assess the performance of different methods considering a specific characteristic of the series, viz., normality or nonnormality, autocorrelation, trend, and shift in the variance. The rank correlation test, Pettitt's test, Buishand's test and the segmentation method with a significance level of 1% (significance level of Scheffé's test) rejected heterogeneity less than 10 series over 100 homogeneous simulated series. On the other hand, the Lee and Heghinian's bayesian method rejected about 40% of the series. This finding suggests that the latter method should be applied only under the hypothesis of heterogeneity. Independent series were simulated by normal, log-normal and Pearson distributions to compare the performance of the methods requiring normality. The results indicated that the normality has no significant impacts on the performance of the methods used. However, the simulation results indicated that the condition of independence of the successive elements of the series is essential to keep the performance constant. Otherwise a trend in the series makes the methods inefficient, except for the rank correlation test for which the alternative is a trend. None of the method were found to be robust against both negative and positive autoregressive dependencies.

Loaiciga et al. (2000) derived climate change scenarios created from scaling factors based on several general circulation models to assess the impacts of aquifer pumping on the water resources of the Edwards Balcones Fault Zone (BFZ) aquifer, Texas, which is one of the largest aquifer systems in the United States. Historical climatic time series for the periods of extreme water shortage (1947-1959), near-average recharge (1978-1989), and above-average recharge (1975-1990) were scaled to $2\times CO_2$ conditions to create aquifer recharge scenarios in a warmer climate. Several pumping scenarios were combined with the $2\times CO_2$ climate scenarios to evaluate the sensitivity of water resources impacts to human-induced stresses on the Edwards BFZ aquifer. The $2\times CO_2$ climate-change scenarios were linked to surface hydrology and used to drive aquifer dynamics with alternative numerical simulation models calibrated for the Edwards BFZ aquifer. The aquifer simulation indicated that given the predicted growth and water demand in the Edwards BFZ aquifer region, the aquifer's groundwater resources seem to be threatened under $2xCO_2$ climate scenarios. It was also found that the $2xCO_2$ climatic conditions could exacerbate negative impacts and water shortages in the Edwards BFZ aquifer even if pumping does not increase above its present average level. Based on the historical evidence and the results of this study, it was concluded that without proper consideration to variations in aquifer recharge and sound pumping strategies, the water resources of the Edwards BFZ aquifer could be severely affected under a warmer climate.

Burn and Elnur (2002) analyzed 18 hydrologic variables for a network of 248 Canadian catchments reflecting natural conditions. The Mann-Kendall test was used to detect trends and a permutation approach was used to estimate the test distribution. The catchments having trends in hydrologic variables were further studied to examine trends in meteorological variables and explore the relationship between hydrologic and meteorological responses to climatic change. It is concluded that a greater number of trends were detected than are expected to occur by chance. There were differences in the geographic location of significant trends in the hydrologic variables, which indicated that the climatic impacts were not spatially uniform.

Yu et al. (2002) investigated the impact of climate change on the water resources of the Kao-Pen Creek basin in southern Taiwan. The historical trends of salient meteorological variables (i.e., mean daily temperature, mean daily precipitation on wet days, monthly wet days, and the transition probabilities of daily precipitation occurrence in each month) were detected using the nonparametric Mann-Kendall test. The trends of these meteorological variables were then employed to generate runoff under future climatic conditions using a continuous rainfall-runoff model. The results indicated that the transition probabilities of daily precipitation occurrence significantly influence the precipitation generation, and the generated runoff under future climatic conditions was found to increase during the wet season and decrease during the dry season.

6.3 Application of Time Series Analysis in Surface Water Hydrology

6.3.1 Streamflow

Cehak (1979) performed the frequency analysis of the flood data from five stations in the east Alps rivers by using Fisher-Tippett II and III distributions. The data of the longest observation periods were fitted well to the Type II distribution. A comparison of the calculations with only a part of the observations showed that the magnitude of the curvature parameter k of the Fisher-Tippett curves increases with increasing observation periods. A trend test indicated an increasing trend in all of the Danube stations after about 1860 and no trend in the flood data from southern rivers. The variance spectra suggested that the series may be considered as realization of autoregressive processes of the first order. Superposed to the red-noise spectra there were waves with periods of 2.4 and 8 years in the case of Danube stations and with periods of 5.7 and 6.7 years in the case of southern rivers.

Lye and Lin (1994) analyzed the peak flow series from 90 Canadian rivers to examine stationarity. The results suggested that although short-term dependence is practically absent for most peak flow series, significant long-term dependence is present for a large number of peak flow series tested. It was demonstrated that the most statistical tests of independence or stationarity are designed to detect only short-term serial correlation. They were found insensitive to the long-term serial correlation structure of flood records, which can be far more important.

Lin and Lye (1994) investigated the suitability of Sen's method (Gilbert, 1987) for modelling hydrologic time series, especially the generation of synthetic flow series. It was found that several problems exist with the proposed method. They are: too many parameters in the model, difficulties in modelling skewed series, and finding a suitable stochastic model for the residuals between the original and fitted cumulative departure curves. On the other hand, it was found that the Sen's method is effective in preserving the Hurst phenomenon and is especially suited for modelling time series with a relatively high Hurst coefficient but a low lag-one serial correlation coefficient. It was also demonstrated that after some modifications in the Sen's method, some of the problems could be overcome to a certain extent.

Knapp (1994) analyzed the long-term streamflow records of Upper Mississippi River Basin to determine trends in streamflows and flooding. Trends in average flow and flooding were found strongly correlated to the coincident increases in average annual precipitation. For many portions of the watershed, the precipitation and streamflows over the last three decades were found higher than any earlier period on record. Outside of the dominant influence of climate variation, only one major change in Mississippi River flood discharges was observed. Further, the flood control reservoirs in the Missouri River watershed appeared to produce a 10% reduction in the average flood peak and flood volume for the Mississippi River at St. Louis, Missouri.

Rao (1995) analyzed the rainfall (1901-1990) and streamflow series (1926-1980) of Mahanadi River, India in relation to the climatic change in the river basin. The analysis of trends in the runoff from the upper catchment suggested a steady decrease in the river flows at Hirakud and Naraj gauging stations during the 55-year period of the study. In order to increase confidence in this result, the moisture indices for the catchment were computed and examined, which also indicated a clear declining trend during the period 1901-80. It was concluded that the climate warming that occurred over the basin resulted in a gradual decrease of river flows of the upper catchment as well as of the entire basin during the period 1926 to 1980.

Lins and Slack (1999) determined secular trends in the streamflows of 395 climate-sensitive streamgaging stations in the conterminous United States by the nonparametric Mann-Kendall test. Trends were calculated for the selected quantiles of discharge [0th] to [100th percentiles] to evaluate the differences between low-, medium-, and high-flow regimes during the twentieth century. Two general patterns emerged: (i) trends are most prevalent in the annual minimum (Q_0) to median (Q_{50}) flow categories and least prevalent in the annual maximum (Q_{100}) category; and (ii) at all but the highest quantiles, streamflow has increased across broad sections of the United States. The decreases in streamflow was found only in parts of the Pacific Northwest and Southeast. Systematic patterns were less apparent in the Q_{100} flow. Hydrologically, these results imply that the conterminous US is getting wetter.

Douglas et al. (2000) evaluated trends in the flood and low streamflows of the US by using a regional average Kendall's S trend test at two spatial scales and over two timeframes. The field significance was assessed following a bootstrap methodology to account for the regional cross-correlation of streamflows. The flood flow series was found trend-free at 5% level of significance, but low streamflows showed upward trends with significant temporal persistence. After removing serial correlation from the series, significant trends in low flows were apparent but were less in numbers. The ignorance of regional cross-correlation resulted in statistically significant trends in all but two of the low flow analyses and in two-thirds of the flood flow analyses. In addition, it was found that the cross-correlation of streamflow records dramatically reduces the effective number of samples required for trend assessment.

Radziejewski et al. (2000) normalized and de-seasonalized the raw series of good quality streamflow data and subsequently transformed to the Fourier spectral domain. Keeping the power spectrum preserved, the phase spectrum was subjected to randomization. After transformation back to the temporal domain, the data were contaminated with trends and step changes in a controlled way. Then the detectability of nonstationarity by particular tests as a quasi-continuous function of magnitude of the contaminating change was evaluated. A method was devised to compare the tests' performance. The analysis of detectability versus the magnitude of change provides a new insight into the properties of the tests.

Zhang et al. (2001) studied the trends computed for past 30-50 years for 11 hydroclimatic variables obtained from the recently created Canadian Reference Hydrometric Basin Network Database. It was found that the annual mean streamflow has generally decreased during the periods, with significant decreases in the southern part of the country. The monthly mean streamflow has also decreased, with the greatest decrease occurring in August and September. However, significant increases in streamflow were observed in March and April. Furthermore, significant increases were identified in lower percentiles of the daily streamflow frequency distribution over northern British

Columbia and the Yukon Territory. On the other hand, in southern Canada, significant decreases were observed in all percentiles of the daily streamflow distribution. The breakup of river ice and the ensuing spring freshet occur significantly especially in British Columbia. The results also suggest earlier freeze-up of rivers, particularly in eastern Canada. The trends observed in the hydroclimatic variables in this study are in agreement with those identified in the climatic variables in other Canadian studies.

Alemaw and Chaoka (2002) investigated possible trends in the annual riverflow of 502 rivers (data from early 1950s to late 1990s) in the region of South Africa by visualization technique. The rescaled adjusted partial sums (RAPS) were used instead of the actual time series plots of runoff. A simulation experiment of the technique was conducted to demonstrate how the plot of RAPS offers a reasonable visualization of the readily apparent mode of underlying trend, which may be hidden in the standard time series plots. The dominantly visualized trends were linear and declining. A subsequent linear trend test by fitting a linear trend model to the annual river discharge series revealed a dominant negative slope ranging from –6.8 to –0.2%; it suggests the existence of declining trends in some rivers of the South African region. Of the 502 time series under study, 137 time series had statistically significant decreasing trend, 96 series had significant increasing trend, and the remaining 269 series had no trend at all.

Birsan et al. (2002) analyzed the mean daily runoff data from undisturbed and independent watersheds in Switzerland for detecting trends by the Mann-Kendall trend test. Based on the seasonal analyses of streamflow quantiles, it was found that: (i) the streamflow has increased in the winter period, especially the winter annual maximum, at about 60% of the stations; and (ii) the streamflow has decreased in the summer period, particularly the low streamflow quantiles. The trends were found to be statistically significant, which indicate a substantial change in the streamflow regime.

Ramesh and Davison (2002) proposed semi-parametric approaches to trend analysis using local likelihood fitting of annual maximum and partial duration series, and demonstrated their application to the exploratory analysis of changes in extremes in sea level and river flow data. Bootstrap methods were used to quantify the variability of estimates.

Adeloye and Montaseri (2002) described three tests for determining consistency, trend, and randomness in hydrological data series. The tests were then applied to monthly streamflow data records from seven sites — three in Iran and four in Yorkshire, England. All hydrological series were found consistent, trend-free and random. Furthermore, few goodness-of-fit tests (i.e., Chi-square, Kolmogorov-Smirnov, probability plot correlation coefficient and moment ratio diagram) for probability distribution are discussed. Based on the results, probability plot correlation coefficient test was found simple to use and this test can be employed even if critical test-statistic values are not known.

Beighley and Moglen (2002) analyzed the trends of nonstationary discharge corresponding to the periods of urbanization by employing three statistical tests: one parametric *t*-test and two nonparametric tests (Kendall's Tau and Spearman Rank Correlation tests) using the annual maximum discharge and annual maximum discharge-precipitation ratios series. It was concluded that the ratios are more effective than the discharges alone for identifying nonstationarity resulting from urbanization. In addition, the relationships between measures of urbanization and the presence/absence of significant trends in the discharge series are presented.

Kahya and Kalayci (2004) presented a trend analysis of 31-year monthly streamflows obtained from 26 basins of Turkey. Four non-parametric trend tests (i.e., the Sen's T, Spearman's Rho, Mann-Kendall, and the Seasonal Kendall), which are popular for detecting linear trends in a hydrological time series were used. The Van Belle and Hughes' basin-wide trend test was also included in the analysis. Homogeneity of trends in monthly streamflows was tested following the method developed by Van Belle and Hughes. Thus, this study presents a complete application of both the Van Belle and Hughes' tests for homogeneity of trends and basin-wide trend tests (originally developed for trend detection in water quality data) in a hydroclimatic variable. The results revealed that the basins located in western Turkey, in general, exhibit downward trend (significant at the 0.05 or lower level), whereas the basins of eastern Turkey have no trends. In most cases, the first four trend tests were found to yield the same conclusion about the trend existence. Furthermore, based on the Van Belle and Hughes' basin-wide trend test, some basins located in southern Turkey were found to exhibit a global trend, which suggests the homogeneity of trends both in seasons and in stations.

Ludwig et al. (2004) presented a detailed portrait of the average hydroclimatic patterns in the Têt River basin using the data from 1980 to 2000, which is a typical Mediterranean river in the south of France where short but violent flash-floods frequently occur. Average temperatures during the selected period were the warmest of the last century. Average spring temperatures in the basin followed a highly significant trend of increasing temperatures. The average autumn temperature, however, decreased and partly counterbalanced the temperature increase in springs. The mean annual runoff was found highly variable, but no clear trend over the investigated period was detected. However, an increasing trend was detected in the flood discharge in the downstream portion of the basin. Also, the mean annual precipitation over the entire basin showed no clear evolution, but the contribution from the upper part of the basin was found decreased, whereas the contribution from the middle and lower parts of the basin was found increased. It was concluded that the humidity of Mediterranean origin is more important for the hydro-climatic functioning of the Têt basin. If the detected trends persist in the future, the flood frequency is most likely to increase.

Yurekli et al. (2004) analyzed daily maximum streamflow data of each month from three gauge stations on Cekerek Stream in Turkey for simulation using stochastic approaches. Initially non-parametric Mann-Kendall (MK) test was used to identify the trend during study period. The two approaches of stochastic modelling, ARIMA and Thomas-Fiering models, were used to simulate monthly maximum data. The error estimates (RMSE and MAE) of predictions from both approaches were compared to identify the most suitable approach for reliable simulation. The MK test suggested no linear trend in monthly maximum data sequences of each mentioned gauge station. The two error estimates calculated for two approaches indicate that ARIMA model appear to be slightly better than Thomas-Fiering. However, both approaches were identified as appropriate method for simulating daily maximum streamflow of Cekerek Stream.

Recent evidence of nonstationary trends in water resources time series as a result of natural and/or anthropogenic climate variability and change, has raised more interest in nonlinear dynamic system modelling methods. Coulibaly and Baldwin (2005) investigated the effectiveness of dynamically driven recurrent neural networks (RNN) for complex time-varying water resources system modelling. An optimal dynamic RNN approach is proposed to directly forecast different nonstationary hydrological time series. The proposed method automatically selects the most optimally trained network in any case. The simulation performance of the dynamic RNN-based model is compared with the results obtained from optimal multivariate adaptive regression splines (MARS) models. It is shown that the dynamically driven RNN model can be a good alternative for the modelling of complex dynamics of a hydrological system, performing better than the MARS model on the three selected hydrological time series, namely the historical storage volumes of the Great Salt Lake, the Saint-Lawrence River flows, and the Nile River flows.

Burn et al. (2008) performed trend analysis on streamflow data in terms of spring and summer runoff volumes, peak flow rates and peak flow occurrences, as well as an annual volume measure, for analysis periods of 1966-2005, 1971-2005, and 1976-2005 for 26 hydrometric gauging stations in Canadian Prairies. The data were analyzed by using Mann-Kendall test. The Mann-Kendall test for trend and bootstrap resampling were used to identify the trends and to determine the field significance of the trends. Partial correlation analysis was used to identify relationships between hydrological variables that exhibit a significant trend and meteorological variables that exhibit a significant trend. The results revealed decreasing trends in the spring snowmelt runoff event volume and peak flow, decreasing trends (earlier occurrence) in the spring snowmelt runoff event peak date and decreasing trends in the seasonal (1 March-31 October) runoff volume. These trends were attributed to a combination of reductions in snowfall and increases in temperatures during the winter months.

Wu et al. (2008) detected spatial and temporal trends in streamflow droughts in terms of frequency, duration and severity in Nebraska. The studies were conducted on three time periods: 1970-2001 (60 stations), 1950-2001 (43 stations), and 1932-2001 (nine stations). The analysis was performed on the drought event parameters by applying correlation between event parameters tests, Hurst coefficients and lag-one coefficients, and trend-free pre-whitening Mann-Kendall (TFPW-MK) tests. The analysis showed that there is no uniform trend on the streamflow drought in the entire state. However, some trends are evident for specific regions. Specifically, it is most likely that droughts in the Republican watershed have become more intense; whereas the drought has become slightly alleviated in the Missouri and nearby watersheds.

Khaliq et al. (2009a) reviewed usefulness of four methods for identification of hydrological trends (Mann-Kendall, Spearman rank correlation, Sen's slope and linear regression tests) in the presence/absence of short-term serial and cross correlations. The ability of the reviewed tests for detecting trends and interpreting their collective results is demonstrated by a case study of annual mean daily flows of Canadian river basins. The results of the case study indicated that failure to incorporate the effects of serial and cross correlations in a trend investigation study could result in erroneous conclusions. It is recommended that the old practice of identifying hydrological trends without the consideration of serial and cross correlations should be avoided and these characteristics should be given adequate attention in all studies on temporal trends.

Khaliq et al. (2009b) investigated trends in annual, summer and winter time series of 30-day low flows occurring in pristine rivers basins of Canada. This study investigated effect of independence (IND), short-term persistence (STP) and long-term persistence (LTP) using the observed low flow data and simulated time series of known STP- and LTP-like structures. Mann-Kendall (MK) test along with two modified versions and a block bootstrap resampling test were used to estimate significance of temporal trends under the assumptions of IND and STP. A semi-parametric and a parametric procedure based on the fractional autoregressive integrated moving average modelling approach and a MK scaling test were used to estimate significance of temporal trends under the assumption of LTP. The results of the study suggested that for a majority of the time series of low flows analyzed, the assumption of IND or STP cannot be ruled out. On the contrary, the fluctuating behaviour of trends revealed in selected time series of low flows, with longer records, using moving window technique favour the LTP hypothesis. In general, the results of trend investigation suggested that the estimates of trend significance are highly sensitive to IND, STP and LTP assumptions, e.g., adopting IND assumption instead of LTP for a given set of hydrological time series exhibiting LTP, could result in incorrect estimation of trend significance. Also, substituting IND assumption for STP would result in incorrect estimates of trend significance. Therefore, for reliable trend investigation, satisfactory identification of STP- or LTP-like behaviour in hydrologic time series, which seldom exceed 100 years, is important and challenging, and must be given adequate attention in all trend investigation studies.

Zhang et al. (2010b) employed repeated monotonic trend tests with varying beginning and ending time for detecting changes in streamflow in tributaries within the Susquehanna River Basin, USA. The method was employed to analyze long-term streamflow trends and detect change for annual minimum, median, and maximum daily streamflow for eight unregulated watersheds within the basin. Monthly baseflow and storm runoff were investigated. The results showed a considerable increase in annual minimum flow for most of the examined watersheds and a noticeable increase in annual median flow for about half of the examined watersheds. Both these streamflow increases were abrupt, with only a few years of transition centered around 1970. The abrupt change in annual minimum and median flows appeared to occur in the summer and fall seasons. The abrupt change in annual minimum and median flows was considered as a result of increased flows in the summer and fall seasons. The results also indicated that there is no long-term significant increasing or decreasing change in annual maximum flow in the examined watersheds.

Zhao et al. (2010) investigated seasonal and annual trends of streamflow and the correlations between streamflow and climatic variables in five sub-basins in Poyang Lake basin in the southeast China over 50-year period. The Theil-Sen approach and the non-parametric Mann-Kendall test were applied to identify the trends in the annual and seasonal streamflow, precipitation and evapotranspiration series. It was found that annual and seasonal streamflow of all the stations had increasing trends except Lijiadu station in wet season. Only 37.5% hydro-stations in annual streamflow increased significantly, while most stations increased at 95% significance level in dry season. Trends in annual and seasonal precipitation during the whole period were generally not as significant as those in evapotranspiration. The correlations between streamflow and climate variables (precipitation and evapotranspiration) were detected by the Pearson's test. The results showed that streamflow in the Poyang Lake basin are more sensitive to changes in precipitation than potential evapotranspiration.

6.3.2 Surface Water Quality

Hirsch et al. (1982) presented techniques for the exploratory analysis of monthly water quality data for monotonic trends. The first procedure is a nonparametric test for trend detection, which is applicable to the datasets having seasonality. The second procedure, the seasonal Kendall slope estimator, is an estimator of trend's magnitude. The third procedure provides a means for testing the temporal change in the relationship between constituent concentration and streamflow.

El-Shaarawi et al. (1983) examined the temporal changes in some water quality parameters (i.e., pH, alkalinity, total phosphorous, and nitrate concentrations) using the 1975-1980 data of the Niagara-on-the-lake in Ontario. The moving averages, Spearman's rank correlation coefficients and regression methods, which model the seasonal cycle, were used in this study. It was found that the pH and alkalinity are decreasing, and nitrate increasing, but these changes were not found in all months.

Van Belle and Hughes (1984) proposed several nonparametric tests for detecting trends in water quality, because the assumptions of classical parametric tests are usually not met by water quality data. Also, the additional peculiarities of data, such as missing values, censored data, and seasonality, compound the problem of analysis.

Harned and Davenport (1990) analyzed trends in the water quality data of 1945-1988 from major streams flowing into the Albemarle-Pamlico estuarine system. The nonparametric seasonal Kendall test indicated a change in the water quality data during the 1945 to 1988 period. The evaluation of water-quality data and more than 50 basin variables indicated 121 significant correlations between 11 basin characteristics and 12 water-quality constituents at 21 estuary locations and seven National Stream Quality Accounting Network stations.

Yu et al. (1993) analyzed the surface water quality data of the Arkansas, Verdigris, Neosho, and Walnut River basins, Kansas to examine trends in 17 major constituents by using four different nonparametric methods. The results indicated that the concentrations of specific conductance, total dissolved solids, calcium, hardness, sodium, potassium, alkalinity, sulfate, chloride, total phosphorus, ammonia plus organic nitrogen, and suspended sediment generally have downward trends. Some of the downward trends were related to the increase in discharge, while the others were attributed to the decrease in pollution sources. The homogeneity tests suggested that both the station-wide trends and basin-wide trends are non-homogeneous.

Robson and Neal (1996) examined the trend of ten years upland stream and bulk deposition water quality data from Plynlimon, mid-Wales by applying the seasonal Kendall test. The plotted data on time scale showed long-term cycles, which relate to the fluctuations in weather patterns at Plynlimon and thus violate the assumptions of common statistical trend tests. Even though the seasonal Kendall test was significant for some determinands, the graphs suggested that many of these trends are unlikely to continue. There was no indication of changing acid deposition inputs or changing acidity within the runoff, despite a decline in the UK sulphur dioxide emissions. The streamwater dissolved organic carbon showed an increase over time, but there was not corresponding decrease in pH as might be expected from the acidification theory. There were cyclic variations in bulk precipitation and in streamwater quality, which indicated that trends cannot be established even with 10 years of data. Therefore, it was strongly recommended that long-term monitoring programmes should continue for several decades. It was also emphasized that graphical analysis greatly enhances data interpretation, and should be considered as an essential component for trend investigation.

Kalayci and Kahya (1996) detected linear trends in the surface water quality of rivers in the Susurluk Basin by employing four nonparametric trend tests, viz., the Sen's T test, Spearman's Rho test, Mann-Kendall test and the Seasonal Kendall test. The linear slopes (change per unit time) of trends were calculated by using a nonparametric estimator. In addition, the homogeneity in monthly trends was tested by the Van Belle and Hughes method. The results of the nonparametric tests indicated that the discharge and sediment concentration have downward trends, while the temperature, EC and the concentrations of sodium, potassium, calcium+magnesium, bicarbonate and chloride have upward trends. In contrast, the concentrations of carbonate, pH, sulfate, organic matter, and boron have no trends.

Harned and McMohan (1997) examined the monotonically increasing or decreasing temporal trends in riverine water quality including the suspended sediment, solids, and nutrients for six stations of the Contentnea Creek Basin by using the seasonal Kendall test. The variation in water quality, because of the variation in streamflow, was also accounted for in cases where streamflow data were available. Nutrient concentrations for Contentnea Creek at Hookerton have declined since 1980. Total nitrogen, nitrate+nitrite, and nitrate concentrations have a significant declined trend, with the greatest reductions occurring from 1980 to 1992. Total ammonia and organic nitrogen concentrations, which were increasing during the 1980s, have declined since around 1990. Total phosphorus, dissolved phosphorus and orthophosphorus, which increased during the 1980, have shown a significant decline since 1988 — the first year of the legislated phosphate detergent ban.

Antonopoulos et al. (2001) analyzed the time series of water quality parameters and the discharge of Strymon River, Greece for the 1980-1997 period. The nonparametric Spearman's criterion was employed to detect the existence of trends for the variables: discharge, ECw, DO, SO_4^{-2}, Na^+, K^+ and NO^{-3} and the evaluation of the best-fitted models were performed by using χ^2-test and the Kolmogorov-Smirnov test. Furthermore, the relationships between concentration and loads of constituents both with the discharge were also examined. According to the correlation coefficient (r) values, the relation between concentrations and discharge is weak ($r < 0.592$), whereas the relation between loads and discharge is very strong ($r > 0.902$).

Stansfield (2001) illustrated the importance of considering detection limits of variables and sampling frequencies by analyzing the trends in water quality time series (part of the Wellington Regional Council's freshwater baseline water quality monitoring programme, New Zealand) using the nonparametric seasonal Kendall test and the Sen Slope estimator. Results indicated that the trends (upward and downward) obtained using a low detection limit are often not discernible when a higher limit is adopted. It was also found that if the sampling frequency was changed from monthly to quarterly, fewer trends were detected. Moreover, the quarterly data exhibited a trend, which was usually of different magnitude (slope) compared to that in the monthly data.

Gangyan et al. (2002) examined the temporal and spatial sediment load characteristics of Asia's longest river, the Yangtze by using the turning point test, Kendall's rank correlation test and the Anderson's correlogram test for randomness and trend identification. The annual sediment load data from 1950 to 1990 and the monthly sediment load data from 1950 to 1969 were used. The periodicity was analysed by harmonic analysis and the stochastic component was modelled by autoregressive model. The analysis of the results indicated that the annual sediment load series is trend-free at 5% significance level and the monthly means and standard deviations of sediment load have periodicity. The month-to-month correlation structure was found nonstationary. Using the AR(1) model for the dependent stochastic component, 100 years of monthly sediment data were generated, and the observed and generated data matched well.

Jassby et al. (2003) developed a time series model of the Secchi depth for Lake Tahoe, USA incorporating a mechanistic understanding of interannual variability. The Secchi depth was found occasionally over 40 m for Lake Tahoe, but the mean annual Secchi depth has declined by about 10 m since 1967, prompting a large-scale restoration programme. The year-to-year variability was found to be extremely high, obscuring restoration actions and compliance with water quality standards. The model focussed on the Secchi depth during summer, when the lake is least transparent and most heavily used. Interannual variability for the summer season was driven largely by precipitation differences. The model offers a tool for determining the compliance with water quality standards when precipitation anomalies may persist for years. It was also demonstrated by means of an ex-post forecast that the increasing Secchi depths during 1999-2001 are simply climate-driven and do not represent a recovery of the lake. The long-term trend for summer is attributed to the accumulation of allochthonous mineral suspension.

Panda et al. (2011) examined trends in sediment load of the tropical (Peninsular) river basins of India and explored influence of climatic and human forcing mechanisms on the land-ocean fluvial systems. Sediment time series of different timescale during the period 1986-87 to 2005-06 from 133 gauging stations was analyzed. Results indicated dramatic reductions in sediment load in the tropical river basins, which is beyond the fold of assignable natural variability. Around 88 and 62% of the total 133 gauging stations showed a decline in sediment loads in the monsoon and non-monsoon seasons, respectively. The significant downward trends outnumbered the corresponding upward trends in high proportions for both the seasons. Striking spatial coherence was observed among the significant trends, suggesting the presence of the cross-correlation among the sediment records. The regional trends, which account the spatial correlation, also indicated a widespread nature of sediment declines. The rainfall, characterized by the non-significant decreasing trends and frequent drought years, was found to be the primary controller of sediment loads for most of the river basins. It was concluded that a little change in rainfall towards the deficit side leads to a significant reduction in the sediment load. Among the tropical rivers, the maximum reduction in sediment flux took place for the Narmada River (-2.07×10^6 t/yr) due to the construction of dam.

6.4 Application of Time Series Analysis in Groundwater Hydrology

6.4.1 Groundwater Flow

Molénat et al. (1999) viewed the catchment as a system that converts the rainfall to the stream discharge through a transfer function (TF). By comparing the observed TF with the simulated TF, the hydrological processes and their time scales were identified. The simulated transfer functions were developed using the Dupuit's assumptions and linear representation of the aquifer. The identification of the TF was based on the stochastic method using a spectral representation of the rainfall and streamflow time series. The novelty of this work is to extend the stochastic approach to the one-order catchment hydrology and to develop a model, which takes into account both the aquifer discharge and rapid flows. The proposed method was applied to three first-order agricultural catchments located in different regions of France. For each site, the obtained results were in good agreement with reality. These results indicated that the streamflow is dominated by the aquifer flow, which is the fast transfer accounting for 3-8% of the total discharge depending on the catchment. The stochastic approach based on the spectral analysis of temporal variations in global observations was emphasized to be useful for extracting significant information about the dominant processes occurring in the catchment and their characteristic time scales.

6.4.2 Groundwater Quality

Chang (1988) developed a modelling technique that includes the homogeneity test of data and the best model selection to fit the water loss series by a stochastic process. The results of this study revealed nonhomogeneities in the annual water loss time series from the Ohio River basin, and hence the adjustments were required before the model fitting by a stochastic process. The best model selected based on the criterion of the parsimony of parameters was successfully used to forecast the regional water losses.

Wilson et al. (1992) established groundwater quality changes caused by anthropogenic activities with the help of a time-series analysis of well water quality data from a 1964-1965 survey. In all cases, Ca^{+2} and Fe^{+3} were found to increase with depth due to the dissolution of Ca minerals as water moves downgradient, and due to a change from oxidizing to reducing conditions downgradient, respectively. NO_3^- and Cl^- concentrations were found higher in the recharge areas possibly due to surface pollution sources. The significant variability of chemical constituents was attributed to the recharge events, aquifer depth, spatial lithological changes, and the anthropogenic activities in recharge areas.

Loftis (1996) reviewed national assessments, agricultural, urban, point source and hazardous waste case studies on regional and localized groundwater quality all over the world, including a few snapshots. Based on this review, the correct meaning of 'trend' was emphasized as a critical step for groundwater quality studies in both temporal and spatial context. Generally, trends are thought of as changes over time at either a regional or local spatial scale and the water quality managers are mostly interested in changes due to artificial activities. It was concluded that although there are many regional groundwater studies, which provide a 'snapshot' water quality description over an area at a time, only few consider temporal changes and still fewer include a statistical analysis of long-term trends.

Lee and Lee (2003) evaluated and quantified the potential of natural attenuation of groundwater at a petroleum-contaminated site in an industrial area of Seoul, Korea. Eight rounds of groundwater sampling and subsequent chemical analyses were performed for a period of three years. Groundwater of the study area was found contaminated by toluene, ethylbenzene and xylene (TEX). TEX concentration was found decreasing with time, with the TEX plume in a quasi-steady state. The trend analysis by the Mann-Kendall Test along with the changes in mass flux and plume area confirmed that the TEX plume reached a quasi-steady state. Furthermore, the proportion of the total attenuation attributable to biodegradation was found decreased during the monitoring period, while the contribution of other attenuation processes (dilution or dispersion) was found increased.

Kim et al. (2005) applied time series analysis to investigate the effect of tide on groundwater quality in a coastal area of Korea. Continuous and regular *in situ* monitored data of electrical conductivity (EC) and groundwater level, and tidal level data measured by the National Oceanographic Research Institute were used for the time series analysis. It was found that EC and groundwater level conspicuously fluctuate with two periodicities (15.4- and 0.52-day), which is very similar to those of the tide. Also, the behaviours of their fluctuations vary in accordance with the tidal period. It was concluded that the groundwater quality is mainly controlled by the tidal level, and the strength of tidal effect on the groundwater quality is different according to the tidal period.

6.5 Time Series Analysis of Irrigation Requirement and Soil Moisture

Gupta and Chauhan (1986) studied the stochastic structure of weekly irrigation requirements of a crop by considering the irrigation requirement time series as an additive model with trend, periodicity and stochasticity as its components. Each component was identified and, if found, removed from the original series. The turning point test and the Kendall's rank correlation test were applied for detecting trends, whereas the correlogram technique was used to detect the periodicity. The harmonic analysis was done for identifying significant harmonics. The series was then tested for stationarity and the dependent part of the stochastic component was found to be expressed well by the second-order autoregressive model. Therefore, the developed model superimposed a periodic-deterministic process and a stochastic component. The adequacy of fit was judged by the insignificant correlation and the normal distribution of obtained residuals. It was concluded that the developed periodic-stochastic model can be used for representing the time-based structure of the irrigation requirement time series for paddy crops.

Wu et al. (1997) evaluated the efficacy of the time series analysis technique to predict average water content in the soil profile and the water content at different soil depths from the measurements made at a single depth. The volumetric water content of a Zimmerman fine sand in Princeton, MN was measured by TDR at six depths during the early 1993 growing season. The time series made up of hourly measurements of soil water content was first-order differenced to obtain stationarity. The differenced data were used to conduct analyses in the frequency domain to evaluate the coherence and cross-amplitude between two soil water content time series and were subsequently fitted to the autoregressive moving average models to obtain coefficients for the transfer function models in the time domain. The transfer function models were then used to predict water contents at 50, 75 and 100 cm depths and the average water content in the top 100-cm soil profile from the measured water content at 25-cm depth. Overall, the predictions were reasonable, with an increased accuracy as the separation distance from the 25-cm depth decreased.

6.6 Concluding Remarks

Time series analysis has been used in a variety of fields in the past, such as hydrology, climatology, geology, ocean engineering, seismology, etc. In this chapter, however, the studies related to only hydrologic and climatologic time series have been reviewed. It is clear from this review that precipitation and streamflow are major hydrologic variables followed by temperature and surface water quality, which attracted the attention of researchers from different parts of the world for applying time series analysis techniques. The application of standard statistical tests and the evaluation of some statistical tests has been a major focus of applied research in this area. Comparatively, less number of studies is reported wherein a new approach is developed or an existing approach is modified to improve overall efficiency of some time series analysis techniques. Furthermore, no study is reported to date which covers all aspects (i.e., basic properties) of hydrologic/hydrogeologic time series analysis. Trend detection by Kendall or seasonal Kendall test has been a major focus of most studies. Unfortunately, remaining trend detection tests and other important properties of the time series (i.e., stationarity, homogeneity, periodicity and persistence) are often ignored. The main reason behind this ignorance seems to be lack of scientists'/researchers' knowledge about the availability of appropriate statistical tests for time series analysis as well as the lack of easy-to-use guidelines/book for their effective application. It is expected that the application of time series analysis to hydrological/hydrogeological variables will expand considerably in the future with gradual advancements in computation technology and increasing availability of software packages for time series analysis. More and more studies encompassing a wide variety of hydrological/hydrogeological variables, together with innovative studies are needed to bring the time series analysis techniques to maturity.

References

Abaurrea, J. and Cebrian, A. (2003). Trend analysis of daily rainfall extremes. http://www.isi-eh.usc.es/resumenes/127_52_abstract.pdf (accessed on 27 July 2003).

Adamowski, K. and Bocci, C. (2001). Geostatistical regional trend detection in river flow data. *Hydrological Processes*, **15:** 3331-3341.

Adamowski, K. and Bougadis, J. (2003). Detection of trends in annual extreme rainfall. *Hydrological Processes*, **17(18):** 3547-3560.

Adeloye, A.J. and Montaseri, M. (2002). Preliminary streamflow data analyses prior to water resources planning study. *Hydrological Sciences Journal*, **47(5):** 679-692.

Alemaw, B.F. and Chaoka, T.R. (2002). Trends in the flow regime of the southern African rivers as visualized from rescaled adjusted partial sums (RAPS). *African Journal of Science and Technology, Science and Engineering Series*, **3(1):** 69-78.

Anderson, P.L., Meerschaert, M.M. and Vecchia, A.V. (1999). Innovations algorithm for periodically stationary time series. *Stochastic Processes and their Applications*, **83(1):** 149-169.

Angel, J.R. and Huff, F.A. (1997). Changes in heavy rainfall in Midwestern United States. *Journal of Water Resources Planning and Management, ASCE*, **123(4):** 246-249.

Anh, V., Lunney, K. and Peiris, S. (1997). Stochastic models for characterisation and prediction of time series with long-range dependence and fractality. *Environmental Modelling and Software*, **12(1):** 67-73.

Antonopoulos, V.Z., Papamichail, D.M. and Mitsiou, K.A. (2001). Statistical and trend analysis of water quality and quantity data for the Strymon River in Greece. *Hydrology and Earth System Sciences*, **5(4):** 679-691.

Astatkie, T., Yiridoe, E.K. and Clark, J.S. (2003). Testing for trend in variability of climate data: Measures and temporal aggregation with applications to Canadian data. *Theoretical and Applied Climatology*, **76(3-4):** 235-247.

Astel, A., Mazerski, J., Polkowska, Ż. and Namieśnik, J. (2004). Application of PCA and time series analysis in studies of precipitation in Tricity (Poland). *Advances in Environmental Research*, **8(3-4):** 337-349.

Aulenbach, B.T., Hooper, R.P. and Bricker, O.P. (1996). Trends in the chemistry of precipitation and surface water in a national network of small watersheds. *Hydrological Processes*, **10(2):** 151-181.

Avinash, A. and Ghanshyam, D. (1988). Time series model of stream flow for a catchment of Ramganga River. *Journal of Institution of Engineers (India), Civil Engineering Division*, **88(Part CI):** 228-230.

Beighley, E. and Moglen, G.E. (2002). Trend assessment in rainfall-runoff behavior in urbanizing watersheds. *Journal of Hydrologic Engineering, ASCE*, **7(1):** 27-34.

Bhuiya, R.K. (1971). Stochastic analysis of periodic hydrologic process. *Journal of the Hydraulics Division, ASCE*, **97(HY 7):** 949-962.

Bîrsan, M.-V., Molnar, P. and Burlando, P. (2002). Streamflow trends in Switzerland. Proceedings of the PHEFRA Workshop, Barcelona, October 16-19, 2002. http://www.ccma.csic.es/dpts/suelos/hidro/phefra/images/chapter_ 38_phefra.pdf (accessed on 27 January 2004).

Boroneant, C., Cazacioc, L. and Gologus, L. (1995). Short time climatic change in precipitation regime. Conference on Climate Dynamics and the Global Change Perspective, Cracow, Poland, October 17-20, 1995. http://www.cyf-kr.edu.pl/~ziniedzw/paper015.html (accessed on 30 January 2004).

Brázdil, R. and Stepánek, P. (1995). Homogenized air temperature and precipitation series of Brno in 1848-1993. Proceedings of the Conference on Climate Dynamics and the Global Change Perspective, Cracow, Poland, October 17-20, 1995. http://www.cyf-kr.edu.pl/~ziniedzw/paper019.html (accessed on 30 January 2004).

Brunetti, M., Buffoni, L., Maugeri, M. and Nanni, T. (2000). Precipitation intensity trends in northern Italy. *International Journal of Climatology*, **20(9):** 1017-1031.

Brunetti, M., Maugeri, M. and Nanni, T. (2001). Changes in total precipitation, rainy days and extreme events in northeastern Italy. *International Journal of Climatology*, **21(7):** 861-871.

Buishand, T.A. (1979). Urbanization and changes in precipitation, a statistical approach. *Journal of Hydrology*, **40(3-4):** 365-375.

Buishand, T.A. (1982). Some methods for testing the homogeneity of rainfall records. *Journal of Hydrology*, **58:** 11-27.

Buishand, T.A. (1984). Tests for detecting a shift in the mean of hydrological time series. *Journal of Hydrology*, **73:** 51-69.

Burn, D.H. (1994). Hydrologic effects of climatic change in west-central Canada. *Journal of Hydrology*, **160(1-4):** 53-70.

Burn, D.H. and Elnur, M.A.H. (2002). Detection of hydrologic trends and variability. *Journal of Hydrology*, **255(1-4):** 107-122.

Burn, D.H., Fan, L. and Bell, G. (2008). Identification and quantification of streamflow trends on the Canadian Prairies. *Hydrological Sciences Journal*, **53(3):** 538-549.

Capodaglio, A.G. and Moisello, U. (1990). Simple stochastic model for annual flows. *Journal of Water Resources Planning and Management, ASCE*, **116(2):** 220-232.

Cayan, D.R., Redmond, K.T. and Riddle, L.G. (1999). ENSO and hydrologic extremes in the western United States. *Journal of Climate*, **12:** 2881-2893.

Cehak, K. (1979). On flood probabilities of East Alpine rivers. *Journal of Hydrology*, **20(1):** 65-82.

Chang, T.J. (1988). Stochastic forecast of water losses. *Journal of Irrigation and Drainage Engineering, ASCE*, **114(3):** 547-558.

Changnon, S.A. and Kunkel, K.E. (1995). Climate-related fluctuation in Midwestern floods during 1921–1985. *Journal of Water Resources Planning and Management, ASCE*, **121(4):** 326-334.

Charvátová, I. (1995). Climatic changes and solar inertial motion. Proceedings of the Conference on Climate Dynamics and the Global Change Perspective, Cracow, Poland, October 17-20, 1995. http://www.cyf-kr.edu.pl/~ziniedzw/paper023.html (accessed on 30 January 2004).

Chen, H.-L. and Rao, A.R. (2002). Testing hydrologic time series for stationarity. *Journal of Hydrologic Engineering, ASCE*, **7(2):** 129-136.

Clarke, R.T. (2002). Fitting and testing the significance of linear trends in Gumbel-distributed data. *Hydrology and Earth System Sciences*, **6(1):** 17-24.

Coulibaly, P. and Baldwin, C.K. (2005). Nonstationary hydrological time series forecasting using nonlinear dynamic methods. *Journal of Hydrology*, **307(1-4):** 164-174.

Cunderlik, J.M. and Burn, D.H. (2002). Local and regional trends in monthly maximum flows in southern British Columbia. *Canadian Water Resources Journal*, **27(2):** 191-212.

Darken , P.F., Holtzman, G.I., Smith, E.P. and Zipper, C.E. (2000). Detecting changes in trends in water quality using modified Kendall's tau. *Environmetrics*, **11(4):** 423-434.

De Luís, M., Raventós, J., González-Hidalgo, J.C., Sánchez, J.R. and Cortina, J. (2000). Spatial analysis of rainfall trends in the region of Valencia (east Spain). *International Journal of Climatology*, **20(12):** 1451-1469.

De Michele, C., Montanari, A. and Rosso, R. (1998). The effects of non-stationarity on the evaluation of critical design storms. *Water Science and Technology*, **37(11):** 187-193.

Douglas, E.M., Vogel, R.M. and Kroll, C.N. (2000). Trends in floods and low flows in the United States: Impact of spatial correlation. *Journal of Hydrology*, **240(1-2):** 90-105.

Ducré-Robitaille, J., Vincent, L.A. and Boulet, G. (2003). Comparison of techniques for detection of discontinuities in temperature series. *International Journal of Climatology*, **23(9):** 1087-1101.

El-Shaarawi, A.H., Esterby, S.R. and Kuntz, K.W. (1983). A statistical evaluation of trends in the water quality of the Niagara River. *Journal of Great Lakes Research*, **9:** 234-240.

Esteban-Parra, M.J. and Castro-Diez, Y. (1995). On the homogeneity of the longest temperature series in Spain: a critical analysis. Proceedings of the Conference on Climate Dynamics and the Global Change Perspective, Cracow, Poland, October 17-20, 1995. http://www.cyf-kr.edu.pl/~ziniedzw/paper034.html (accessed on 30 January 2004).

Fanta, B., Zaake, B.T. and Kachroo, R.K. (2001). A study of variability of annual river flow of the southern African region. *Hydrological Sciences Journal*, **46(4):** 513-524.

Feng, X., Zhang, G. and Yin, X. (2011). Hydrological responses to climate change in Nenjiang river basin, northeastern China. *Water Resources Management*, **25:** 677-689.

Fortuniak, K. (1995). Periodicity of temperature and precipitation in Poland in the period 1956-1990. Conference on Climate Dynamics and the Global Change

Perspective, Cracow, Poland, October 17-20, 1995. http://www.cyf-kr.edu.pl/~ziniedzw/paper037.html (accessed on 30 January 2004).

Gangyan, Z., Goel, N.K. and Bhatt, V.K. (2002). Stochastic modelling of the sediment load of the upper Yangtze River (China). *Hydrological Sciences Journal*, **47(S):** S93-S105.

Giakoumakis, S.G. and Baloutsos, G. (1997). Investigation of trend in hydrological time series of the Evinos River basin. *Hydrological Sciences Journal,* **42(1):** 81-88.

Grew, H. and Werrity, A. (1995). Changes in flood frequency and magnitude in Scotland. Proceedings of the BHS Fifth National Hydrology Symposium, Edinburgh, pp. 3.1-3.9.

Gupta, R.K. and Chauhan, H.S. (1986). Stochastic modeling of irrigation requirements. *Journal of Irrigation and Drainage Engineering, ASCE*, **112(1):** 65-76.

Hamed, K.H. and Rao, A.R. (1998). A modified Mann-Kendall trend test for autocorrelated data. *Journal of Hydrology*, **204(1-4):** 182-196.

Hameed, T., Marino, M.A., DeVries, J.J. and Tracy J.C. (1997). Method for trend detection in climatological variables. *Journal of Hydrologic Engineering, ASCE,* **2(4):** 154-160.

Harned, D.A. and Davenport, M.S. (1990). Water-quality Trends and Basin Activities and Characteristics for the Albemarle-Pamlico Estuarine System, North Carolina and Virginia. U.S. Geological Survey Open-File Report 90-398, 164 pp. http://nc.water.usgs.gov /albe/pubs/APEStrends.html (accessed on 25 January 2004).

Harned, D.A. and McMahon, G. (1997). Trends in surface water quality for the Contentnea Creek Basin, 1980-1996. Proceedings of the Albemarle-Pamlico Estuarine Study Comprehensive Conservation and Management Plan Implementation Forum, June 5-6. http://nc.water.usgs.gov/albe/pubs/APEScont.html (accessed on 25 January 2004).

Haywood, J. and Wilson, G.T. (2000). Selection and estimation of component models for seasonal time series. *Journal of Forecasting*, **19(5):** 393-417.

Henderson, R.J. (1989). Rainfall time series for storm overflow assessment. *Water Science and Technology*, **21:** 1789-1791.

Higashino, M., Kanda, T. and Michioku, K. (1999). Time series analysis and transformation of water quality in an eutrophic reservoir. *Mizu Kankyo Gakkaishi*, **22(8):** 668-676.

Hirsch, R.M. and Slack, J.R. (1984). Nonparametric trend test for seasonal data with serial dependence. *Water Resources Research*, **20(6):** 727-732.

Hirsch, R.M., Slack, J.R. and Smith, R.A. (1982). Techniques of trend analysis for monthly water quality data. *Water Resources Research*, **18(1):** 107-121.

Hobbins, M.T., Ramirez, J.A. and Brown, T.C. (2001). Trends in regional evapotranspiration across the United States under the complementary relationship hypothesis. Hydrology Days 2001, pp. 106-121.

Jain, S. and Lall, U. (2001). Floods in a changing climate: Does the past represent the future? *Water Resources Research*, **37(12):** 3193-3205.

Janos, B., Lucien, D. and Omar, H.R. (1988). Practical generation of synthetic rainfall event time series in a semi-arid climatic zone. *Journal of Hydrology*, **103:** 357-373.

Jassby, A.D., Reuter, J.E. and Goldman, C.R. (2003). Determining long-term water quality change in the presence of climate variability: Lake Tahoe (USA). *Canadian Journal of Fisheries and Aquatic Sciences*, **60(12):** 1452-1461.

Jayawardena, A.W. and Lai, F. (1989). Time series analysis of water quality data in Pearl River, China. *Journal of Environmental Engineering, ASCE*, **115(3):** 590-607.

Johann, G., Papadakis, I. and Pfister, A. (1998). Historical precipitation time series for applications in urban hydrology. *Water Science and Technology*, **37(11):** 147-153.

Kadioglu, M. (1997). Trends in surface air temperature data over Turkey. *International Journal of Climatology*, **17(5):** 511-520.

Kahya, E. and Kalayci, S. (2004). Trend analysis of streamflow in Turkey. *Journal of Hydrology*, **289(1-4):** 128-144.

Kalayci, S. and Kahya, E. (1996). Detection of water quality trends in the rivers of the Susurluk Basin. *Turkish Journal of Engineering and Environmental Sciences*, **22(6):** 503-514.

Keiser, D.T. and Griffiths, J.F. (1998). Problems associated with homogeneity testing in climate variation studies: A case study of temperature in the northern Great Plains, USA. *International Journal of Climatology*, **17(5):** 497-510.

Khaliq, M.N., Ouarda, T.B.M.J. and Gachon, P. (2009b). Identification of temporal trends in annual and seasonal low flows occurring in Canadian rivers: The effect of short- and long-term persistence. *Journal of Hydrology*, **369 (1-2):** 183-197.

Khaliq, M.N., Ouarda, T.B.M.J., Gachon, P., Sushama, L. and St-Hilaire, A. (2009a). Identification of hydrological trends in the presence of serial and cross correlations: Review of selected methods and their application to annual flow regimes of Canadian rivers. *Journal of Hydrology*, **368(1-4):** 117-130.

Khan, A.R. (2001). Analysis of hydro-meteorological time series in the upper Indus basin: Searching evidence for climatic change. International Water Management Institute (IWMI), Working Paper 23, Pakistan Country Series Number 7, Colombo, Sri Lanka.

Kim, J.-H., Lee, J., Cheong, T.-J., Kim, R.-H., Koh, D.-C., Ryu, J.-S. and Chang, H.-W. (2005). Use of time series analysis for the identification of tidal effect on groundwater in the coastal area of Kimje, Korea. *Journal of Hydrology*, **300(1-4):** 188-198.

Kite, G. (1989). Use of time series analyses to detect climatic change. *Journal of Hydrology*, **111:** 259-279.

Knapp, H.V. (1994). Hydrologic trends in the Upper Mississippi River Basin. *Water International*, **19(4):** 199-206.

Kothyari, U.C., Singh, V.P. and Aravamuthan, V. (1997). An investigation of changes in rainfall and temperature regimes of the Ganga basin in India. *Water Resources Management*, **11**(1): 17-34.

Kripalani, R.H. and Kulkarni, A. (2001). Monsoon rainfall variations and teleconnections over South and East Asia. *International Journal of Climatology*, **21(5):** 603-616.

Kristev, L. and Koleva, E. (1995). Variation of the snow cover characteristics in mountain region of Bulgaria. Proceedings of the Conference on Climate Dynamics and the Global Change Perspective, Cracow, Poland, October 17-20, 1995. http://www.cyf-kr.edu.pl/~ziniedzw/paper056.html (accessed on 30 January 2004).

Kumar, V., Jain, S.K. and Singh, Y. (2010). Analysis of long-term rainfall trends in India. *Hydrological Sciences Journal*, **55(4):** 484-496.

Lee, J. and Lee, K. (2003). Viability of natural attenuation in a petroleum-contaminated shallow sandy aquifer. *Environmental Pollution*, **126(2):** 201-212.

Lin, Y. and Lye, L.M. (1994). Modelling long-term dependence based on cumulative departures of annual flow series. *Journal of Hydrology*, **160(1-4):** 105-121.

Lins, H.F. and Slack, J.R. (1999). Streamflow trends in the United States. *Geophysical Research Letters*, **26(2):** 227-230.

Loaiciga, H.A., Maidment, D.R. and Valdes, J.B. (2000). Climate-change impacts in a regional karst aquifer, Texas, USA. *Journal of Hydrology*, **227(1):** 173-194.

Loftis, J.C. (1996). Trends in groundwater quality. *Hydrological Processes*, **10(2):** 335-355.

Loukas, A. and Quick, M.C. (1996). Effect of climate change on hydrologic regime of two climatically different watersheds. *Journal of Hydrologic Engineering, ASCE*, **1(2):** 77-87.

Lubes-Niel, H., Masson, J.M., Paturel, J.E. and Servat, E. (1998). Climatic variability and statistics: A simulation approach for estimating power and robustness of tests of stationarity. *Journal of Water Science*, **11(3):** 383-408 (in French).

Ludwig, W., Serrat, P., Cesmat, L. and Garcia-Esteves, J. (2004). Evaluating the impact of the recent temperature increase on the hydrology of the Têt River (Southern France). *Journal of Hydrology*, **289(1-4):** 204-221.

Lye, L.M. and Lin, Y. (1994). Long-term dependence in annual peak flows of Canadian rivers. *Journal of Hydrology*, **160(1-4):** 89-103.

Machiwal, D. and Jha, M.K. (2006). Time series analysis of hydrologic data for water resources planning and management: A review. *Journal of Hydrology and Hydromechanics*, **54(3):** 237-257.

Machiwal, D. and Jha, M.K. (2008). Comparative evaluation of statistical tests for time series analysis: Application to hydrological time series. *Hydrological Sciences Journal*, **53(2):** 353-366.

Maidment, D.R. and Parzen, E. (1984). Time patterns of water use in six Texas cities. *Journal of Water Resources Planning and Management, ASCE*, **110(1):** 90-106.

Mirza, M.Q., Warrick, R.A., Ericksen, N.J. and Kenny, G.J. (1998). Trends and persistence in precipitation in the Ganges, Brahmaputra and Meghna river basins. *Hydrological Sciences Journal*, **43(6):** 845-858.

Molénat, J., Davy, P., Gascuel-Odoux, C. and Durand, P. (1999). Study of three subsurface hydrologic systems based on spectral and cross-spectral analysis of time series. *Journal of Hydrology*, **222(1-4):** 152-164.

Molénat, J., Davy, P., Gascuel-Odoux, C. and Durand, P. (2000). Spectral and cross-spectral analysis of three hydrological systems. *Physics and Chemistry of the Earth, Part B: Hydrology, Oceans and Atmosphere*, **25(4):** 391-397.

Moraes, J.M., Pellegrino, G.Q., Ballester, M.V., Martinelli, L.A., Victoria, R.L. and Krusche, A.V. (1998). Trends in hydrological parameters of a southern Brazilian watershed and its relation to human induced changes. *Water Resources Management*, **12(4):** 295-311.

Nieplová, E. (1995). Climate changes and variability monitoring and homogenization of observation series in Slovakia. Conference on Climate Dynamics and the Global Change Perspective, Cracow, Poland, October 17-20, 1995, http://www.cyf-kr.edu.pl/~ziniedzw/paper072.html (accessed on 30 January 2004).

Oguntunde, P.G., Friesen, J., van de Giesen, N. and Savenije, H.H.G. (2006). Hydroclimatology of the Volta River Basin in West Africa: Trends and variability from 1901 to 2002. *Physics and Chemistry of the Earth*, **31:** 1180-1188.

Pagliara, S., Viti, C., Gozzini, B., Meneguzzo, F. and Crisci, A. (1998). Uncertainties and trends in extreme rainfall series in Tuscany, Italy: Effects on urban drainage networks design. *Water Science and Technology*, **37(11):** 195-202.

Panda, D.K., Kumar, A. and Mohanty, S. (2011). Recent trends in sediment load of the tropical (Peninsular) river basins of India. *Global and Planetary Change*, **75(3-4):** 108-118.

Perreault, L., Bernier, J., Bobee, B. and Parent, E. (2000). Bayesian change-point analysis in hydrometeorological time series, Part 1: The normal model revisited. *Journal of Hydrology*, **235(3):** 221-241.

Pugacheva, G., Gusev, A., Martin, I., Schuch, N. and Pankov, V. (2003). 22-year periodicity in rainfalls in littoral Brazil. Geophysical Research Abstracts, EGS-AGU-EUG Joint Assembly, Abstracts from the meeting held in Nice, France, April 6-11, 2003, p. 6797.

Radziejewski, M., Bardossy, A. and Kundzewicz, Z.W. (2000). Detection of change in river flow using phase randomization. *Hydrological Sciences Journal*, **45(4):** 547-558.

Raghuwanshi, N.S. and Wallender, W.W. (1997). Field measured evapotranspiration as a stochastic process. *Agricultural Water Management*, **32:** 111-129.

Ramesh, N.I. and Davison, A.C. (2002). Local models for exploratory analysis of hydrological extremes. *Journal of Hydrology*, **256(1-2):** 106-119.

Rao, G.P. (1995). Effect of climate change on streamflows in the Mahanadi River Basin, India. *Water International*, **20(4):** 205-212.

Reed, D.W., Jakob, D. and Robson, A.J. (1999). Statistical procedures for flood frequency estimation. *In:* A.J. Robson and D.W. Reed (editors), Flood Estimation Handbook, Vol. 3, Institute of Hydrology, 338 pp.

Robson, A.J. and Neal, C. (1996). Water quality trends at an upland site in Wales, UK, 1983-1993. *Hydrological Processes*, **10**(2): 183-203.

Robson, A.J., Jones, T.K., Reed, D.W. and Bayliss, A.C. (1998). A study of national trend and variation in UK floods. *International Journal of Climatology*, **18:** 165-182.

Sahin, S. and Cigizoglu, H.K. (2010). Homogeneity analysis of Turkish meteorological data set. *Hydrological Processes*, **24(8):** 981-992.

Schwankl, L.J., Raghuwanshi, N.S. and Wallender, W.W. (2000). Time series modeling for predicting spatially variable infiltration. *Journal of Irrigation and Drainage Engineering, ASCE*, **126(5):** 283-287.

Sen, P.K. (1968). Estimates of the regression coefficient based on Kendall's tau. *Journal of the American Statistical Association*, **63(324):** 1379-1389.

Serra, C., Burgueño, A. and Lana, X. (2001). Analysis of maximum and minimum daily temperatures recorded at Fabra Observatory (Barcelona, NE Spain) in the period 1917-1998. *International Journal of Climatology*, **21(5):** 617-636.

Sharma, K.P., Moore, B. and Vorosmarty, C.J. (2000). Anthropogenic, climatic, and hydrologic trends in the Kosi Basin, Himalaya. *Climatic Change*, **47(1-2):** 141-165.

Singh, P., Kumar, V., Thomas, T. and Arora, M. (2008). Basin-wide assessment of temperature trends in northwest and central India. *Hydrological Sciences Journal*, **53(2):** 421-433.

Stansfield, B. (2001). Effects of sampling frequency and laboratory detection limits on the determination of time series water quality trends. *New Zealand Journal of*

Marine and Freshwater Research, **35(5)**, http://www.rsnz.govt.nz/publish/nzjmfr/2001/91.php (accessed on 26 January 2005).

Tarhule, A. and Woo, M. (1998). Changes in rainfall characteristics in northern Nigeria. *International Journal of Climatology*, **18(11):** 1261-1271.

Tayanç, M. and Toros, H. (1997). Urbanization effects on regional climate change in the case of four large cities of Turkey. *Climatic Change*, **35(4):** 501-524.

Tayanç, M., Dalfes, H.N., Karaca, M. and Yenigün, O. (1998). A comparative assessment of different methods for detecting inhomogeneities in Turkish temperature data set. *International Journal of Climatology*, **18(5):** 561-578.

Tsakalias, G. and Koutsoyiannis, D. (1999). A comprehensive system for the exploration and analysis of hydrological data. *Water Resources Management*, **13(4):** 269-302.

Van Belle, G. and Hughes, J.P. (1984). Nonparametric tests for trend in water quality. *Water Resources Research*, **20(1):** 127-136.

Vassilev, I. and Georgiev, B.N. (1996). River runoff changes and recent climatic fluctuations in Bulgaria. *GeoJournal*, **40(4):** 379-385.

Vincent, L.A. and Gullett, D.W. (1999). Canadian historical and homogeneous temperature datasets for climate change analyses. *International Journal of Climatology*, **19(12):** 1375-1388.

Walanus-Gliwice, A. (1995). The problem of periodicity in the hydrometeorology. Proceedings of the Conference on Climate Dynamics and the Global Change Perspective, Cracow, Poland, October 17-20, 1995. http://www.cyf-kr.edu.pl/~ziniedzw/paper101.html (accessed on 30 January 2004).

Webb, B.W. (1996). Trends in stream and river temperature. *Hydrological Processes*, **10(2):** 205-226.

Westmacott, J.R. and Burn, D.H. (1997). Climate change effects on the hydrologic regime within the Churchill-Nelson River Basin. *Journal of Hydrology*, **202(1-4):** 263-279.

Wilson, T.M., Ogden, A.E. and Mills, H.H., III (1992). Time-series analysis of groundwater chemistry in the west Tennessee sand aquifers. *Journal of the Tennessee Academy of Science*, **67(3):** 29-33.

Wu, H., Soh, L.-K., Samal, A. and Chen, X.H. (2008). Trend analysis of streamflow drought events in Nebraska. *Water Resources Management*, **22(2):** 145-164.

Wu, L., Jury, W.A., Chang, A.C. and Allmaras, R.R. (1997). Time series analysis of field-measured water content of a sandy soil. *Soil Science Society of America Journal*, **61(3):** 736-742.

Xu, Z.X., Takeuchi, K. and Ishidaira, H. (2003). Monotonic trend and step changes in Japanese precipitation. *Journal of Hydrology*, **279(1-4):** 144-150.

Yu, P.-S., Yang, T.-C. and Kuo, C.-C. (2006). Evaluating long-term trends in annual and seasonal precipitation in Taiwan. *Water Resources Management*, **20:** 1007-1023.

Yu, P.-S., Yang, T.-C. and Wu, C.-K. (2002). Impact of climate change on water resources in southern Taiwan. *Journal of Hydrology*, **260(1-4):** 161-175.

Yu, Y-S., Zou, S. and Whittemore, D. (1993). Non-parametric trend analysis of water quality data of rivers in Kansas. *Journal of Hydrology*, **150(1):** 61-80.

Yue, S. and Wang, C.Y. (2004). The Mann-Kendall test modified by effective sample size to detect trend in serially correlated hydrological series. *Water Resources Management*, **18(3):** 201-218.

Yue, S., Pilon, P. and Cavadias, G. (2002b). Power of the Mann-Kendall and Spearman's rho tests for detecting monotonic trends in hydrological series. *Journal of Hydrology*, **259(1-4):** 254-271.

Yue, S., Pilon, P. and Phinney, B. (2003). Canadian streamflow trend detection: Impacts of serial and cross-correlation. *Hydrological Sciences Journal*, **48(1):** 51-64.

Yue, S., Pilon, P., Phinney, B. and Cavadias, G. (2002a). The influence of autocorrelation on the ability to detect trend in hydrological series. *Hydrological Processes*, **16(9):** 1807-1829.

Yurekli, K., Kurunc, A. and Simsek, H. (2004). Prediction of daily maximum streamflow based on stochastic approaches. *Journal of Spatial Hydrology*, **4(2):** 1-12.

Zaninovic, K. and Gajic-Capka, M. (2000). Changes in components of the water balance in the Croatian lowlands. *Theoretical and Applied Climatology*, **65(1-2):** 111-117.

Zhang, X., Harvey, K.D., Hogg, W.D. and Yuzyk, T.R. (2001). Trends in Canadian streamflow. *Water Resources Research*, **37(4):** 987-998.

Zhang, Z., Dehoff, A.A. and Pody, R.D. (2010a). A new approach to identify trend pattern of streamflows. *Journal of Hydrologic Engineering, ASCE*, **15(3):** 244-248.

Zhang, Z., Dehoff, A.A., Pody, R.D. and Balay, J.W. (2010b). Detection of streamflow change in the Susquehanna River Basin. *Water Resources Management*, **24(10):** 1947-1964.

Zhao, G., Hörmann, G., Fohrer, N., Zhang, Z. and Zhai, J. (2010). Streamflow trends and climate variability impacts in Poyang lake basin, China. *Water Resources Management*, **24(4):** 689-706.

Part II

Salient Case Studies

7

Efficacy of Time Series Tests: A Critical Assessment

7.1 Introduction

The application of statistical hydrology in earlier days was restricted to surface water problems only, especially for analyzing the hydrologic extremes such as floods and droughts (McCuen, 2003). However, during past three decades, the statistical domain of hydrology has broadened to encompass the problems related to both surface water and groundwater resources (Shahin et al., 1993; Machiwal and Jha, 2006). With such a broad domain, time series analysis has emerged as a powerful tool for the efficient planning and management of scarce freshwater resources.

Time series analysis has a vast scope in geology, ocean technology, seismology, etc. Time series analysis has also been applied to many hydrological and climatological situations in the past. For instance, time series studies have been carried out for analyzing the historic rainfall data (e.g., Henderson, 1989; De Michele et al., 1998; Mirza et al., 1998; Pagliara et al., 1998; Abaurrea and Cebrian, 2003; Pugacheva et al., 2003; Astel et al., 2004), streamflow data (Avinash and Ghanshyam, 1988; Capodaglio and Moisello, 1990; Radziejewski et al., 2000; Fanta et al., 2001; Adeloye and Montaseri, 2002; Chen and Rao, 2002), flood data (Grew and Werrity, 1995; Changnon and Kunkel, 1995; Westmacott and Burn, 1997; Robson et al., 1998; Reed et al., 1999; Lins and Slack, 1999; Loukas and Quick, 1996, 1999; Cayan et al., 1999; Jain and Lall, 2001; Douglas et al., 2000; Adamowski and Bocci, 2001; Zhang et al., 2001; Cunderlik and Burn, 2002), infiltration data (Schwankl et al., 2000), and surface water quality data (Jayawardena and Lai, 1989; Higashino et al., 1999) as well as for generating synthetic rainfall data in semi-arid regions (Janos et al., 1988), determining water consumption patterns (Maidment and Parzen, 1984), detecting trends in evapotranspiration and wind speed (Hameed et al., 1997; Raghuwanshi and Wallender, 1997), and for detecting climatic changes (Kite, 1989; Khan, 2001).

Natural time series, including hydrologic, climatic and environmental time series, which satisfy the assumptions of normality, homogeneity, randomness, non-periodic, non-persistence and stationarity, seem to be the exception rather than the rule (Rao et al., 2003). In fact, for all water resources studies involving the use of hydrologic time series data, preliminary statistical analyses must always be carried out (Adeloye and Montaseri, 2002). Nevertheless, most time series analysis is performed using standard methods after relaxing the required conditions one way or another in the hope that the departure from these assumptions is not large enough to affect the result of the analysis (Rao et al., 2003). A comprehensive survey of the past studies on the hydrologic time series analysis (Machiwal and Jha, 2006) revealed that no studies to date have considered all the aspects of time series analysis. Major work is reported dealing with only linear trend analysis. However, other equally important characteristics of the hydrologic time series, i.e., normality, homogeneity, stationarity, periodicity and persistence, have been ignored. Further, most past studies report only Regression and/or Kendall's Rank Correlation tests for trend detection. Two of the past studies (Esterby, 1996; Hess et al., 2001) report an overview of some selected trend tests. Thus, no studies are reported to date, which deal with a complete and extensive analysis of normality, homogeneity, stationarity, periodicity, persistence and stochastic component in the hydrologic time series (Machiwal and Jha, 2006).

As mentioned in Chapter 4, several statistical tests are available for determining a particular time series characteristic. Some of the available time series tests are more powerful than others. However, the use of a specific statistical test is still dependent on the user's familiarity with the test rather than on the strength of the test. It has been found that the results of two different tests may be dissimilar in characterizing the same characteristic of a hydrologic time series (Machiwal and Jha, 2008). Therefore, the goal of this chapter is to demonstrate the efficacy of various time series tests for detecting particular characteristics through a case study on the annual and salient consecutive days' maximum rainfall series of Kharagpur, West Bengal, eastern India. This chapter draws significantly from Machiwal and Jha (2008).

7.2 Methodology

In this case study, annual rainfall series of 46 years (1957-2002) and the six consecutive days' maximum rainfall series of 47 years (1956-2002) in Kharagpur, West Bengal, India have been analyzed. Annual rainfall for a year is the total rainfall occurring in that year. The consecutive days' maximum rainfall denotes the maximum rainfall, which occurs during given consecutive days in a particular year. For instance, consecutive 2-day maximum rainfall denotes the amount of maximum rainfall that occurs in any of the two consecutive days in a particular year. Needless to mention that the maximum rainfall series comprising one-day, consecutive two, three, four, five and six days maximum rainfall are equally important for estimating the total annual runoff and maximum runoff produced during a rainfall event. The daily rainfall data of Kharagpur for the 1956-2002 period were obtained from the Physics and Meteorology Department of Indian Institute of Technology Kharagpur, West Bengal, India. These data were checked for the anomalies, and the rainfall records were found free from anomalies, with no missing data in the series. It is worth mentioning that the length of the daily rainfall records is large enough to be used successfully for demonstrating the proper application of various time series tests.

Three tests (i.e., Geary's test, Kolmogorov-Smirnov test and D'Agostino-Pearson Omnibus test) have been applied for testing normality, seven tests (i.e., The von Neumann test, Cumulative Deviations tests, Bayesian test, Tukey test, Link-Wallace test, Bartlett test, and Hartley test) for testing homogeneity, three tests (i.e., Student's *t*-test, Simple *t*-test and Mann-Whitney test) for examining stationarity, and twelve tests (i.e., Regression test, Spearman Rank Order Correlation test, Turning Point test, Kendall's Phase test, Wald-Wolfowitz Total Number of Runs test, Sum of Squared Lengths test, Adjacency test, Difference Sign test, Run test on Successive Differences, Inversions test, Kendall's Rank Correlation test and Mann-Kendall test) for detecting trend. Additionally, periodicity and persistence have been examined through harmonic analysis and autocorrelation analysis, respectively. These time series tests are described in Chapters 3 and 4.

7.3 Graphical Interpretation

It is always a good practice to present a time series in the form of a simple x-y plot prior to the application of statistical techniques. Seven such plots (i.e., total annual, one-day, 2-, 3-, 4-, 5- and 6-day maximum rainfall) showing mean and range of the rainfall time series in this study were drawn and two plots for annual and one-day maximum rainfall are shown in Figs 7.1(a-b) as an example. It is apparent from Fig. 7.1(a) that the time series plot of annual rainfall does not depict any temporal trend. Similarly, the plots of maximum rainfalls have no overall trends [Fig. 7.1(b)]. One significant observation is discernible from time plots of maximum rainfall series that the time pattern of maximum rainfalls is similar for all the series regardless of the consecutive days (i.e., one-day, 2-, 3-, 4-, 5- and 6-day). It is also obvious that the increment in rainfall with an increase in the number of consecutive days is not considerable compared to the amount of maximum rainfall, which is a major reason for the similar time patterns in all the maximum rainfall series.

In addition to the time plots, box plots (Fig. 7.2) were drawn to compare all the rainfall time series under investigation which provide an excellent summary of five important aspects (lowest value, 25th percentile, median, 75th

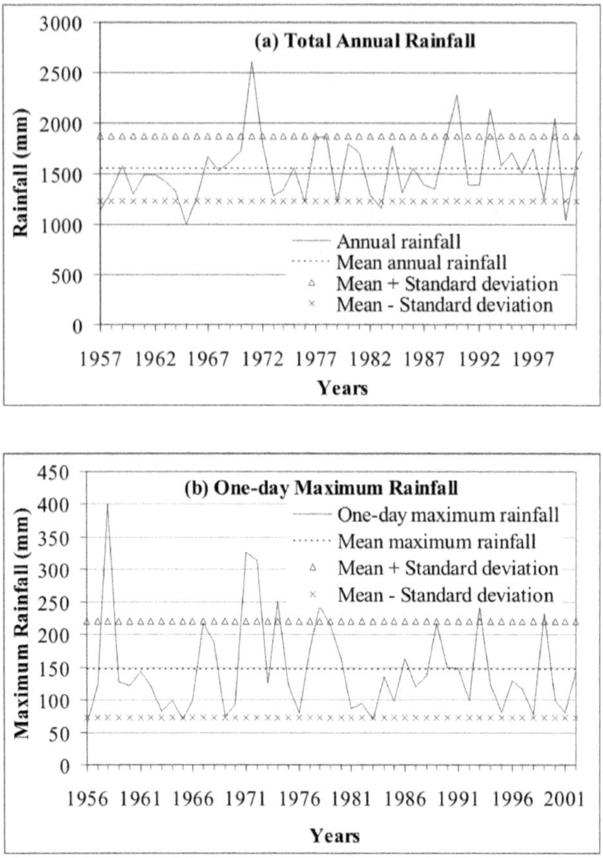

Fig. 7.1. Time series plots of (a) annual and (b) 1-day maximum rainfall time series.

percentile and highest value) of the distribution of rainfall data together with the identification of 'outliers'. The bottom and top horizontal lines in the box in a box plot indicate the 25th and 75th percentile, respectively, of the statistics computed from the observed data. The square within the box represents the median. The whiskers are lines extending from each end of the box to show the extent of the rest of the data. The whisker extends to the most extreme data value within 1.5 times the interquartile range of the data. The values beyond the ends of the whiskers are called 'outliers'. Further details about the box plot can be found in Chapter 2 of this book and USEPA (1998).

It is apparent from Figs 7.2(a, b) that for the annual rainfall series, median is at the centre in between upper and lower quartiles with the upper adjacent value larger than the lower one. Only a single mild outlier is found in the annual series beyond the upper adjacent value. This figure also reveals that the distribution of annual rainfall data below the median value is more condensed than that above the median. A comparison of the box plots of maximum rainfall series reveals that more data are lying in the lower half of the range in all the series (i.e., one-day, 2-, 3-, 4-, 5- and 6-day). In other words, the lower half range of the maximum rainfall is heavily weighted than its upper half range. The medians of the one-day, 2- and 3-day maximum rainfall series are below the centre of the rectangle, whereas the medians of the 4-, 5- and 6-day maximum rainfall series are at the centre of the rectangle. Therefore, it can be concluded that the maximum rainfall time series becomes more uniform in distribution by increasing the number of consecutive days. Furthermore, mild outliers can be seen in the one-day and consecutive 2-day maximum rainfall time series, which disappear completely in the consecutive 3-, 4-, 5- and 6-day maximum rainfall time series. It should be noted that in all the maximum rainfall box plots [Fig. 7.2(b)], the upper whisker (i.e., distance of upper adjacent value from the upper quartile) is longer than the lower whisker, which indicates the less density of data in the upper half range compared to the density in the lower half range.

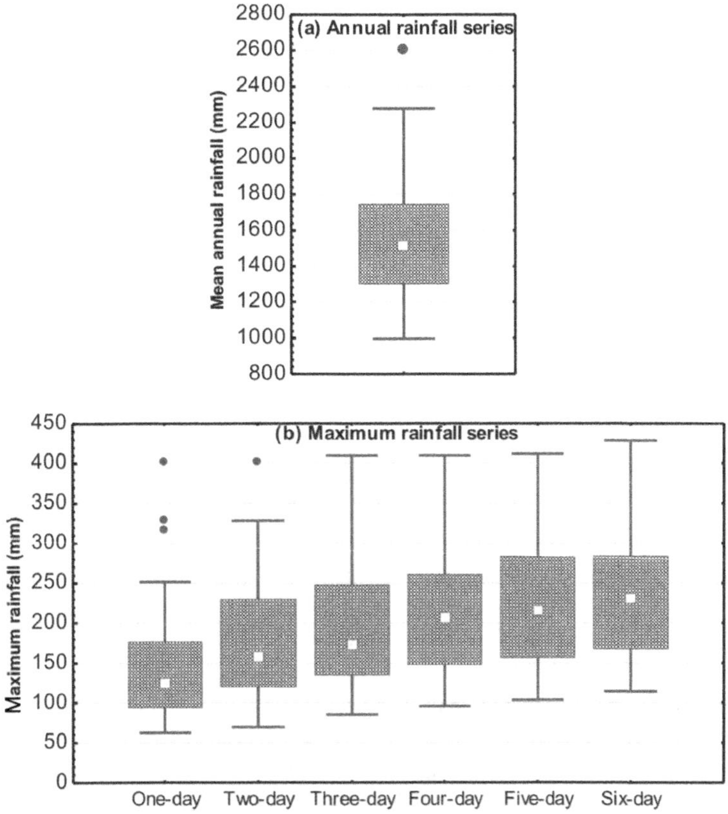

Fig. 7.2. Box plots of the (a) annual and (b) maximum rainfall time series of Kharagpur.

One of the most significant findings of box plot analysis, is the presence of mild outliers in the total annual rainfall, one-day and consecutive 2-day maximum rainfall time series. Therefore, normal probability plots [Figs 7.3(a-g)] were drawn for testing the significance of mild outliers and normality of the rainfall data. The normal probability plot is reported to be the single-most valuable graphical aid in diagnosing how a population distribution appears to differ from a normal distribution (PROPHET StatGuide, 2007). These plots revealed that for the one-day and consecutive 2- and 3-day maximum rainfall, a straight line cannot be obtained on normal probability plots, whereas the data of the remaining rainfall series do not deviate significantly from the straight line. The non-normality in one-day and consecutive 2-day maximum rainfall time series could be attributed to the presence of few mild outliers as mentioned earlier.

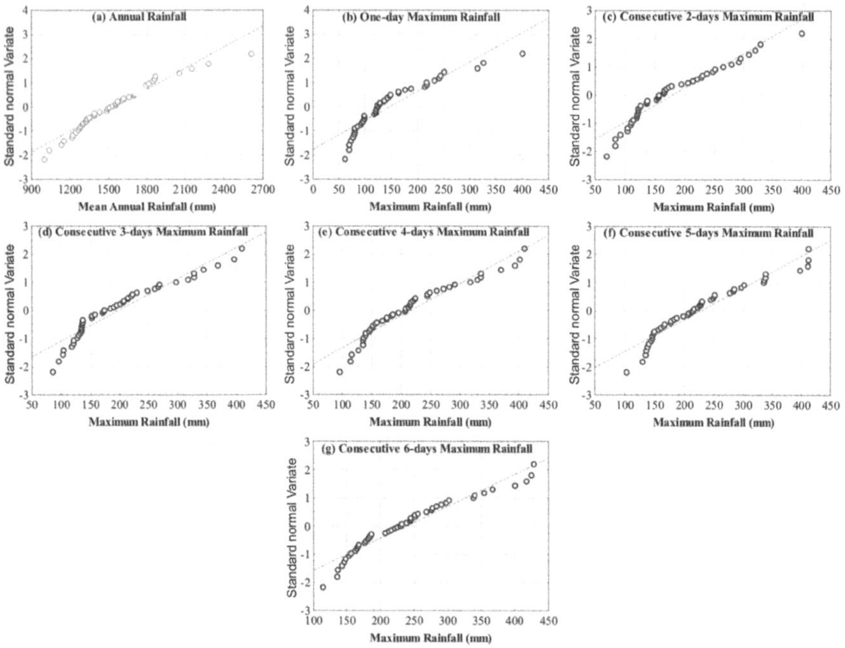

Fig. 7.3. Normal probability plots of the (a) annual and (b-g) salient consecutive days' maximum rainfall time series of Kharagpur. Dotted straight lines represent regression lines.

7.4 Checking Normality

Three most-widely used statistical tests, i.e., Geary's Test (Walpole and Myers, 1989), Kolmogorov-Smirnov test (NIST/SEMATECH, 2007) and D'Agostino-Pearson Omnibus test (D'Agostino, 1986), were applied to examine the normality of the rainfall series under study. Results of these normality tests are presented in Table 7.1. If normality is present in a time series, the Geary's

Table 7.1. Observed test-statistics and the results of three normality tests

Test-Statistic	Annual Rainfall	One-day max.	2-day max.	3-day max.	4-day max.	5-day max.	6-day max.
(a) Geary's Test							
Geary's test-statistic	0.981	0.951	1.036	1.017	0.994	1.007	1.003
Normality	**Yes**	**Yes**	**Yes**	**Yes**	**Yes**	**Yes**	**Yes**
(b) Kolmogorov-Smirnov (KS) Test							
KS test-statistic	0.1053	0.1877	0.1504	0.1499	0.1446	0.1283	0.1185
P-value	>0.1	0.0003	0.0094	0.0099	0.0152	0.0509	0.0964
Normality	**Yes**	No	No	No	No	**Yes**	**Yes**
(c) D'Agostino-Pearson Omnibus Test							
K^2 test-statistic	9.788	19.11	6.367	6.661	5.475	4.907	4.624
P-value	0.0075	<0.0001	0.0414	0.0358	0.0647	0.0860	0.0991
Normality	No	No	No	No	**Yes**	**Yes**	**Yes**
(d) D'Agostino-Pearson Omnibus Test after removing mild outliers							
Number of removed outliers	–	3	1	3	–	–	–
K^2 test-statistic	–	5.475	4.603	5.480	–	–	–
P-value	–	0.0647	0.1001	0.0646	–	–	–
Normality	–	**Yes**	**Yes**	**Yes**	–	–	–

Note: max.: maximum.

test-statistic value approaches one (Walpole and Myers, 1989). Given this criterion, all the seven rainfall time series can be considered normal based on the Geary's test (Table 7.1). The results of the Kolmogorov-Smirnov test and the D'Agostino-Pearson Omnibus test can be interpreted by comparing observed *P*-values with 0.05. If the *P*-value is more than 0.05, the null hypothesis of normality cannot be rejected. It can be seen from Table 7.1 that observed *P*-values for the annual rainfall and 5- and 6-day maximum rainfalls are greater than 0.05 for the Kolmogorov-Smirnov test. Similarly, observed P-values are greater than 0.05 for the 4-, 5- and 6-day maximum rainfall time series for the D'Agostino-Pearson Omnibus test. Thus, based on the box plots, normal probability plots and three normality tests, the rainfall series under study can be considered to be normal, though one-day and consecutive 2-day maximum rainfall time series have few mild outliers that causes deviation from normality. After removing these mild outliers and then applying the D'Agostino-Pearson Omnibus test, which is reported as powerful normality test (e.g., DeCarlo, 1997; Öztuna et al., 2006), it was found that observed *P*-value is not significant (Table 7.1). Thus, all the seven rainfall series under study could be considered to be normal.

7.5 Checking Homogeneity

In this study, three homogeneity tests (i.e., von Neumann test, Cumulative Deviations test and Bayesian test) were employed to examine the homogeneity in the annual and maximum rainfall series. The results of the three homogeneity tests are presented in Table 7.2. The von Neumann ratio is a statistic that has an expected value of 2 for a homogeneous series, but it tends to be less than 2 for a non-homogeneous series. It is apparent from Table 7.2 that the von Neumann ratio (N) becomes smaller than 2 for the one-day and consecutive 2-day maximum rainfall series, which suggests non-homogeneity in these rainfall series of Kharagpur. However, N reaches close to 2 for the total annual, consecutive three-, four-, five- and six-day maximum rainfall series. In case of Cumulative Deviations and Bayesian tests, all the applied test-statistics (i.e., Q, R, U and A) have the smaller values compared to their critical values (Buishand, 1982) at 5% significance level for the total annual and the maximum rainfall series, which indicate that all the rainfall time series are homogenous and belong to the same population. Furthermore, as mentioned in Chapter 4, non-homogeneity arises due to changes in the method of data collection and/or the environment in which it is done. Environmental or physical factors are type, height and exposure of the raingauge, which may affect homogeneity (Buishand, 1982). For the present study, location of the raingauge station and the method of recording rainfall did not change since its establishment. Also, the surrounding environment of the raingauge station and/or physical factors has not altered over the years. Therefore, the physical factors affecting homogeneity did not change for the raingauge station under study for the entire period of collected data. Hence, it is believed that the rainfall series could be homogeneous. Based on the results of the Cumulative Deviations and the Bayesian tests, the rainfall series in this study are found homogeneous. Thus, based on the available history of the raingauge station, the Cumulative Deviations and the Bayesian tests seem to be superior to the classical von Neumann test. Similar finding has also been reported by Buishand (1982) based on the data generation method.

Table 7.2. Observed and critical test-statistics of the three homogeneity tests for the annual and maximum rainfall time series of Kharagpur

Rainfall	*von Neumann*	*Cumulative deviations*		*Bayesian test*	
time series	*test (N)*	$Q/\sqrt{n}$	$R/\sqrt{n}$	*U*	*A*
1	*2*	*3*	*4*	*5*	*6*
Annual	1.908	0.990 (1.265)	0.990 (1.540)	0.249 (0.450)	1.471 (2.460)
1-day max.	1.784	0.752 (1.265)	0.745 (1.540)	0.113 (0.450)	0.651 (2.460)
2-day max.	1.873	0.637 (1.265)	0.631 (1.540)	0.067 (0.450)	0.407 (2.460)
3-day max.	1.922	0.603 (1.265)	0.602 (1.540)	0.076 (0.450)	0.462 (2.460)
4-day max.	2.008	0.670 (1.265)	0.657 (1.540)	0.087 (0.450)	0.504 (2.460)
5-day max.	2.120	0.613 (1.265)	0.611 (1.540)	0.100 (0.450)	0.584 (2.460)
6-day max.	2.022	0.575 (1.265)	0.571 (1.540)	0.099 (0.450)	0.601 (2.460)

Note: max.: maximum; Bracketed figures in Columns 3 to 6 are critical values.

Moreover, four tests for multiple comparisons of homogeneity were also applied to all the seven rainfall time series after fragmenting these in two ways: (a) two fragmentations each comprising half of the entire series, and (b) three fragmentations each consisting of one-third of the entire series. The computed values of all the homogeneity test-statistics for multiple comparisons were compared with their critical values (Table 7.3). It is clear from this table that for the Tukey test, the calculated difference in the means of different possible combinations of two subseries in two individual fragmentations are always less than the critical limit. Therefore, the null hypotheses of homogeneity cannot be rejected for annual, one-day, 2-, 3-, 4-, 5- and 6-day maximum rainfall time series, which also have equal variances. Similarly, the observed value of the test-statistic K_L is found less than its critical value in case of Link-Wallace test, which suggests the presence of homogeneity in the annual rainfall

Table 7.3. Observed and critical test-statistics of the homogeneity tests for multiple comparisons for the annual and maximum rainfall time series of Kharagpur

Fragmentation	*Tukey test*		*Link-Wallace test*		*Bartlett test*		*Hartley test*	
	Observed difference	*Critical limit, W*	K_L	$K_{critical}$	$B_{computed}$	$B_{critical}$	F_{max}	$F_{critical}$
(a) Annual rainfall time series								
First	89.68 15.90	194.44	0.721	1.78	0.06	3.84	1.108	2.382
Second	132.54 116.64	239.31	Not applicable		1.43	5.99	1.881	3.54
(b) 1-day maximum rainfall time series								
First	24.99 32.06	43.48	Not applicable		**7.75**	3.84	**3.35**	2.34
Second	2.70 29.36	53.61			3.65	5.99	2.65	2.86
(c) Consecutive 2-day maximum rainfall time series								
First	23.90 36.93	45.04	Not applicable		3.10	3.84	2.12	2.34
Second	12.50 24.43	55.38			2.93	5.99	2.49	2.86

(d) Consecutive 3-day maximum rainfall time series							
First	25.17 44.73	49.32	Not applicable	**5.44**	3.84	**2.73**	2.34
Second	12.37 32.36	60.24		4.41	5.99	**3.01**	2.86
(e) Consecutive 4-day maximum rainfall time series							
First	29.35 39.56	48.00	Not applicable	**5.09**	3.84	**2.64**	2.34
Second	2.51 37.05	58.97		3.90	5.99	**2.88**	2.86
(f) Consecutive 5-day maximum rainfall time series							
First	25.99 48.19	48.84	Not applicable	**5.21**	3.84	**2.67**	2.34
Second	1.76 46.43	58.99		4.23	5.99	**2.89**	2.86
(g) Consecutive 6-day maximum rainfall time series							
First	24.88 49.00	48.71	Not applicable	**4.89**	3.84	**2.59**	2.34
Second	3.40 45.60	58.76		4.16	5.99	**2.83**	2.86

Note: Figures in bold face indicate the rejection of the null hypothesis of homogeneity.

series of Kharagpur. The Link-Wallace test could not be applied to the second fragmentation of annual rainfall time series and to both the fragmentations (i.e., first and second) of maximum rainfall time series because of their unequal sample sizes. The results of the Bartlett test (Table 7.3) indicate that the homogeneity is associated with the annual rainfall time series for both the fragmentations and with the second fragmentation of all maximum rainfall time series (i.e., 1-, 2-, 3-, 4-, 5- and 6-day). However, the first fragmentation of one-day, consecutive 3-, 4-, 5- and 6-day maximum rainfall time series are found non-homogeneous based on the Bartlett test. Table 7.3 also reveals that the results of the Hartley test are exactly similar to that of the Bartlett test for the first fragmentation of all the rainfall time series under study. However, the second fragmentation of consecutive 3-, 4- and 5-day maximum rainfall can be declared non-homogeneous based on the Hartley test. Furthermore, based on the historical information of the raingauge station, the Tukey test is found to be more powerful in identifying the homogeneity than the Bartlett and Hartley tests.

Based on the above results of three homogeneity tests and available historical information for the raingauge station, it can be concluded that the Cumulative Deviations and Bayesian tests are superior to the von Neumann test. Similarly, Tukey test is better than Bartlett, Hartley and Link-Wallace tests for multiple comparisons. All these superior tests indicated that annual and maximum rainfall time series of Kharagpur are homogeneous. Here, it is emphasized that adequate number of tests should be applied and the results of all the applied tests should be critically analyzed to arrive at a reliable decision about the characteristics of a hydrologic time series.

7.6 Checking Stationarity

The entire time series of annual and maximum rainfalls were divided into five subseries according to the first-half, second-half, first one-third, second one-third, and last one-third of the entire period of rainfall records, and then the stationary tests were applied to examine whether the means of the five subseries are significantly different from that of the entire series. The salient statistical parameters of the entire and subseries, i.e., annual and maximum rainfalls, are summarized in Table 7.4. It is obvious from Table 7.4 that the annual and maximum rainfall series have been derived from positively skewed distributions with wide variations from the mean (standard deviation ≥18% for annual rainfall and standard deviation ≥25% for all maximum rainfall series). It is also evident that the skewness coefficients for the first one-third rainfall subseries are comparatively high, which indicates that the subseries contains more low values than high values. This finding supports the earlier made inferences based on the box plots of time series.

Table 7.4. Summary of the results of the Student's t-test for stationarity, together with the statistical characteristics of entire series and subseries of the annual and maximum rainfall time series

Time series	*Mean (mm)*	*Standard deviation (mm)*	*Skewness coefficient*	*Kurtosis*	*Degrees of freedom*	*Student's t-statistic*	
						Calculated	*Critical***
(a) Annual rainfall time series							
Entire series	1547.65	326.56	0.954	1.461			
First half	1502.81	335.29	1.610	4.406	22	-0.644	1.717
Second half	1592.49	318.59	0.374	-0.287	22	0.644	1.717
First one-third	1496.37	366.76	2.006	6.170	14	-0.588	1.761
Second one-third	1512.27	267.44	0.074	-1.850	14	-0.405	1.761
Last one-third	1628.91	341.60	0.334	-0.411	15	0.964	1.753
(b) 1-day maximum rainfall time series							
Entire series	146.31	74.22	1.513	2.306			
First half	159.07	92.32	1.262	0.846	22	0.806	1.717
Second half	134.09	50.47	0.842	-0.133	23	-0.790	1.714
First one-third	134.48	84.90	2.458	7.009	14	-0.597	1.761
Second one-third	166.54	82.32	0.765	-0.471	15	1.056	1.753
Last one-third	137.18	52.20	0.985	0.124	15	-0.477	1.753

(Contd.)

Table 7.4 (*Contd.*)

Time series	*Mean* (*mm*)	*Standard deviation* (*mm*)	*Skewness coefficient*	*Kurtosis*	*Degrees of freedom*	*Student's t-statistic*	
						Calculated	*Critical***
(c) Consecutive 2-day maximum rainfall time series							
Entire series	178.83	76.73	0.901	0.183			
First half	191.04	89.68	0.795	-0.356	22	0.746	1.717
Second half	167.14	61.57	0.607	-0.559	23	-0.731	1.714
First one-third	162.01	80.44	2.002	5.029	14	-0.821	1.761
Second one-third	198.94	89.72	0.126	-1.588	15	1.015	1.753
Last one-third	174.51	56.85	0.807	-0.429	15	-0.218	1.753
(d) Consecutive 3-day maximum rainfall time series							
Entire series	198.57	83.94	0.940	0.047			
First half	211.42	102.02	0.729	-0.826	22	0.718	1.717
Second half	186.25	61.75	0.675	-0.303	23	-0.704	1.714
First one-third	179.13	90.96	1.619	2.270	14	-0.866	1.761
Second one-third	223.86	97.93	0.447	-1.257	15	1.167	1.753
Last one-third	191.50	56.41	0.384	-1.019	15	-0.326	1.753
(e) Consecutive 4-day maximum rainfall time series							
Entire series	216.32	82.10	0.842	-0.085			
First half	231.31	98.82	0.565	-0.936	22	0.856	1.717
Second half	201.96	60.84	0.757	0.094	23	-0.839	1.714
First one-third	202.00	96.72	1.244	0.703	14	-0.653	1.761
Second one-third	241.56	87.49	0.391	-0.942	15	1.191	1.753
Last one-third	204.51	57.04	0.786	0.197	15	-0.557	1.753

		(f) Consecutive 5-day maximum rainfall time series					
Entire series	227.71	83.20	0.790	-0.179			
First half	240.99	100.70	0.592	-0.965	22	0.748	1.717
Second half	215.00	61.64	0.476	-0.627	23	-0.733	1.714
First one-third	210.71	94.64	1.206	0.621	14	-0.765	1.761
Second one-third	258.90	90.53	0.278	-0.936	15	1.452	1.753
Last one-third	212.47	55.66	0.707	0.110	15	-0.710	1.753
		(g) Consecutive 6-day maximum rainfall time series					
Entire series	239.09	82.90	0.764	-0.131			
First half	251.79	99.99	0.576	-0.897	22	0.719	1.717
Second half	226.91	62.17	0.487	-0.342	23	-0.704	1.714
First one-third	221.25	93.49	1.173	0.657	14	-0.805	1.761
Second one-third	270.25	90.82	0.339	-0.915	15	1.456	1.753
Last one-third	224.65	55.61	0.378	-0.462	15	-0.675	1.753

** Critical values are at 5% significance level.

Moreover, two parametric tests (i.e., Student's *t*-test and Simple *t*-test) and one non-parametric Mann-Whitney test were used for testing stationarity in the rainfall series. The values of the Student's *t*-statistics for annual and maximum (one-day, consecutive 2-, 3-, 4-, 5- and 6-day) rainfall series are less than their critical values (Table 7.4), and hence the null hypothesis cannot be rejected at 5% significance level. Here, the choice of 5% significance level (or 95% confidence level) is arbitrary, but it seems to be a reasonable limit for the type of data under study (Jayawardena and Lai, 1989). Critical values for the *t*-statistic were taken from standard statistical texts (e.g., Shahin et al., 1993). Further, the parametric Simple *t*-test was applied to the entire rainfall series after dividing them into two half-subseries viz., annual rainfall series of 46 years into two 23-year series and the maximum rainfall series of 47 years into two series (i.e., 23-year series and 24-year series). A comparison between the calculated and the critical values of the test-statistics is shown in Table 7.5. It is clear from this table that the stationarity exists in the annual series as well as one-day, 2-, 3-, 4-, 5- and 6-day maximum rainfall time series of Kharagpur. Also, the calculated values of the Mann-Whitney test-statistic are less than their critical values (±1.96) for all the seven rainfall time series, which indicates stationarity in the rainfall time series of Kharagpur. Thus, based on the results of three stationarity tests, the annual and maximum rainfall time series under study are undoubtedly stationary.

7.7 Checking Trend

7.7.1 Application of Trend Tests

The annual and the maximum rainfall time series of Kharagpur were examined for the presence of a linear trend by applying the trend detection tests mentioned in methodology. It can be seen that most of the available trend tests are of parametric character, which necessitates that series should follow a normal statistical distribution. Hence, the normality of the Kharagpur rainfall time series data was tested using normal probability plots, and it was found that they can be considered to follow a normal distribution as discussed earlier in Section on *Graphical Interpretation*. The results of the twelve trend detection tests are presented in Tables 7.5 and 7.6, together with the critical values of the test-statistics. Critical values for the test-statistics of Regression test, Spearman Rank Order Correlation test, Wald-Wolfowitz Total Number of Runs test, Sum of Squares Lengths test, Adjacency test, Difference Sign test and Runs test on Successive Differences were taken from Shahin et al. (1993). Similarly, the critical values for the Mann-Kendall test were obtained from Salas (1993). In the Kendall's Phase Lengths test, the phase length and observed number of each phase length was counted in annual and maximum rainfall (i.e., one-day, 2-, 3-, 4-, 5- and 6-day) time series, and then the expected number of phase lengths was computed. For the one-day and consecutive

Table 7.5. Comparison of observed and critical test-statistics of the Simple *t*-test, Mann-Whitney test and Kendall's phase lengths test results

Rainfall	*Simple t-test*		*Mann-Whitney test*		*Kendall's phase length test*			
time series	$ts_{computed}$	$ts_{critical}$	u_c	$u_{critical}$	*1*	2	3	*4*
Annual	0.93	1.30	-1.22	±1.96	17 (17.92)	7 (7.70)	2 (2.16)	1 (0.46)
One-day max.	1.16	1.30	0.40	±1.96	**13** (18.33)	**12** (7.88)	2 (2.22)	-
2-day max.	1.07	1.30	0.66	±1.96	**13** (18.33)	**9** (7.88)	**4** (2.22)	-
3-day max.	1.03	1.30	0.38	±1.96	**22** (18.33)	8 (7.88)	0 (2.22)	1 (0.47)
4-day max.	1.23	1.30	0.64	±1.96	19 (18.33)	8 (7.88)	0 (2.22)	**2** (0.47)
5-day max.	1.07	1.30	0.53	±1.96	**24** (18.33)	**9** (7.88)	0 (2.22)	0 (0.47)
6-day max.	1.03	1.30	0.55	±1.96	**21** (18.33)	**9** (7.88)	1 (2.22)	0 (0.47)

Note: Bracketed figures in last four columns are critical values at 5% significance level.
Figures in bold face indicate the rejection of the null hypothesis of no trend.
max. = maximum.

2-day maximum rainfall series, three phases of lengths 1, 2 and 3 were encountered, whereas in case of annual and 3-, 4-, 5- and 6-day maximum rainfall time series, phases with lengths of 1, 2, 3 and 4 were found. The observed and expected numbers of phase lengths are given in Table 7.5. It is apparent from Table 7.5 that the differences between the observed and expected numbers of individual phases are small for the annual rainfall time series, whereas the differences are significant for the maximum rainfall series of all durations. Thus, the Kendall's Phase test indicates absence of trend in the annual rainfall time series and presence of trend in all the maximum rainfall time series of Kharagpur.

The results of the linear model or Regression test are summarized in Table 7.6. It is clear from this table that the computed values of the test-statistic are less than their critical values (±1.96) for all the rainfall time series under study, which suggests that the annual and maximum rainfalls of Kharagpur do not have any linear trends. However, the computed test-statistic value (1.78) for the nonparametric Spearman Rank Order Correlation (SROC) test is greater than its critical value (1.681) for the annual rainfall time series (Table 7.6), and hence the null hypothesis of no trend is rejected. Thus, the annual rainfall series of Kharagpur has a trend based on the SROC test. On the other hand, results of the SROC test indicate that one-day, 2-, 3-, 4-, 5- and 6-day maximum rainfalls of Kharagpur are trend-free. Similarly, the annual and maximum rainfall series were found not to have any trend based on the Turning Point test (Table 7.6). On the contrary, the results of the Wald-Wolfowitz Total Number of Runs test indicated trends for all the seven rainfall time series. Furthermore, it can also be seen from Table 7.6 that the computed values of the Sum of Squared Lengths test-statistic vary from 49 to 67 for all the seven rainfall time series, which are considerably less than its critical value (>118). This suggests no trend in any of the rainfall series. The computed and critical values of the test-statistic of the Adjacency and the Difference Sign tests are also compared in Table 7.6, which reveals that all the rainfall series are free from any kind of trend. However, based on the Difference Sign test, the 4-day maximum rainfall series has some trend, but the remaining six rainfall series are trend-free.

Finally, a comparison between the computed and critical values of the test-statistics of Run test on Successive Differences (Table 7.6) reveals that the annual rainfall series and all the maximum rainfall time series of Kharagpur are trend-free at 5% significance level. Thus, the results of the four tests (i.e., Run test on Successive Differences, Inversions test, Kendall's Rank Correlation test and Mann-Kendall test) are similar to each other (Table 7.6). It can also be seen that the computed test-statistic values of the Kendall's Rank Correlation test are closely related to that of the Mann-Kendall test, which emphasizes the equal power of both the tests in detecting a trend in hydrologic time series. This might be the reason that of these four tests, the Kendall's Rank Correlation

Table 7.6. Observed and critical test-statistic values of the 11 trend tests for the annual and the maximum rainfall time series

Trend test	*Annual rainfall time series*	*Consecutive days maximum rainfall time series*					
		One-day	*2-day*	*3-day*	*4-day*	*5-day*	*6-day*
1. Linear model or Regression test	1.49	-0.72	-0.18	-0.06	-0.38	-0.40	-0.32
	(±1.96)	(±1.96)	(±1.96)	(±1.96)	(±1.96)	(±1.96)	(±1.96)
2. SROC test	**1.78**	0.02	0.25	0.47	0.16	0.02	0.11
	(1.681)	(1.680)	(1.680)	(1.680)	(1.680)	(1.680)	(1.680)
3. Turning Point test	0.48	0.71	1.06	0.71	0.00	1.41	1.71
	(±1.96)	(±1.96)	(±1.96)	(±1.96)	(±1.96)	(±1.96)	(±1.96)
4. Wald-Wolfowitz Total Number of Run test	**3.43**	**3.54**	**3.54**	**3.54**	**3.24**	**3.24**	**3.24**
	(±1.96)	(±1.96)	(±1.96)	(±1.96)	(±1.96)	(±1.96)	(±1.96)
5. Sum of Squared Lengths test	67	63	55	55	49	53	59
	(>118)	(>118)	(>118)	(>118)	(>118)	(>118)	(>118)
6. Adjacency test	0.046	0.108	0.063	0.039	-0.004	-0.060	-0.011
	(±1.96)	(±1.96)	(±1.96)	(±1.96)	(±1.96)	(±1.96)	(±1.96)
7. Difference Sign test	0.505	1.750	1.250	1.750	**2.750**	1.250	0.250
	(±1.96)	(±1.96)	(±1.96)	(±1.96)	(±1.96)	(±1.96)	(±1.96)
8. Run test on Successive Differences	0.654	1.235	1.235	0.529	0.176	1.235	0.529
	(±1.96)	(±1.96)	(±1.96)	(±1.96)	(±1.96)	(±1.96)	(±1.96)
9. Inversions test	1.81	0.12	0.16	0.30	0.03	0.19	0.01
	(±1.96)	(±1.96)	(±1.96)	(±1.96)	(±1.96)	(±1.96)	(±1.96)
10. Kendall's Rank Correlation test	1.808	-0.174	0.156	0.284	0.009	-0.193	-0.009
	(±1.96)	(±1.96)	(±1.96)	(±1.96)	(±1.96)	(±1.96)	(±1.96)
11. Mann-Kendall test	1.76	0.17	-0.15	-0.28	0	0.18	0.02
	(±1.96)	(±1.96)	(±1.96)	(±1.96)	(±1.96)	(±1.96)	(±1.96)

Note: Bracketed figures in rows 2 and 5 show critical values, while in rows 1, 3, 4 and 6-11 show critical ranges at 5% significance level. Figures in bold face indicate the rejection of the null hypothesis of no trend.

and Mann-Kendall tests have been used widely in recent hydrological studies as well as in studies related to soil, plant and meteorology.

7.7.2 Assessment of Trend Tests

Based on the results of various trend detection tests (Tables 7.5 and 7.6), a comparative evaluation could be done. The only test which indicates presence of trend in all the seven rainfall time series under study (i.e., annual rainfall and one-day, 2-, 3-, 4-, 5- and 6-day maximum rainfalls) is Wald-Wolfowitz test. However, this test is reported to be neither very powerful nor efficient by Himmelblau (1969) because this test does not take into account the phase length for computing the test-statistic and thus some useful information is ignored. The Kendall's Phase test indicates a trend in all the maximum rainfall time series, whereas it indicated no trends in the annual rainfall time series. On the contrary, the SROC test, which is recommended by the World Health Organization for hydrologic time series analysis, suggests presence of trend of the annual rainfall time series, but it suggests no trend in all the maximum rainfall time series. It is worth mentioning that the Kendall's Phase test is currently outdated due to the availability of sound and robust trend detection tests (Shahin et al., 1993). Therefore, results of SROC test can be considered superior to the Kendall's Phase test. A trend is also detected in the 4-day maximum rainfall time series by the Difference Sign test. Apart from the above-mentioned tests, the remaining nine trend tests did not reveal any trends in the annual as well as the one-day, 2-, 3-, 4-, 5- and 6-day maximum rainfall time series of Kharagpur. Thus, bearing in mind the fact that more the applied tests, more is the chance to reject a true null hypothesis, it can be inferred that the annual rainfall series (46 years) of Kharagpur does not have any trends (i.e., series is stable) and similarly the one-day, consecutive 2-, 3-, 4-, 5- and 6-day maximum rainfall series (47 years) are free from any linear trends. Based on the above results, it can be concluded that the SROC test, Kendall's Rank Correlation test and Mann-Kendall test are more reliable or powerful than the Wald-Wolfowitz test, Kendall's Phase test and Difference Sign test.

It is a usual practice to consider a hydrologic time series to be stationary if it is homogeneous and trend-free. Therefore, all the rainfall time series of Kharagpur can also be considered stationary. This finding is in agreement with that based on the three specific stationary tests (mentioned in Section on *Checking Stationarity*). It is also well discernible from the results of different trend tests that if more tests are applied for the same objective, it is usually difficult to arrive at a common conclusion. This is due to the fact that on increasing the number of tests for analyzing a time series, the probability that at least one test rejects the null hypothesis of being true increases. Therefore, it is strongly recommended that the null hypothesis should not be rejected on the basis of only one or two test results (Brockwell and Davis, 1991).

7.8 Investigating Periodicity

The harmonic analysis was performed to examine the periodic/cyclic nature of the annual rainfall series for past 46 years (1957-2002) and maximum rainfall (one-day maximum, consecutive 2-, 3-, 4-, 5- and 6-day maximum) series for past 47 years (1956-2002) of Kharagpur. The harmonic analysis was performed up to 23 maximum number of harmonics in all these rainfall series. It was found that the periodicity effect is not apparent in the seven rainfall series and about 75% variation in all the rainfall series is caused by 15 harmonics cumulatively. Furthermore, since the periodicity in a time series is generally introduced due to the earth's rotation around the sun (Kite, 1989), it is expected to be inherent in the time series having a period of less than one year (e.g., monthly and/or seasonal) rather than annual or annual maximum rainfall series. In this context, the average monthly rainfall time series of Kharagpur were also subjected to the harmonic analysis, which revealed six-month and one-year periodicities in the rainfall series.

7.9 Investigating Persistence

The persistence in the total annual and the selected consecutive days' maximum rainfall series was tested by using the autocorrelation technique. Separate autocorrelograms were prepared for all the rainfall time series as shown in Figs 7.4(a-g). According to the time series lengths of 46 and 47 years, the autocorrelation functions/coefficients were determined up to a maximum order of 12 years. The upper and lower bounds for defining the non-critical (acceptable) range were computed by the Anderson's test (Anderson, 1942) for large samples. It is apparent from these figures that for all the rainfall time series, the autocorrelograms have a zigzag path closer to zero, and the autocorrelation coefficients are more or less same for a given time lag. Further, all the autocorrelograms have almost the similar profile over the time lag scale.

Moreover, for the annual rainfall time series, the autocorrelation function at nine years time lag falls outside the acceptable region [Fig. 7.4(a)]. The appearance of autocorrelation coefficient (r_k) beyond the upper critical limit indicates persistency in the annual rainfall time series of Kharagpur with nine years time lag. For the remaining time lags, the autocorrelation coefficients were found within the acceptable range. Thus, it can be safely inferred that the annual rainfall time series of Kharagpur is slightly persistent in nature with a nine-year time lag. This means that the annual rainfall of a tth year has some relationship with the annual rainfall of $(t+9)^{th}$ year, though it is not significant in this study as the deviation of r_k from the critical limit (i.e., 0.14) may not be considered very significant for practical purposes. However, persistency from the annual rainfall series was removed through transformation of the series by removing autocorrelation for nine years time lag prior to the

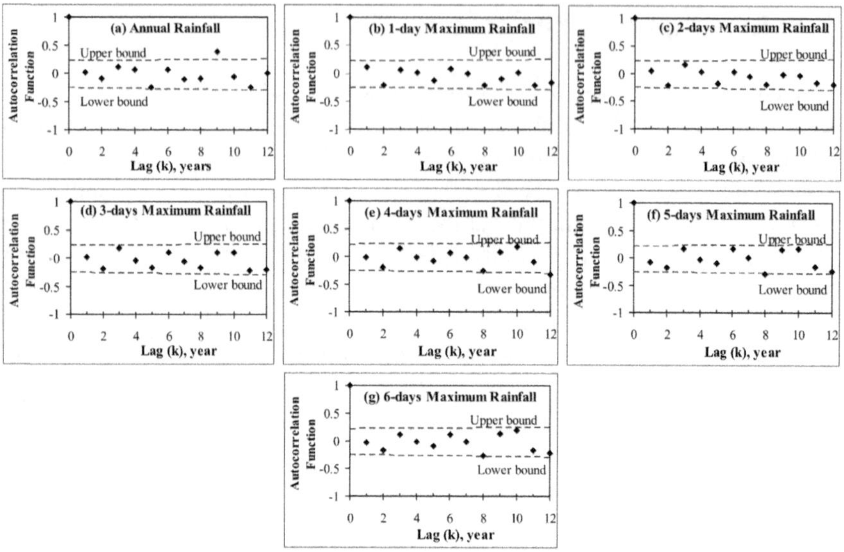

Fig. 7.4. Autocorrelograms for the (a) annual and (b-g) maximum rainfall series of Kharagpur.

subsequent stochastic analysis. On the other hand, it is obvious from Figs 7.4(b-g) that the autocorrelograms for the one-day, consecutive 2-, 3-, 4-, 5- and 6-day maximum rainfall time series are non-persistent, i.e., they are independent of each other.

7.10 Conclusions

A critical assessment of various time series tests has been carried out to demonstrate the efficacy of these tests for analyzing hydrologic time series through a case study on the annual and salient consecutive days' maximum rainfall series of Kharagpur, India. The time series plots, box plots and normal probability plots of the one annual and six maximum rainfall time series revealed no trends. The box plots indicated no severe outliers in both types of rainfall series (i.e., annual rainfall and consecutive days maximum rainfalls), except for one or two mild outliers in the 1-day and 2-day maximum rainfall time series but they were also not found significant. The normal probability plots indicated normality in the annual rainfall series as well as in the consecutive 3-, 4-, 5- and 6-day maximum rainfall series, with slight non-normality in the one-day and consecutive 2-day maximum rainfall series. However, after removing two-three mild outliers, the D'Agostino-Pearson Omnibus test suggested normality in all the seven rainfall series. Therefore, the annual and maximum rainfall time series under investigation are considered to be normally distributed.

Analysis of the results of three homogeneity tests and four multiple comparisons tests indicated that the annual and maximum rainfall time series are homogeneous. Based on the physical parameters affecting homogeneity, the Cumulative Deviations test and the Bayesian test were found superior to the classical von Neumann test. The performance of the Tukey test was found excellent among all the multiple comparisons tests used in this study. The results of the Bartlett test were found to be almost similar to that of the Hartley test. The applicability of the Link-Wallace test, however, is limited due to the basic assumption of equal sample size. Here, it is emphasized that the history of data-recording stations should always be associated with the time series records in order to assess the performance of different types of homogeneity tests.

The rainfall time series of Kharagpur were found stationary at 5% significance level based on the two parametric *t*-tests and one nonparametric Mann-Whitney test. Out of the twelve trend-detecting tests applied, nine tests revealed randomness (i.e., no trends) in all the seven rainfall time series. Though the Wald-Wolfowitz Total Number of Runs test and the Spearman Rank Order Correlation test reject the randomness in the annual rainfall time series, this series is still considered random based on the results of other ten tests (some of which such as Kendall's Rank Correlation and Mann-Kendall tests are equally powerful in detecting randomness in the hydrologic time series). Besides the results of three specific tests for stationarity, the results of homogeneity and randomness tests also suggested stationarity in the annual and maximum rainfall time series of Kharagpur. Furthermore, the Fourier series analysis did not indicate apparent periodicity in any of the seven rainfall series. The autocorrelation analysis indicated persistence in the annual rainfall series with a time lag of nine years, though the deviation from the critical value may not be considered significant for practical purpose. This observed persistency in the annual rainfall series can be removed through transformations (Machiwal and Jha, 2008).

Finally, it is concluded that the application of several statistical tests for the same purpose in a time series analysis increases the chance for rejecting a true null hypothesis. Therefore, the decision about the rejection of null hypothesis should be made by critically analyzing the results of adequate number of statistical tests (at least more than two tests). Such an approach for time series analysis is essential to ensure efficient application of time series tests in analyzing hydrologic time series, thereby enhancing the reliability of time series tests in scientific decision making.

References

Abaurrea, J. and Cebrian, A. (2003). Trend analysis of daily rainfall extremes. http://www.isi-eh.usc.es/resumenes/127_52_abstract.pdf (accessed on 27 July 2003).

Adamowski, K. and Bocci, C. (2001). Geostatistical regional trend detection in river flow data. *Hydrological Processes*, **15:** 3331-3341.

Adeloye, A.J. and Montaseri, M. (2002). Preliminary streamflow data analyses prior to water resources planning study. *Hydrological Sciences Journal*, **47(5):** 679-692.

Anderson, R.L. (1942). Distribution of the serial correlation coefficient. *Annals of Mathematical Statistics*, **13:** 1-13.

Astel, A., Mazerski, J., Polkowska, Ż. and Namieśnik, J. (2004). Application of PCA and time series analysis in studies of precipitation in Tricity (Poland). *Advances in Environmental Research*, **8(3-4):** 337-349.

Avinash, A. and Ghanshyam, D. (1988). Time series model of stream flow for a catchment of Ramganga River. *Journal of Institution of Engineers (India), Civil Engineering Division*, **88 (Part CI):** 228-230.

Brockwell, P.J. and Davis, R.A. (1991). Time Series: Theory and Methods. Springer Series in Statistics. Springer, New York.

Buishand, T.A. (1982). Some methods for testing the homogeneity of rainfall records. *Journal of Hydrology*, **58:** 11-27.

Capodaglio, A.G. and Moisello, U. (1990). Simple stochastic model for annual flows. *Journal of Water Resources Planning and Management, ASCE*, **116(2):** 220-232.

Cayan, D.R., Redmond, K.T. and Riddle, L.G. (1999). ENSO and hydrologic extremes in the western United States. *Journal of Climate*, **12:** 2881-2893.

Changnon, S.A. and Kunkel, K.E. (1995). Climate-related fluctuation in Midwestern floods during 1921–1985. *Journal of Water Resources Planning and Management, ASCE*, **121(4):** 326-334.

Chen, H.-L. and Rao, A.R. (2002). Testing hydrologic time series for stationarity. *Journal of Hydrologic Engineering, ASCE*, **7(2):** 129-136.

Cunderlik, J.M. and Burn, D.H. (2002). Local and regional trends in monthly maximum flows in southern British Columbia. *Canadian Water Resources Journal*, **27(2):** 191-212.

D'Agostino, R.B. (1986). Tests for the normal distribution. *In:* R.B. D'Agostino and M.A. Stephens (editors), Goodness of Fit Techniques, Marcel Decker Inc., pp. 367-419.

De Michele, C., Montanari, A. and Rosso, R. (1998). The effects of non-stationarity on the evaluation of critical design storms. *Water Science and Technology*, **37(11):** 187-193.

DeCarlo, L.T. (1997). On the meaning and use of kurtosis. *Psychological Methods, American Psychological Association, Inc.*, **2(3):** 292-307.

Douglas, E.M., Vogel, R.M. and Kroll, C.N. (2000). Trends in floods and low flows in the United States: Impact of spatial correlation. *Journal of Hydrology*, **240(1-2):** 90-105.

Esterby, S.R. (1996). Review of methods for the detection and estimation of trends with emphasis on water quality applications. *Hydrological Processes*, **10(2):** 127-149.

Fanta, B., Zaake, B.T. and Kachroo, R.K. (2001). A study of variability of annual river flow of the southern African region. *Hydrological Sciences Journal*, **46(4):** 513-524.

Grew, H. and Werrity, A. (1995). Changes in flood frequency and magnitude in Scotland. Proceedings of the BHS Fifth National Hydrology Symposium, Edinburgh, pp. 3.1-3.9.

Hameed, T., Marino, M.A., DeVries, J.J. and Tracy, J.C. (1997). Method for trend detection in climatological variables. *Journal of Hydrologic Engineering, ASCE*, **2(4):** 154-160.

Henderson, R.J. (1989). Rainfall time series for storm overflow assessment. *Water Science and Technology*, **21:** 1789-1791.

Hess, A., Iyer, H. and Malm, W. (2001). Linear trend analysis: A comparison of methods. *Atmospheric Environment*, **35(30):** 5211-5222.

Higashino, M., Kanda, T. and Michioku, K. (1999). Time series analysis and transformation of water quality in an eutrophic reservoir. *Mizu Kankyo Gakkaishi*, **22(8):** 668-676.

Himmelblau, D.M. (1969). Process Analysis by Statistical Methods. John Wiley and Sons, Inc., New York.

Jain, S. and Lall, U. (2001). Floods in a changing climate: Does the past represent the future? *Water Resources Research*, **37(12):** 3193-3205.

Janos, B., Lucien, D. and Omar, H.R. (1988). Practical generation of synthetic rainfall event time series in a semi-arid climatic zone. *Journal of Hydrology*, **103:** 357-373.

Jayawardena, A.W. and Lai, F. (1989). Time series analysis of water quality data in Pearl River, China. *Journal of Environmental Engineering, ASCE*, **115(3):** 590-607.

Khan, A.R. (2001). Analysis of hydro-meteorological time series in the upper Indus basin: Searching evidence for climatic change. International Water Management Institute (IWMI), Working Paper 23, Pakistan Country Series Number 7, Colombo, Sri Lanka.

Kite, G. (1989). Use of time series analyses to detect climatic change. *Journal of Hydrology*, **111:** 259-279.

Lins, H.F. and Slack, J.R. (1999). Streamflow trends in the United States. *Geophysical Research Letters*, **26(2):** 227-230.

Loukas, A. and Quick, M.C. (1996). Effect of climate change on hydrologic regime of two climatically different watersheds. *Journal of Hydrologic Engineering, ASCE*, **1(2):** 77-87.

Loukas, A. and Quick, M.C. (1999). The effect of climate change on floods in British Columbia. *Nordic Hydrology*, **30:** 231-256.

Machiwal, D. and Jha, M.K. (2006). Time series analysis of hydrologic data for water resources planning and management: A review. *Journal of Hydrology and Hydromechanics*, **54(3):** 237-257.

Machiwal, D. and Jha, M.K. (2008). Comparative evaluation of statistical tests for time series analysis: Application to hydrological time series. *Hydrological Sciences Journal*, **53(2):** 353-366.

Maidment, D.R. and Parzen, E. (1984). Time patterns of water use in six Texas cities. *Journal of Water Resources Planning and Management, ASCE*, **110(1):** 90-106.

McCuen, R.H. (2003). Modeling Hydrologic Change: Statistical Methods. CRC Press LLC, Florida.

Mirza, M.Q., Warrick, R.A., Ericksen, N.J. and Kenny, G.J. (1998). Trends and persistence in precipitation in the Ganges, Brahmaputra and Meghna river basins. *Hydrological Sciences Journal*, **43(6):** 845-858.

NIST/SEMATECH (2007). e-Handbook of Statistical Methods. www.itl.nist.gov/div898/handbook/eda/section3/eda35g.htm. (accessed on 26 January 2007).

Öztuna, D., Elhan, A.H. and Tüccar, E. (2006) Investigation of four different normality tests in terms of type 1 error rate and power under different distributions. *Turkish Journal of Medical Sciences*, **36(3):** 171-176.

Pagliara, S., Viti, C., Gozzini, B., Meneguzzo, F. and Crisci, A. (1998). Uncertainties and trends in extreme rainfall series in Tuscany, Italy: Effects on urban drainage networks design. *Water Science and Technology*, **37(11):** 195-202.

PROPHET StatGuide. (2007) Examining normality test results. www.basic.northwestern.edu/statguidefiles/n-dist_exam_res.html (accessed on 26 January 2007).

Pugacheva, G., Gusev, A., Martin, I., Schuch, N. and Pankov, V. (2003). 22-year periodicity in rainfalls in littoral Brazil. Geophysical Research Abstracts, EGS-AGU-EUG Joint Assembly, Abstracts from the meeting held in Nice, France, April 6-11, 2003, p. 6797.

Radziejewski, M., Bardossy, A. and Kundzewicz, Z.W. (2000). Detection of change in river flow using phase randomization. *Hydrological Sciences Journal*, **45(4):** 547-558.

Raghuwanshi, N.S. and Wallender, W.W. (1997). Field measured evapotranspiration as a stochastic process. *Agricultural Water Management*, **32:** 111-129.

Rao, A.R., Hamed, K.H. and Chen, H.-L. (2003). Nonstationarities in Hydrologic and Environmental Time Series. Water Science and Technology Library, 45, 392 pp.

Reed, D.W., Jakob, D. and Robson, A.J. (1999). Statistical procedures for flood frequency estimation. *In:* A.J. Robson and D.W. Reed (editors), Flood Estimation Handbook, Vol. 3. Institute of Hydrology, 338 pp.

Robson, A.J., Jones, T.K., Reed, D.W. and Bayliss, A.C. (1998). A study of national trend and variation in UK floods. *International Journal of Climatology*, **18:** 165-182.

Salas, J.D. (1993). Analysis and Modeling of Hydrologic Time Series. *In:* D.R. Maidment (editor-in-chief), Handbook of Hydrology. McGraw-Hill, Inc., USA, pp. 19.1-19.72.

Schwankl, L.J., Raghuwanshi, N.S. and Wallender, W.W. (2000). Time series modeling for predicting spatially variable infiltration. *Journal of Irrigation and Drainage Engineering, ASCE*, **126(5):** 283-287.

Shahin, M., Van Oorschot, H.J.L. and De Lange, S.J. (1993). Statistical Analysis in Water Resources Engineering. A.A. Balkema, Rotterdam, The Netherlands, 394 pp.

USEPA (1998). Guidance for data quality assessment: Practical methods for data analysis. United States Environmental Protection Agency (USEPA), Quality Assurance Division, EPA QA/G-9, QA97 Version, pp. 2.3-3 to 2.3-5.

Walpole, R.E. and Myers, R.H. (1989). Probability and Statistics for Engineers and Scientists. 4th Edition, Macmillan Publishing Company, New York.

Westmacott, J.R. and Burn, D.H. (1997). Climate change effects on the hydrologic regime within the Churchill-Nelson River Basin. *Journal of Hydrology*, **202(1-4):** 263-279.

Zhang, X., Harvey, K.D., Hogg, W.D. and Yuzyk, T.R. (2001). Trends in Canadian streamflow. *Water Resources Research*, **37(4):** 987-998.

8

Trend and Homogeneity in Subsurface Hydrologic Variables: Case Study in a Hard-Rock Aquifer of Western India

8.1 Introduction

A comprehensive review on the applications of time series analysis in surface water hydrology, climatology and groundwater hydrology (Machiwal and Jha, 2006) revealed that although several studies deal with the application of time series analysis in surface water hydrology, the application of time series analysis in subsurface hydrology is greatly limited. In subsurface hydrology, time series analysis has been mostly used for detecting trends in groundwater quality (Loftis, 1996; Broers and van der Grift, 2004; Chang, 2008; Visser et al., 2009).

Trend and homogeneity are the two most important characteristics of hydrologic time series, which have been investigated in most studies (e.g., Esterby, 1996; Loftis, 1996; Hess et al., 2001; Machiwal and Jha, 2006). A time series is said to have trends, if there is a significant correlation (positive or negative) between the observations and time. Trends and shifts in hydrologic time series are usually introduced due to natural or artificial changes (Salas, 1993). The trend in a time series can be expressed by a suitable linear or nonlinear model. However, the linear models are more widely used in hydrology than the nonlinear ones (Shahin et al., 1993). Various parametric and nonparametric statistical tests have been reported in the literature for detecting the trend in the hydrologic time series, viz., turning point test, the Kendall's phase test, the Kendall's rank correlation test, regression test, the Wald-Wolfowitz total number of runs test, sum of squared lengths test, adjacency test, difference sign test, the run test on successive differences, inversion test (Shahin et al., 1993), the Spearman rank order correlation test, the Mann-

Kendall test for a linear and/or nonlinear trend (Salas, 1993), the Hotelling-Pabst test (Conover, 1971), and the Sen test (Gilbert, 1987). Except one, all these trend detection tests are discussed in Chapter 4 of this book. A few more rank correlation tests have also been suggested by Kanji (2001).

On the other hand, homogeneity implies that the data in the series belong to one population, and hence have a time invariant mean. Non-homogeneity arises due to changes in the method of data collection and/or the environment in which it is done (Fernando and Jayawardena, 1994). Three homogeneity tests, viz., the von Neumann test, Cumulative Deviations test and the Bayesian test (Buishand, 1982) and four multiple comparison tests, viz., Tukey test, Link-Wallace test, Bartlett test and Hartley test (Kanji, 2001) are most widely used for exploring homogeneity in hydrologic time series. All these homogeneity tests are described in Chapter 4 of this book.

The main intent of this chapter is to demonstrate the application of time series analysis to subsurface hydrologic time series viz., groundwater level and net recharge time series. Trends and homogeneity have been examined in these time series as well as in rainfall time series using the data of Udaipur district, Rajasthan, western India. Udaipur district (study area) is situated in the hard-rock hilly terrains of Aravalli Range in Rajasthan and suffers from frequent droughts due to poor and delayed monsoon, low rainfall, abnormally high summer-temperature and inadequate water resources (Bhuiyan et al., 2006). Among consecutive five drought years (1998-2002), the 2002 drought was one of the severest droughts in Rajasthan as well as in the history of India, which affected 56% of the geographical area and the livelihoods of 300 million people in 18 states (Samra, 2004). Thus, the study area is severely affected by water scarcity, which has a direct impact on the livelihood, health and hygiene of the inhabitants. Considering growing water scarcity and global climate change, it is essential to examine trends and homogeneity in rainfall, groundwater level and net recharge time series for the efficient management of water resources in the study area. This case study also demonstrates the use of geographical information system (GIS) in presenting the results of trend and homogeneity tests.

8.2 Study Area and Data

The study area (Udaipur district) is situated in the southern part of the largest and driest state (Rajasthan) of India (Fig. 8.1). It lies between 23°45′ and 25°10′ North latitude and 73°0′ and 74°35′ East longitude encompassing a geographical area of about 12,698 km^2. It consists of 11 blocks (viz., Badgaon, Bhinder, Dhariawad, Girwa, Gogunda, Jhadol, Kherwara, Kotra, Mavli, Salumber and Sarada). It is worth mentioning that for the administration purpose, a state in India is divided into districts, districts into blocks and blocks into *Gram Panchayats*; each *Gram Panchayat* consists of several villages. The Udaipur district is bounded by Rajsamand district in the north, Pali district in the northwest, Dungarpur and partly Banswara districts in the south and Chittaurgarh district in the east.

The climate of Udaipur is tropical, semi-arid with mercury staying between a maximum of 42.3 °C and a minimum of 28.8 °C during summers. Winters are a little cold with the maximum temperature rising to 28.8 °C and the minimum dipping to 2.5 °C. January is the coldest month and May is the hottest month. The mean annual rainfall is 625 mm, precipitating more than 80% during June through September. The rainy season (i.e., wet season) usually starts from mid-June and lasts for about four months up to the end of October. November to May can be characterized as the dry period.

There are ten rainfall gauging stations in Udaipur district for recording rainfall on a regular basis. The locations of the rainfall stations are shown in Fig. 8.1. Monthly rainfall data of these ten standard rainfall stations for the

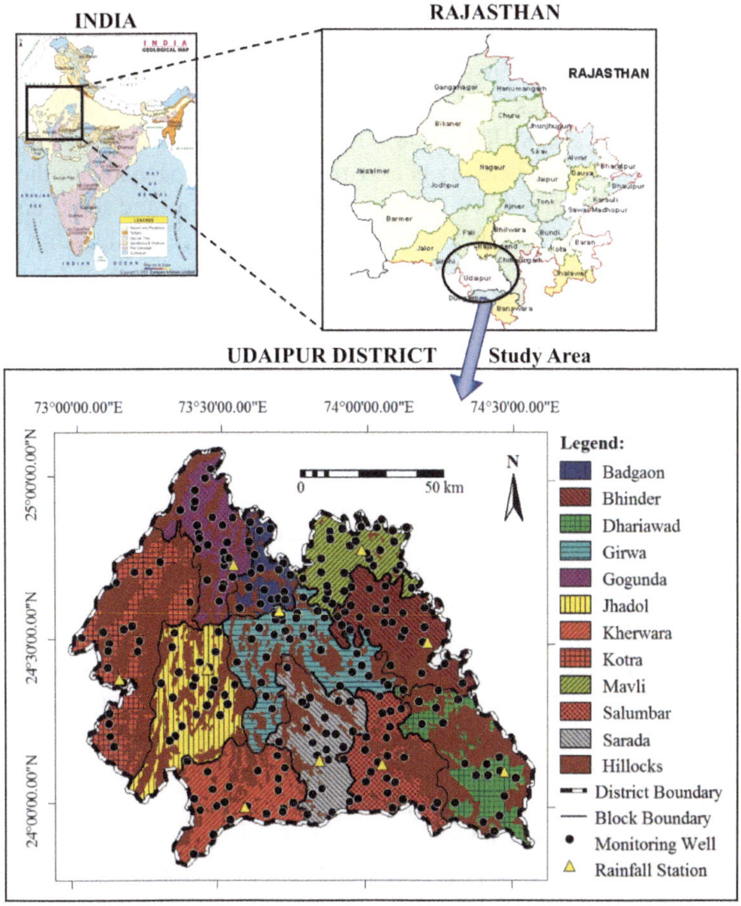

Fig. 8.1. Location map of the study area.

period 1965-2006 were collected from the Land Record Section of Collectorate, Udaipur, Rajasthan. The collected rainfall data of the ten rainfall stations were processed to prepare annual rainfall time series at each station. Pre- and post-monsoon groundwater level data of 251 monitoring wells over the study area for the 19-year period (1988-2006) were collected from the Ground Water Department, Udaipur, Rajasthan; the location of sites are shown in Fig. 8.1. All the collected groundwater level data were screened to check anomalies and 140 monitoring wells were identified with continuous and unambiguous records of pre- and post-monsoon groundwater levels for a period of 16 years (1991-2006). Using 1991-2006 years data of groundwater levels, the annual net recharge was calculated at 140 sites by water table fluctuation method.

8.3 Application of Time Series Tests

In this study, both graphical and statistical methods of time series analysis were applied to determine spatial and temporal patterns in rainfall, pre- and post-monsoon groundwater levels and net recharge time series. Trend and homogeneity are two most important statistical characteristics of the hydrological time series, which reveal temporal variability of hydrologic variables (Machiwal and Jha, 2008). The GIS technique is used to present spatial patterns of presence/absence of trend and homogeneity in the study area. It is quite common to use only one or two tests for time series analysis, which facilitates easy decision making. However, Machiwal and Jha (2008) recommended that an adequate number of statistical tests must be applied for detecting a particular time series characteristic and the results should be analyzed critically to arrive at a reliable decision. Therefore, unlike the customary approach, adequate/multiple statistical tests were applied in this study in order to ensure realistic decisions about the time series analysis.

Firstly, spatial and temporal variations in annual rainfalls of ten rainfall stations were analyzed using 43 years (1965-1997) data. Box and whisker plots of annual rainfalls, which provide a summary of five statistical properties, were drawn for ten rainfall stations and for 43 individual years. In addition, annual rainfall time series of ten rainfall stations for the 43 years (1965-2007) were analyzed for detecting trends by applying three most powerful statistical tests, namely Mann-Kendall test, Spearman Rank Order Correlation (SROC) test and Kendall Rank Correlation test. The details of these tests are given in Chapter 4 (Section 4.3). Spatial homogeneity of the annual rainfall time series was also examined by applying Levene's Analysis of Variance (ANOVA) test and Levene's Median test. To carry out these time series analyses, spreadsheet programs were developed using MS-Excel software.

Moreover, trend in the seasonal (pre-monsoon and post-monsoon) groundwater level time series was examined by using above-mentioned statistical tests. Homogeneity of the seasonal groundwater level time series was investigated by using seven statistical tests, namely Hartley test, Link-

Wallace test, Bartlett test, Tukey test, von Neumann test, Cumulative Deviation test and Bayesian test. Theoretical backgrounds of these statistical tests are provided in Chapter 4 (Section 4.1). The trend and homogeneity tests were applied separately for pre- and post-monsoon seasons using 16 years (1991-2006) groundwater level time series data of 140 sites. Thus, in total, ten time series tests were applied to 280 individual groundwater level time series. Finally, trend and homogeneity were examined in the annual net recharge time series of 16 years (1991-2006) at 140 sites over the study area by applying three trend tests and seven homogeneity tests. To accomplish these tasks, spreadsheet programs were developed using MS-Excel software.

8.4 Spatial and Temporal Variations of Annual Rainfall

8.4.1 Annual Rainfall Pattern

The long-term variation of rainfall is also of prime importance for the efficient management of water resources. The variation of annual rainfalls over the study area along with the 43-year mean annual rainfall is shown with the help of box and whisker plots in Fig. 8.2. It is evident from this figure that the temporal variation of annual rainfall is the highest for Kotra rainfall station and the median rainfalls of Dhariawad and Kotra stations are more than the 43-year mean annual rainfall (662 mm) in the area. It is also revealed from the box and whisker plots (Fig. 8.2) that some outliers and/or extreme rainfall events occurred during the 43-year period at almost all the ten rainfall stations in the study area. The outliers and extremes were detected by the STATISTICA software, which considers a data point to be an outlier if the data point is

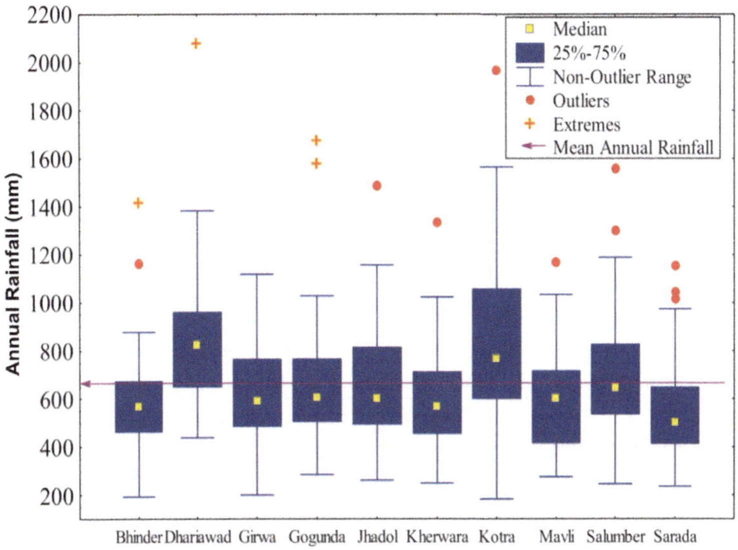

Fig. 8.2. Spatial variation of 43-year period annual rainfalls.

outside the 1.5 times box length range from the upper and lower values of the box. An extreme value is that which is outside the three times box length range from the upper and lower values of the box (Tukey, 1977).

Moreover, long-term temporal variation of the annual rainfall for the 43-year period is shown in Fig. 8.3. Clearly, no overall trend in the annual rainfall is visible, but the presence of outliers indicates that the annual rainfalls in some of the years at Dhariawad and Kotra stations are substantially higher than the rainfalls at other stations. Of the total 43-year period, the annual rainfalls at all the rainfall stations exceeded the mean annual rainfall in five years (1973, 1983, 1990, 1994 and 2006), whereas they were below the mean annual rainfall in eleven years (1966, 1972, 1974, 1982, 1986, 1987, 1988, 1995, 1999, 2000 and 2002). Thus, the rainfall in the study area exhibits considerable variations with space and time.

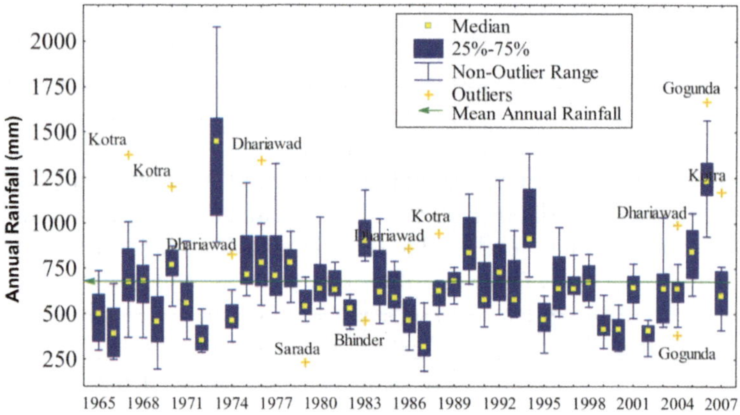

Fig. 8.3. Spatio-temporal variations of 43 years annual rainfalls.

8.4.2 Trend and Homogeneity in the Annual Rainfall Time Series

Results of the Mann-Kendall, SROC, and Kendall Rank Correlation tests are summarized in Table 8.1. It can be seen from this table that the calculated test-statistic values of all three trend tests is more than their critical values at 5% level of significance ($\alpha = 0.05$) for Mavli rainfall station, which indicates presence of trend in the annual rainfall time series of Mavli station. However, the calculated test-statistic values of all trend tests for the remaining nine rainfall stations are less than their critical values, and hence there is no trend in the rainfall series of these stations. Negative test-statistic value of the Mann-Kendall test suggests declining trend in the annual rainfall time series of Mavli rainfall station.

Furthermore, the results of the Levene's ANOVA test (Table 8.2) revealed that the calculated test-statistic value (0.014) is less than its critical value ($\approx$1.96) for 9 degrees of freedom in the numerator and 420 degrees of freedom in the denominator. Hence, the annual rainfalls at the ten rainfall stations do

Table 8.1. Calculated and critical test-statistic values of trend and homogeneity tests for the rainfall time series

Rainfall gauging station	*Levene's median test*	*Mann-Kendall test*		*Spearman Rank Order Correlation test*		*Kendall Rank Correlation test*	
		Calculated	*Critical*[a]	*Calculated*	*Critical*[a]	*Calculated*	*Critical*[a]
Bhinder	0.11	0.10	±1.96	0.06	1.683	-0.115	±1.96
Dhariawad	0.11	0.36	±1.96	0.40	1.683	-0.366	±1.96
Girwa	0.11	0.44	±1.96	0.47	1.683	-0.450	±1.96
Gogunda	0.12	-0.04	±1.96	0.05	1.683	0.052	±1.96
Jhadol	0.12	-0.38	±1.96	0.57	1.683	0.387	±1.96
Kherwara	0.13	-0.48	±1.96	0.46	1.683	0.492	±1.96
Kotra	0.15	-0.17	±1.96	0.16	1.683	0.178	±1.96
Mavli	0.12	-2.09*	±1.96	2.25*	1.683	2.104*	±1.96
Salumber	0.12	-0.77	±1.96	0.78	1.683	0.785	±1.96
Sarada	0.12	-0.36	±1.96	0.45	1.683	0.366	±1.96

Note: * $p < 0.05$; [a]Critical values are at $\alpha = 0.05$.

Table 8.2. Results of Levene's analysis of variance test

Source	*Sum of Squares (SS)*	*Degree of freedom*	*F-ratio*
SS between rainfall stations	268,66,147	9	0.014
Residual SS	268,58,056	420	–

not have significant variance at $\alpha = 0.05$. It is apparent from Table 8.1 that the calculated test-statistic values of the Levene's median test do not vary significantly for the ten rainfall stations. Thus, the results of the Levene's median test are in agreement with those of the Levene's ANOVA test. This finding suggests that the annual rainfalls over the area are spatially homogeneous.

8.5 Trend and Homogeneity in Seasonal Groundwater Levels

8.5.1 Results of Trend Tests

Results of the three trend tests indicating number of sites with presence/ absence of the trends in pre- and post-monsoon groundwater levels are shown in Figs 8.4 and 8.5, respectively. It is apparent that the number of sites with significant trends of increasing groundwater level at 5% significance level ($\alpha = 0.05$) is approximately same for Mann-Kendall test (49% and 7% of the sites in pre- and post-monsoon seasons, respectively) and Kendall Rank Correlation test (51% and 8% of the sites in pre- and post-monsoon seasons, respectively). However, Spearman Rank Order Correlation test does not indicate nature of trend (i.e., increasing/decreasing) and results in relatively large number of sites with significant trends (i.e., 61% in pre-monsoon and 20% in post-monsoon) at $\alpha = 0.05$ compared to two earlier trend tests.

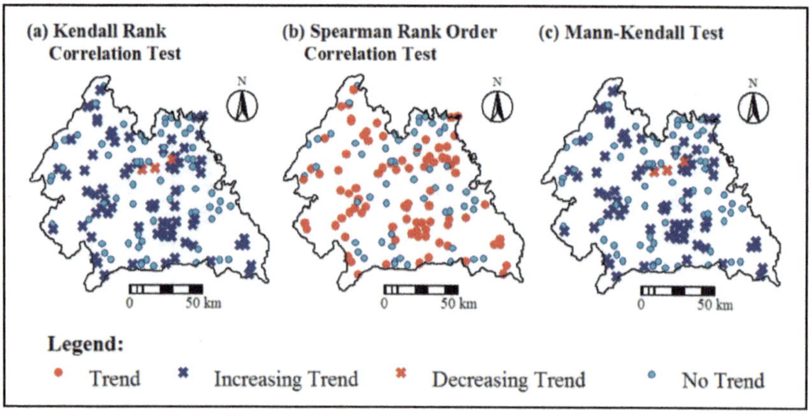

Fig. 8.4. Trends in pre-monsoon groundwater levels based on the three trend tests.

It is apparent from Figs 8.4 and 8.5 that the number of sites with significant trends of increasing groundwater levels (at $\alpha = 0.05$) is relatively more in the pre-monsoon season than in the post-monsoon season. Presence of the increasing trends of post-monsoon groundwater level at relatively small number of sites compared to that in the pre-monsoon groundwater level is due to the rise in groundwater levels during monsoon season, which results in increased hydraulic connectivity among the sites.

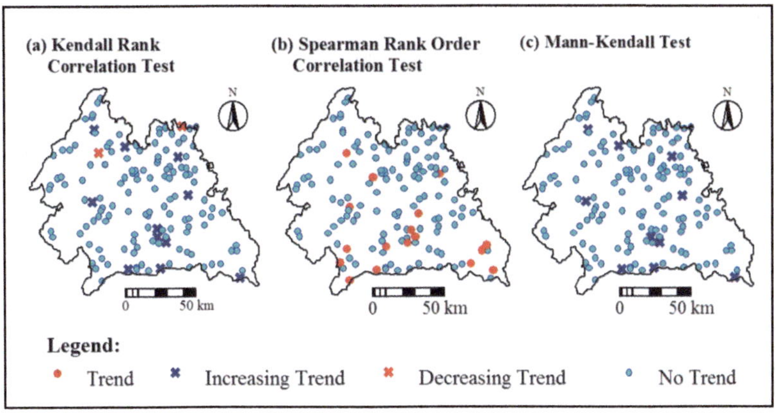

Fig. 8.5. Trends in post-monsoon groundwater levels based on the three trend tests.

It is also evident from Fig. 8.4 that the sites with significant trends ($p < 0.05$) appear to be in three major clusters during pre-monsoon season over the study area: (i) in northeast portion, (ii) in southwest portion, and (iii) in south portion of the area. Of the three clusters, the first cluster falls in the residential area, second in hillocks, and the third cluster falls in the cultivated command area. Based on the type of land use/land cover in significant trend clusters, the significant trends in the pre-monsoon groundwater levels of the first cluster may be due to unsystematic and uncontrolled seasonal domestic groundwater withdrawals. Similarly, the significant trends of the third cluster may be attributed to the non-systematic and uncontrolled groundwater withdrawals for irrigation during pre-monsoon season when surface water supply is not adequate to meet the crop water requirements in the command area. Thus, the significant trends in pre- and post-monsoon groundwater levels in the first and third clusters are due to anthropogenic factors. However, the significant trends in the second cluster could be attributed to geogenic factors (e.g., natural geologic processes). It can be seen from Figs 8.4 and 8.5 that significant increasing trends in post-monsoon groundwater level time series exist at less number of sites than the number of sites having increasing trends in pre-monsoon groundwater level time series.

8.5.2 Results of Homogeneity Tests

Results of the seven homogeneity tests indicating sites with presence/absence of homogeneity in the time series of pre- and post-monsoon groundwater levels are shown in Figs 8.6 and 8.7. Clearly, the number of sites with homogeneity and non-homogeneity differs for different homogeneity tests in both pre- and post-monsoon groundwater levels. For example, according to

Hartley test, the pre-monsoon groundwater levels at about 96% of the sites are homogeneous, while von Neumann test indicates that only 11% of the sites have homogeneity in pre-monsoon groundwater levels (Fig. 8.6). The results of the Tukey and Link-Wallace tests suggest that homogeneity is associated with 31% of the sites in pre-monsoon groundwater levels. On the other hand, three test-statistics (Q of Cumulative Deviations test and U and A of Bayesian test) indicate that 47, 44 and 46% of the sites have homogeneity in pre-monsoon groundwater level time series, respectively, whereas the R test-

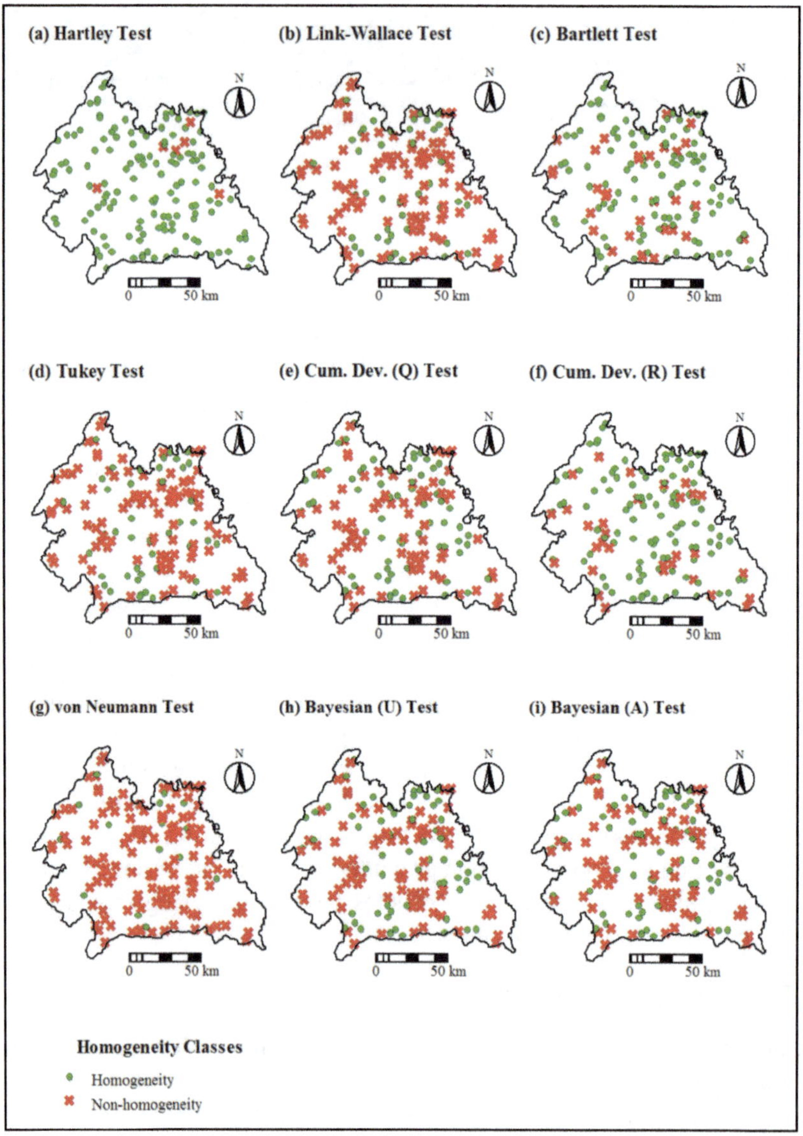

Fig. 8.6. Homogeneous and non-homogeneous pre-monsoon groundwater levels over the study area based on the seven homogeneity tests.

statistics of Cumulative Deviation test reveals that 78% of the sites have homogeneity (Fig. 8.6).

As far as the post-monsoon groundwater level time series is concerned, the results of the Hartley test (Fig. 8.7) suggest presence of homogeneity in all 140 post-monsoon groundwater level time series. The Tukey, Bartlett and the Link-Wallace tests suggest that 66, 66 and 59% of the sites have homogeneous post-monsoon groundwater level time series. Three test-statistics (i.e., *Q*, *U*

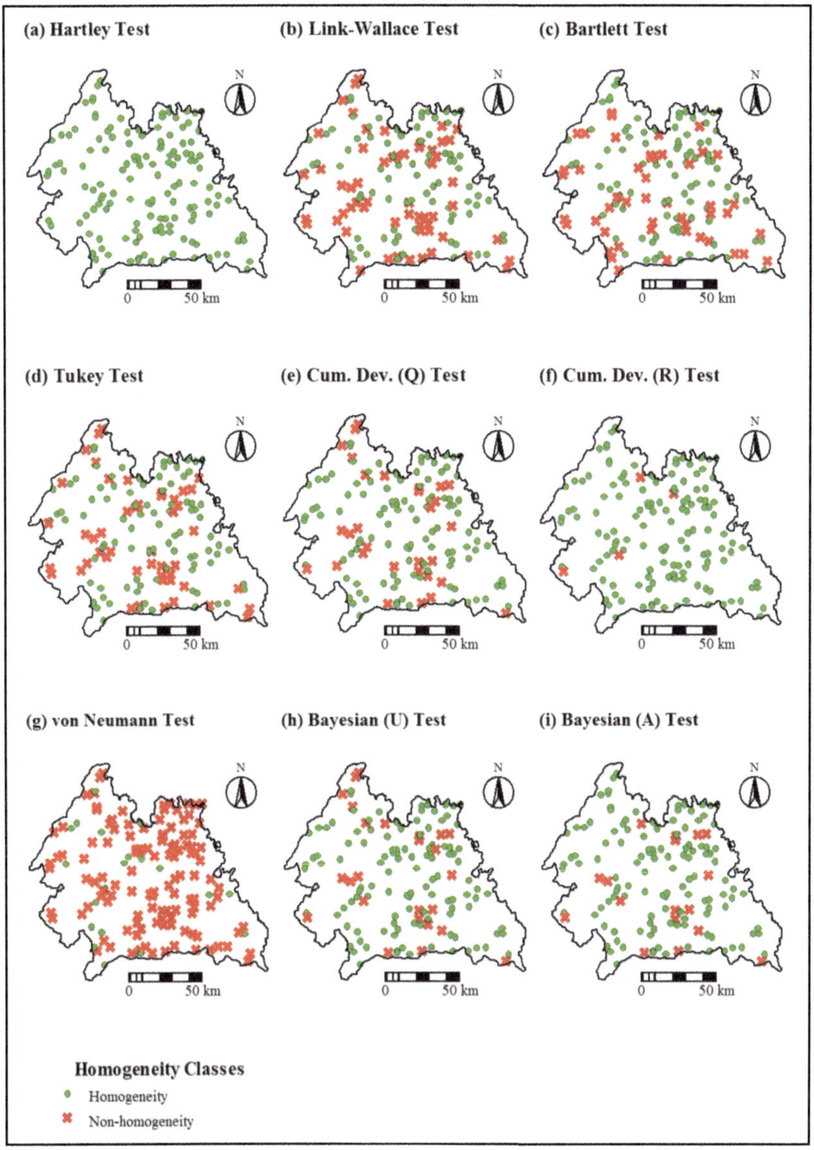

Fig. 8.7. Homogeneous and non-homogeneous post-monsoon groundwater levels over the study area based on the seven homogeneity tests.

and A) of the Cumulative Deviations and the Bayesian tests indicate that 79, 82 and 88% of the sites have homogeneity in post-monsoon groundwater level time series, respectively, but the R-statistic of the Cumulative Deviations test shows that homogeneity is present in 97% of the post-monsoon groundwater level time series.

It is apparent from Figs 8.6 and 8.7 that except for Bartlett and von Neumann tests, all other tests indicate more number of sites with homogeneous groundwater levels in post-monsoon season than in pre-monsoon season. The large homogeneity in post-monsoon groundwater levels seems to be logical because of limited human stress on the aquifer during monsoon/post-monsoon season.

The homogeneity tests (Fig. 8.6) suggest that non-homogeneous sites appear in three major clusters during pre-monsoon season over the study area almost similar to the clusters found in case of trend tests. The first cluster is in the northeast portion, second in southwest portion, and third in south portion of the area. The factors responsible for non-homogeneities ($p < 0.05$) in these clusters could be explained based on the type of land use/land cover in the area as discussed in Section 8.5.1. It should be noted that non-homogeneities in the first and third clusters are due to non-systematic variation of groundwater withdrawals for domestic and irrigation purposes, respectively (or anthropogenic sources) since the clusters exist at the same location where the clusters of significant increasing trends exist. However, non-homogeneities in the second cluster may be attributed to geogenic factors. Almost similar types of three non-homogeneity clusters are also discernible in post-monsoon groundwater levels (Fig. 8.7) for the Link-Wallace test, Tukey test, Cumulative Deviations (Q-statistic) test, and Bayesian test (both U and A test-statistics). However, these clusters are relatively less dense for the post-monsoon groundwater levels compared to the pre-monsoon groundwater levels. The lesser number of non-homogeneous groundwater level sites in post-monsoon season is reasonable because of the fact that sufficient surface water is available for domestic and agricultural purposes as well as groundwater supply is augmented due to natural recharge, and hence, the underlying groundwater system is almost free from artificial stress.

8.6 Trend and Homogeneity in Annual Net Recharge

8.6.1 Trends in Annual Net Recharge

Results of the three trend tests indicating number of sites with presence/absence of trends in the annual net recharge are depicted in Fig. 8.8. This figure shows that the number of sites having significant trends of increasing net recharge (at $\alpha = 0.05$) is almost same for the Mann-Kendall test (5%) and the Kendall Rank Correlation test (6%). However, the Spearman Rank Order Correlation test suggests that the annual net recharge has trends at 12% sites.

The three clusters of increasing trend sites are not clearly visible in the case of net recharge due to relatively less number of sites having increasing trend compared to the pre- and post-monsoon groundwater level time series.

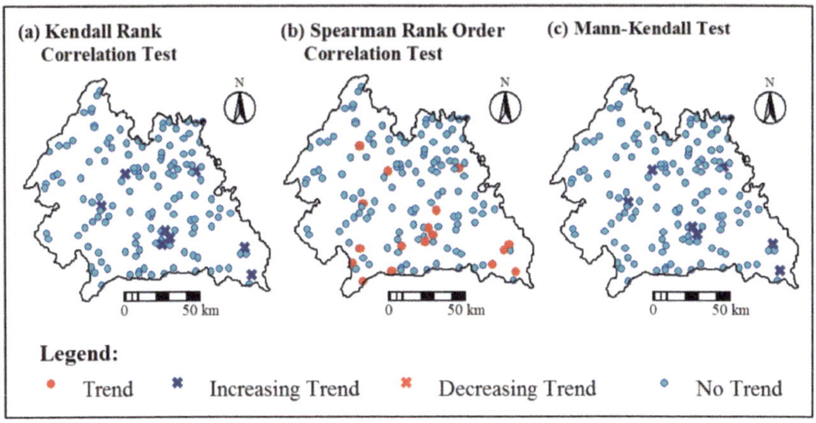

Fig. 8.8. Trends and no trends in the annual net recharge based on the three trend tests.

8.6.2 Homogeneity/Non-homogeneity of Annual Net Recharge

Results of the seven homogeneity tests indicating number of sites with presence/absence of the homogeneity in the annual net recharge time series are shown in Fig. 8.9.

Figure 8.9 shows that the number of sites having homogeneity or non-homogeneity differs for different homogeneity tests. The Hartley test suggests homogeneity in the annual net recharge at about 98% of the sites, whereas the von Neumann test indicates homogeneity only at 34% sites. The results of the Link-Wallace and both the test-statistics (U and A) of the Bayesian tests show that annual net recharge time series is homogeneous at 93% sites. In case of Cumulative Deviations tests, the applied Q and R test-statistics indicate homogeneity in 88 and 99% sites, respectively (Fig. 8.9). The Tukey test reveals that 96% sites have homogeneous annual net recharge time series. Thus, six of the nine applied test-statistics suggest homogeneity in annual net recharge at more than 90% sites.

A comparison of the homogeneity results for annual net recharge time series with those of seasonal groundwater level time series reveals that the annual net recharge time series is relatively more homogeneous than pre- and post-monsoon groundwater levels (Figs 8.6 and 8.7). It is worth mentioning that unlike pre- and post-monsoon groundwater levels non-homogeneous annual net recharge sites are not very dense and hence, no cluster could be delineated for the net recharge.

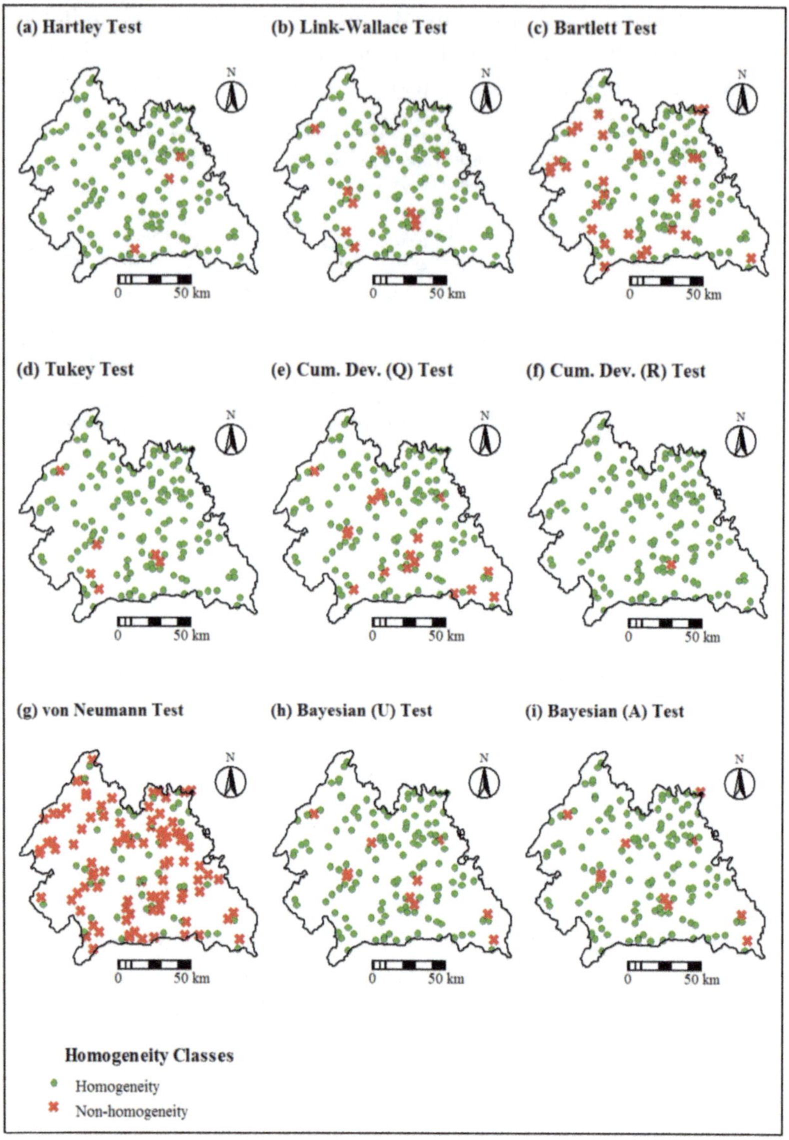

Fig. 8.9. Homogeneous and non-homogeneous annual net recharge in the study area.

8.7 Conclusions

This study demonstrates the application of statistical tests to determine trend and homogeneity in two subsurface hydrologic variables (groundwater levels and net recharge) as well as in rainfall time series of Udaipur district, Rajasthan, western India. Both graphical and statistical methods are used along with GIS technique to determine spatial and temporal patterns in rainfall, pre- and post-monsoon groundwater levels and net recharge time series. The box and whisker plots for the mean annual rainfalls reveal that long-term variation in the rainfall is the highest at Kotra rainfall station. The annual rainfalls in some of the years at Dhariawad and Kotra stations are substantially higher than the rainfalls at other stations. The results of trend tests indicate the presence of significant declining trend at Mavli rainfall station. The results of the Levene's ANOVA test suggest that the rainfall over the study area is spatially homogeneous. Overall, the spatial and temporal variations of annual rainfall in the study area are not statistically significant. Also, no trend was detected in the annual rainfall series except at single rainfall station.

The results of the seven homogeneity tests and the three trend tests for groundwater level time series in both pre- and post-monsoon seasons indicate that non-homogeneous sites appear in three dense clusters over the study area: (i) in northeast portion (dominated by residential area), (ii) in southwest portion (dominated by hillocks), and (iii) in south portion (dominated by cultivated command). The non-homogeneity in groundwater levels in the first and third clusters is attributed to anthropogenic factors, and that in the second cluster is due to geogenic factors. It is also found that the non-homogeneity and trend are present at relatively large number of sites in the pre-monsoon season compared to that in the post-monsoon season.

The results of two homogeneity tests, viz., Link-Wallace test and Bayesian test for the annual net recharge time series are almost similar. It is also found that the annual net recharge time series are relatively more homogeneous compared to the pre- and post-monsoon groundwater level time series. The results of the Mann-Kendall test and Kendall Rank Correlation test for trend detection are almost similar. Homogeneity and trend tests suggest that the annual net recharge time series is homogeneous without any trends in a major portion (more than 90% sites) of the study area.

Overall, it is concluded that the Tukey, Link-Wallace, Bayesian tests and Q-statistic of the Cumulative Deviations test are superior to other homogeneity tests. Further, it is recommended to use either Mann-Kendall test or Kendall Rank Correlation test for exploring trends in similar future studies.

References

Bhuiyan, C., Singh, R.P. and Kogan, F.N. (2006). Monitoring drought dynamics in the Aravalli region (India) using different indices based on ground and remote sensing data. *International Journal of Applied Earth Observation and Geoinformation*, **8(4):** 289-302.

Broers, H.P. and van der Grift, B. (2004). Regional monitoring of temporal changes in groundwater quality. *Journal of Hydrology*, **296:** 192-220.

Buishand, T.A. (1982). Some methods for testing the homogeneity of rainfall records. *Journal of Hydrology*, **58:** 11-27.

Chang, H. (2008). Spatial analysis of water quality trends in the Han River basin, South Korea. *Water Research*, **42:** 3285-3304.

Conover, W.J. (1971). Practical Non-Parametric Statistics. Wiley, New York.

Esterby, S.R. (1996). Review of methods for the detection and estimation of trends with emphasis on water quality applications. *Hydrological Processes*, **10:** 127-149.

Fernando, D.A.K. and Jayawardena, A.W. (1994). Generation and forecasting of monsoon rainfall data. Proceedings of the 20th WEDC Conference on Affordable Water Supply and Sanitation, Colombo, Sri Lanka, pp. 310-313, http://wedc.lboro.ac.uk/conferences/pdfs/20/Fernandd.pdf (accessed on 27 February 2008).

Gilbert, R.O. (1987). Statistical Methods for Environmental Pollution Monitoring. Van Nostrand Reinhold, New York.

Hess, A., Iyer, H. and Malm, W. (2001). Linear trend analysis: A comparison of methods. *Atmospheric Environment*, **35(30):** 5211-5222.

Kanji, G.K. (2001). 100 Statistical Tests. Sage Publication, New Delhi, India, 215 pp.

Loftis, J.C. (1996). Trends in groundwater quality. *Hydrological Processes*, **10:** 335-355.

Machiwal, D. and Jha, M.K. (2006). Time series analysis of hydrologic data for water resources planning and management: A review. *Journal of Hydrology and Hydromechanics*, **54(3):** 237-257.

Machiwal, D. and Jha, M.K. (2008). Comparative evaluation of statistical tests for time series analysis: Application to hydrological time series. *Hydrological Sciences Journal*, **53(2):** 353-366.

Salas, J.D. (1993). Analysis and modeling of hydrologic time series. *In:* D.R. Maidment (editor), Handbook of Hydrology, McGraw-Hill, Inc., New York, pp. 19.1-19.72.

Samra, J.S. (2004). Review and Analysis of Drought Monitoring, Declaration and Management in India. Working Paper 84, International Water Management Institute (IWMI), Colombo, Sri Lanka.

Shahin, M., Van Oorschot, H.J.L. and De Lange, S.J. (1993). Statistical Analysis in Water Resources Engineering. A.A. Balkema, Rotterdam, The Netherlands, 394 pp.

Tukey, J.W. (1977). Exploratory Data Analysis. Addison-Wesley, MA.

Visser, A., Broers, H.P., Heerdink, R. and Bierkens, M.F.P. (2009). Trends in pollutant concentrations in relation to time of recharge and reactive transport at the groundwater body scale. *Journal of Hydrology*, **369:** 427-439.

9

Analysis of Streamflow Trend in the Susquehanna River Basin, USA

9.1 Introduction

Streamflow statistics are extensively employed for the management and development of water resources. The magnitude and frequency of streamflows in the Susquehanna River Basin (SRB) are often used by the Susquehanna River Basin Commission (SRBC) and other agencies for the purposes of water resources planning and management (SRBC, 2006). For example, a wide range of streamflow statistics are used for consumptive water use mitigation, reservoir operation, and minimum release management. Water resources engineers and managers often implicitly assume that streamflow series are stationary over time when using streamflow data and statistics (SRBC, 2006; Zhang and Kroll, 2007a,b; Milly et al., 2008). This assumption may not be valid if the watershed under consideration is sensitive to human disturbance and/or climate change. More generally, climate variability, and change in population, land use and water use are implicated in the non-stationarity of streamflow series (Koutsoyiannis et al., 2009; Lins and Stakhiv, 1998; Milly et al., 2008). In a review of its consumptive use mitigation strategy, the SRBC examined the frequency and duration of consumptive use compensation releases from reservoirs located in the upper reaches of the SRB. It was evident that the number and frequency of 7-day-10-year low flow ($Q_{7,10}$) events had dropped substantially since around 1970. This suggests that the assumption of stationarity in the basin might be invalid. Therefore, an investigation of the assumption of streamflow stationarity in the SRB was of interest.

Invited contribution by Zhenxing Zhang, Robert D. Pody, Andrew D. Dehoff and John W. Balay – Susquehanna River Basin Commission, 1721 N. Front Street, Harrisburg, PA, USA.

The changes of means and the variability of streamflows are the major factors that have contributed to the end of streamflow stationarity. To accurately characterize streamflows, it is of interest to detect these changes. Trend analyses are commonly used in literature to detect changes in streamflow time series. Linear and nonlinear models are often employed to express the trend in a time series in hydrology literature (Shahin et al., 1993). Student's *t*-test is commonly used to detect linear trend. However, this method assumes that the time series is normally distributed. For non-normal data, a nonparametric test such as the Mann-Kendall test (Mann, 1945; Kendall, 1962) is preferred (Hirsch and Slack, 1984; Helsel and Hirsch, 1992). The serial correlation in a time series will impact the ability to evaluate the significance of the test (Kulkarni and von Storch, 1995; von Storch, 1995; Yue et al., 2002b). Lettenmaier et al. (1994) found strong upward trends in about half the 1009 investigated streams in the continental United States for the months of November through April using the streamflow records from 1948 through 1988. Lins and Slack (1999) found similar results with lower magnitude streamflow quantiles.

Noticing that many studies did not consider the regional cross-correlation of streamflow, Douglas et al. (2000) proposed a bootstrap approach to account for it and detected upward trends in low flows in the Midwestern U.S. An increase in high streamflows in the conterminous US has been reported by Groisman et al. (2001). Zhu and Day (2005) reported downward trends in 47 streams across Pennsylvania for the 1971-2001 period. Kalra et al. (2008) documented increased streamflow in the Mississippi and Missouri regions for the period 1951-2002. Different trends in streamflows in varying months in the Colorado River Basin have been documented by Miller and Piechota (2008). Wu et al. (2008) found that there is no uniform trend in droughts in Nebraska. The detection and characterization of trends should be studied in a framework that recognizes and characterizes the dependence structure of hydroclimatic records (Koutsoyiannis and Montanari, 2007). Bhutiyani et al. (2008) reported changing streamflow patterns in the rivers of northwestern Himalaya during the 20th century. Milly et al. (2005) demonstrated an ensemble of 12 climate models to simulate patterns of changes in global streamflows. Changnon and Demissie (1996) examined streamflow changes in the Midwest region of the United States and investigated the effects of land use and climate fluctuations. Machiwal and Jha (2006) provided an excellent review of trend analysis on hydrologic time series, together with the application of other time series analysis techniques in hydrology and climatology.

Traditionally, a period of time is pre-specified and a trend test is conducted using the data within the selected period. However, this approach cannot demonstrate the pattern of change. To effectively and efficiently manage water resources, water resources engineers and managers need to know not only if there are trends but also if the trends are abrupt or gradual. The magnitude of the change and length of the period during which the change occurred have considerably different implications. For a gradual trend, the change occurs during a relatively long period of time and is expected to continue into the future. If the trend is abrupt, the change is a level shift that occurs over a relatively short time. Once the level shift is completed, the streamflows remain stationary at the new level until another change occurs (McCabe and Wolock, 2002). The hydrologic literature has so far devoted very limited attention to the characterization of trend patterns. Kalra et al. (2008) studied the step change in 639 U.S. streamflows over period of 1951-2002 using the Rank Sum test and Student's *t*-test. However, in their study, a prior hypothesis of a time of change was required for detecting the step change. Miller and Piechota (2008) investigated the step change in monthly hydroclimatic variables around the Colorado River Basin. They divided the time series into two sub-series for the step change analysis which requires the prior knowledge of the time of change. McCabe and Wolock (2002) employed multiple nonparametric statistical trend tests with various record lengths to detect change of streamflow across the conterminous United States. Changes in streamflow could be detected by examining the number of stream sites with significant trends. To employ the method suggested by McCabe and Wolock (2002), a group of stream gauges are required. Hydrology literature is often focussed on large spatial scale (national or regional) flow change analyses with a large number of stream gauges. Zhang et al. (2010b) found that their coarse spatial resolution limits the practical application to specific watersheds in the SRB. To detect a change in streamflow for single streams, Zhang et al. (2010a) proposed to use the nonparametric trend test with varying record lengths. Zhang et al. (2010b) applied this approach as a screening tool to identify potential changes in streamflow of watersheds within the SRB.

This study expands on the work of Zhang et al. (2010b). The method proposed by Zhang et al. (2010a) is employed to detect trend patterns in streamflow in the SRB. Both annual and monthly streamflow characteristics are investigated. The two components of streamflow, i.e. 'baseflow' and 'storm runoff', have been examined. Baseflow and storm runoff are estimated by employing the baseflow separation programme, BFI which is available at the U.S. Bureau of Reclamation website (http://www.usbr.gov/pmts/hydraulics_lab/twahl/bfi/). Three questions have been addressed in this study: (i) Are there changes in the streamflow in the basin? (ii) If so, what is the pattern of the change, i.e., are the flow changes increasing or decreasing and are they gradual or abrupt? (iii) If there are changes, in which month(s) do they occur?

9.2 Study Area

The area of the Susquehanna River Basin (SRB) is 71,250 km^2, which includes portions of the states of Pennsylvania, New York and Maryland. It is the largest tributary of the Chesapeake Bay, comprising 43% of the bay's drainage area and providing 50% of its freshwater. The length of the Susquehanna River is about 715 km and it is the largest river within the United States that drains into the Atlantic Ocean. The SRB extends from the Atlantic coastal plain inland approximately 260 km, crossing several physiographic regions (Fig. 9.1) of the Appalachian Highlands. These regions generally run northeast-southwest, approximately parallel to the Atlantic coastline. The physiographic provinces covered by the basin include Appalachian Plateau, Valley and Ridge, Blue Ridge and Piedmont (Marsh and Peirce, 1995; Pennsylvania Department of Environmental Protection, 2009).

The northern and western portions of the basin are located in the Appalachian Plateau. This region is characterized by sedimentary rocks of Devonian, Mississippian and Pennsylvanian age that are flat lying or very gently folded. The rocks are dominantly siliciclastics. The Pennsylvanian formations contain coal beds that are widely mined. The streams generally flow in deep, steep-sided valleys with 150 to 300 m of local relief. Dissection of the plateaus is variable, with some portions having extensive summit areas while other portions are nearly all in slope. The Appalachian Plateau is largely forested. The population is concentrated in widely scattered towns with a population of a few hundreds to a few thousands. The Valley and Ridge section occupies the central portion of the basin. This region is characterized by strongly folded and faulted sedimentary rocks of Paleozoic age. Erosion of this terrain has produced even-crested parallel ridges of Paleozoic sandstone with intervening valleys formed on less resistant rock types. The ridges typically have 300-450 m of local relief. The valleys are generally hilly. The ridges are primarily forested while the valleys are primarily comprised of agricultural lands. The population is concentrated in towns of a few hundred to several thousand people. The larger streams flow parallel to the ridges, while the trunk streams cross the ridges in deeply cut water gaps. The Susquehanna River and its major tributaries actually flow across this structural and topographic grain. The Blue Ridge section is a broad anticlinorium of largely Lower Cambrian metasediments with a core of late pre-Cambrian igneous and meta-volcanic rocks. The Blue Ridge is about 10 km in width, and ends abruptly about 24 km west of the Susquehanna River. Summit areas are gently rounded, with extensive flat areas. The Piedmont section is located seaward of the Blue Ridge. It is approximately 60 km wide where it crosses the SRB. Landuse is a mix of forestry and agriculture. The Piedmont consists largely of uplands underlain by meta-sedimentary and meta-volcanic rocks, with minor lowlands underlain by Mesozoic age sediments and igneous intrusions. The structural grain of the Piedmont is parallel to that in the neighbouring physiographic sections: northeast-southwest. Local relief is typically 30-90 m. The Susquehanna River flows through a rugged bedrock valley as it crosses the Piedmont.

The climate of the SRB is humid continental, and reflects both an alternation and interplay between oceanic and continental air masses, and their associated weather. The Appalachian Plateau section of the SRB is characterized by average annual precipitation in the range of 90-100 cm,

Table 9.1. Salient characteristics of the watersheds investigated

Sl. No.	*Station name*	*Drainage area* km^2	*Channel slope m/km*	*Channel length km*	*Mean basin elevation m*	*Annual precipitation cm*
1.	Wapwallopen Creek near Wapwallopen, Pennsylvania	114	9.9	35.6	405	106
2.	Fishing Creek near Bloomsburg, Pennsylvania	710	7.5	52.0	363	111
3.	West Branch Susquehanna River at Bower, Pennsylvania	816	1.7	56.8	368	115
4.	Sinnemahoning Creek at Sinnemahoning, Pennsylvania	1774	1.7	74.2	512	113
5.	Pine Creek at Cedar Run, Pennsylvania	1564	4.3	71.6	579	102
6.	Penns Creek, Pennsylvania	780	3.6	72.7	405	107
7.	East Mahantango Creek near Dalmatia, Pennsylvania	420	2.1	70.3	262	108
8.	Frankstown Branch Juniata River at Williamsburg, Pennsylvania	754	2.8	53.3	254	109
9.	Dunning Creek at Belden, Pennsylvania	445	8.1	46.6	475	109
10.	Sherman Creek at Shermans Dale, Pennsylvania	518	1.4	69.2	360	108
11.	Deer Creek at Rocks, Maryland	243	3.4	46.0	200	113

average minimum January temperatures of –10 °C and average maximum July temperatures of 27 °C (Pennsylvania Department of Environmental Protection, 2009). The Valley and Ridge section is characterized by average annual precipitation in the range of 84-102 cm in the west, and 107-130 cm in the east. Average minimum January temperatures range from –8 °C in the valleys to –10 °C in the mountains. Average maximum July temperatures range from 29 °C in the valleys to 24 °C in the mountains. The Blue Ridge and Piedmont sections are characterized by average annual precipitation in the range of 102-119 cm. Average minimum January temperature is about –6 °C.

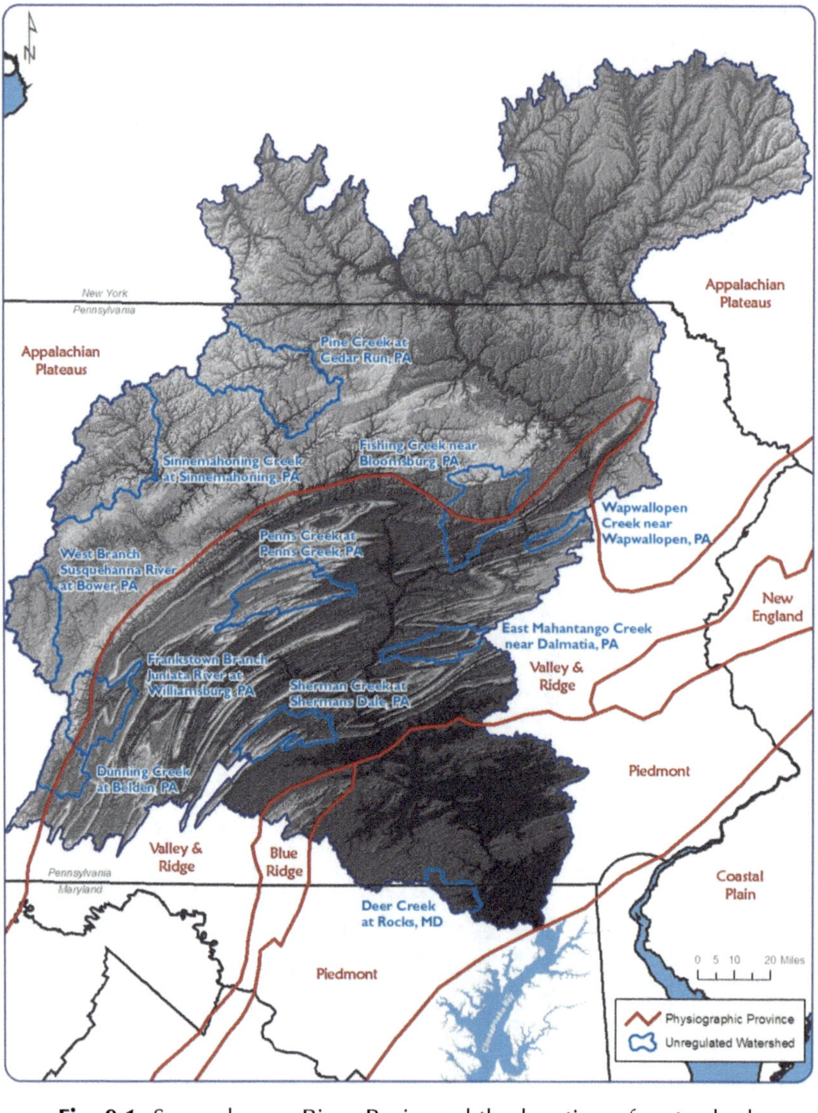

Fig. 9.1. Susquehanna River Basin and the location of watersheds under investigation.

Average maximum July temperatures range from 30 °C in the valleys to 28 °C in the mountains.

To investigate the potential streamflow changes and their character, streamflow measurements at unregulated streams are warranted as they are minimally impacted by flow regulations or other human activities. Eleven long-term gauges distributed across the SRB were identified by SRBC staff as unregulated streams which are suitable for the purpose of this study (Fig. 9.1). Salient characteristics of the identified gauge stations are provided in Table 9.1. Most of these streams were also listed as such in the USGS Hydro-Climatic Data Network (HCDN) (Slack and Landwehr, 1992). In the HCDN datasets, about 1,600 unregulated gauged watersheds distributed throughout the United States were identified as being potentially suitable for study of potential impact of climate change on hydrology. Hydrology literature extensively employed the HCDN dataset for streamflow trend analyses in the United States (McCabe and Wolock, 2002; Vogel et al., 1997; Kroll et al., 2004; Zhang and Kroll 2007, a, b). Daily streamflow data from 1940 to 2006 were downloaded from the USGS website (http://waterdata.usgs.gov/nwis/sw). Annual daily minimum, median and maximum flows are used for the detection of streamflow changes. The monthly average baseflow, storm runoff, total streamflow, and monthly minimum total streamflow are used for detection of changes as well.

9.3 Methodology

When the hydrologic time series records being investigated are in two non-overlapping periods with a lengthy gap between them or a known event which is expected to change the hydrologic time series has occurred at a time during the record, the step change in hydrologic record can be tested by dividing the record into two periods and using techniques such as the rank-sum test, two-sample *t*-test, and analysis of covariance (Helsel and Hirsch, 1992). These techniques require a highly specific situation or prior hypothesis of a time of change. Otherwise, monotonic trend tests such as the Mann-Kendall test are appropriate. In hydrologic practice, hydrologists and water resources managers often do not have the prior information or a known event to identify the time. Thus, many studies of changes in hydrologic time series have focussed on the monotonic changes over time, which can be tested with the Mann-Kendall test. However, the Mann-Kendall test with a specified period of record is not capable of showing the trend pattern, i.e., whether the change is gradual or abrupt (McCabe and Wolock, 2002).

Zhang et al. (2010a) suggested apply multiple Mann-Kendall tests with varying beginning and ending times to detect trend patterns in hydrologic time series. The step-by-step procedure is presented here.

Step 1: Rank the data set according to time T.
Step 2: Set the beginning time $T_b = 1$.

Step 3: Set the ending time $T_e = T_b + 9$ so that the minimal data points in the selected period is 10.

Step 4: Select the subset with the determined beginning time and ending time and conduct the Mann-Kendall test for the selected subset.

Step 5: Set the ending time $T_e = 11, 12, \ldots, n$, and repeat step 4.

Step 6: Set the beginning time $T_b = 2, 3, \ldots, n - 9$, and repeat steps 3, 4, and 5.

Step 7: The resulting p-values of each Mann-Kendall test are plotted with graduated colour (or shades) against beginning and ending times of each subset. The graduated colour (or shades) denotes the underlying *p*-values of individual tests. Therefore, the single Mann-Kendall test with beginning time of T_b and ending time of T_e will have the coordinate (T_b, T_e) on the plot with the graduated colour (or shades) determined by the resulting *p*-values of the test. The significant increasing trends at significance level of 5% are marked with upward triangles and the significant decreasing trends at the same significance level are denoted by downward triangles.

Step 8: The 45-degree line that is parallel to the 1:1 line represents the results with the same record length with different beginning or ending times. To investigate the impact of record lengths, the lines denoting record lengths of 10, 30 and 50 years are plotted.

In the procedure, 10 or more data points are used for the Mann-Kendall test as at least 10 data points are needed so that the test-statistic follows a standard normal distribution (Kendall, 1962; Douglas et al., 2000). Yue et al. (2002a) suggested that the more data points available, the more powerful the test. The multiple Mann-Kendall test with all data series of at least 10 data points and the visualization of the test results can be used to examine the impact of record length, beginning time and ending time. The lines representing varying record lengths can be used to examine the impact of record length. The pattern of various beginning time and fixed ending time demonstrates the impact of the beginning time. The pattern of changing ending time and a specified beginning time shows the impact of the ending time. Trend patterns can be demonstrated by the sensitivity of the multiple trend tests to the beginning time, ending time and record length. For example, Fig. 9.2 shows the raw data and Fig. 9.3 shows the results of multiple Mann-Kendall tests. It is obvious from Fig. 9.2 that the annual minimum flows in the study area abruptly changed over a short period around 1970. From Fig. 9.3, it can be seen that the monotonic trend tests are often significant if the beginning year is before 1970 and the ending year is after 1970. If either the beginning year is after 1970 or the ending year is before 1970, the monotonic tests are usually insignificant. This suggests that the change is abrupt over a short period around 1970. The impact of record length can be explored by Fig. 9.3, which reveals that most long-term trends in the annual minimum flows are significant. However, the short-term trends are usually insignificant unless the records include the period around the year of 1970.

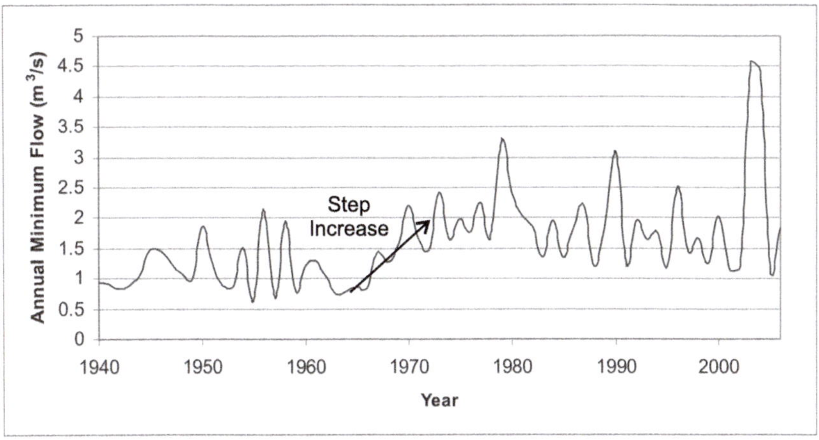

Fig. 9.2. Annual minimum flows of the West Branch Susquehanna River at Bower, Pennsylvania (modified from Zhang et al., 2010a).

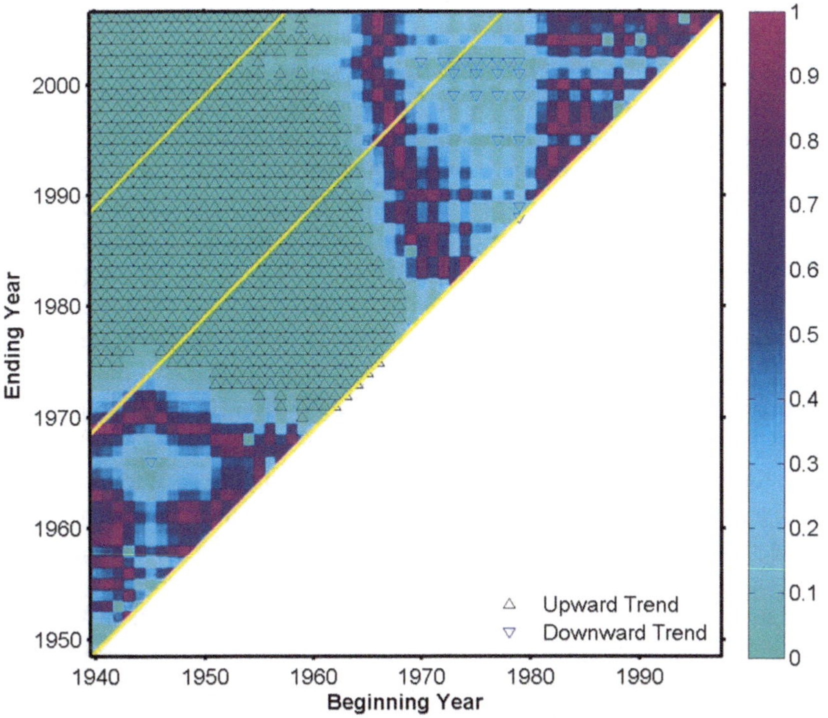

Fig. 9.3. Results of the trend pattern detection in the annual minimum streamflows of the West Branch Susquehanna River at Bower, Pennsylvania. Upward triangles indicate significant increasing trend at the significance level of 0.05 and downward triangles indicate significant decreasing trend at the significance level of 0.05 (modified from Zhang et al., 2010a).

9.4 Results and Discussion

9.4.1 Annual Streamflow Time Series

The annual minimum, median and maximum streamflows over the period of 1940 to 2006 are analyzed in this study and the results are based on the detailed analysis of Sinnemahoning Creek at Sinnemanhoning, Pennsylvania, and summaries of the detailed analysis of the other watersheds.

Figure 9.4 shows the annual minimum streamflows at Sinnemahoning Creek and Fig. 9.5 demonstrates the test results. The individual trend tests are usually significant if the beginning year is in the 1960s or before and the ending year is in the 1970s or later. The long-term trends are usually significant. As discussed in 'Methodology', this finding suggests that a long-term change in annual minimum streamflows at Sinnemahoning happened during a short period around 1970. For the trends of shorter periods, there are often no significant trends if both beginning and ending years are before 1970. If the beginning year is in the mid-1970s to mid-1980s, there are often significantly decreasing trends, which show that the annual minimum streamflows in Sinnemahoning are relatively high in that period.

Other than Deer Creek at Rock, Maryland, and Dunning Creek at Belden, Pennsylvania, all the remaining watersheds show sensitive annual minimum streamflows as what have been seen in Sinnemahoning Creek and West Branch Susquehanna River at Bower, Pennsylvania. These watersheds are distributed in the western and middle portions of the Susquehanna River Basin (Fig. 9.6). On the other hand, Deer Creek and Dunning Creek do not show the abrupt increased annual minimum flows which occurred around 1970.

Moreover, the annual median streamflows at Sinnemahoning Creek are shown in Fig. 9.7 and the results of the trend pattern identification are shown in Fig. 9.8. While only some of the trend tests are significant if the beginning year is before the mid-1960s and the ending year is after the mid-1970s, those tests often have p-values of less than 0.10. If the significance level of 0.10 is adopted, the pattern is very like the one seen in the annual minimum streamflow. This indicates that the abrupt increase in annual median streamflows in Sinnemahoning Creek is of a lesser magnitude than the abrupt increase in annual minimum streamflows. Pine Creek at Cedar Run, Pennsylvania, West Branch Susquehanna River at Bower, Pennsylvania, and Frankstown Branch Juniata River at Williamsburg, Pennsylvania, share similar trend patterns with Sinnemahoning Creek in that all have relatively sensitive annual median streamflows. The watersheds with relatively sensitive annual median streamflows are located in the western portion of the Susquehanna River Basin (Fig. 9.9).

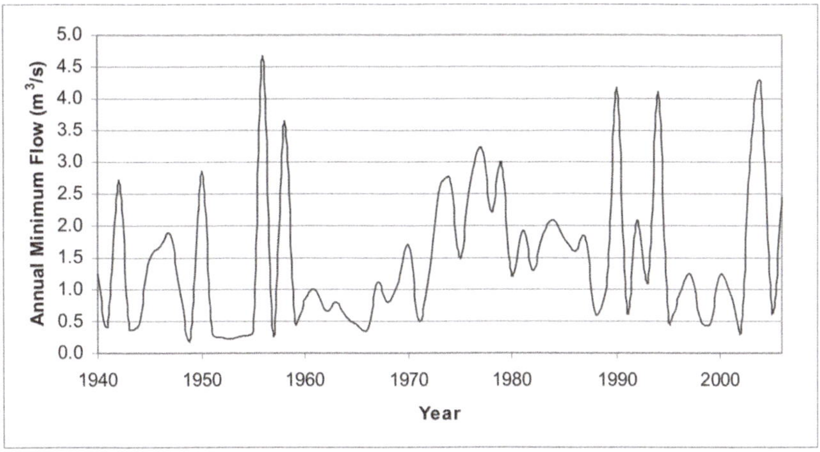

Fig. 9.4. Annual minimum streamflows of the Sinnemahoning Creek at Sinnemahoning, Pennsylvania.

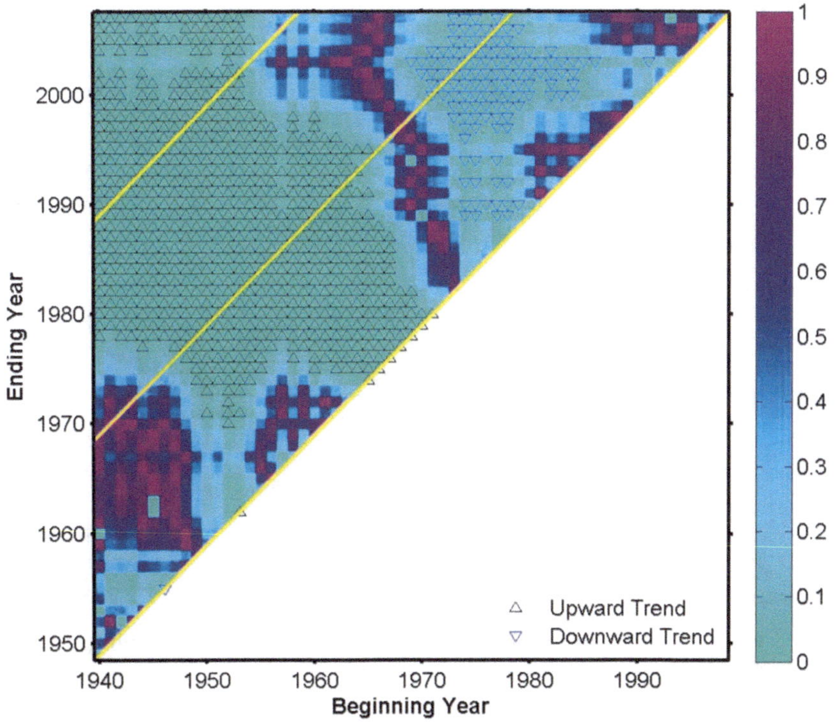

Fig. 9.5. Results of the trend pattern detection in the annual minimum streamflows of the Sinnemahoning Creek at Sinnemahoning, Pennsylvania. Upward triangles indicate significant increasing trend at the significance level of 0.05 and downward triangles indicate significant decreasing trend at the significance level of 0.05.

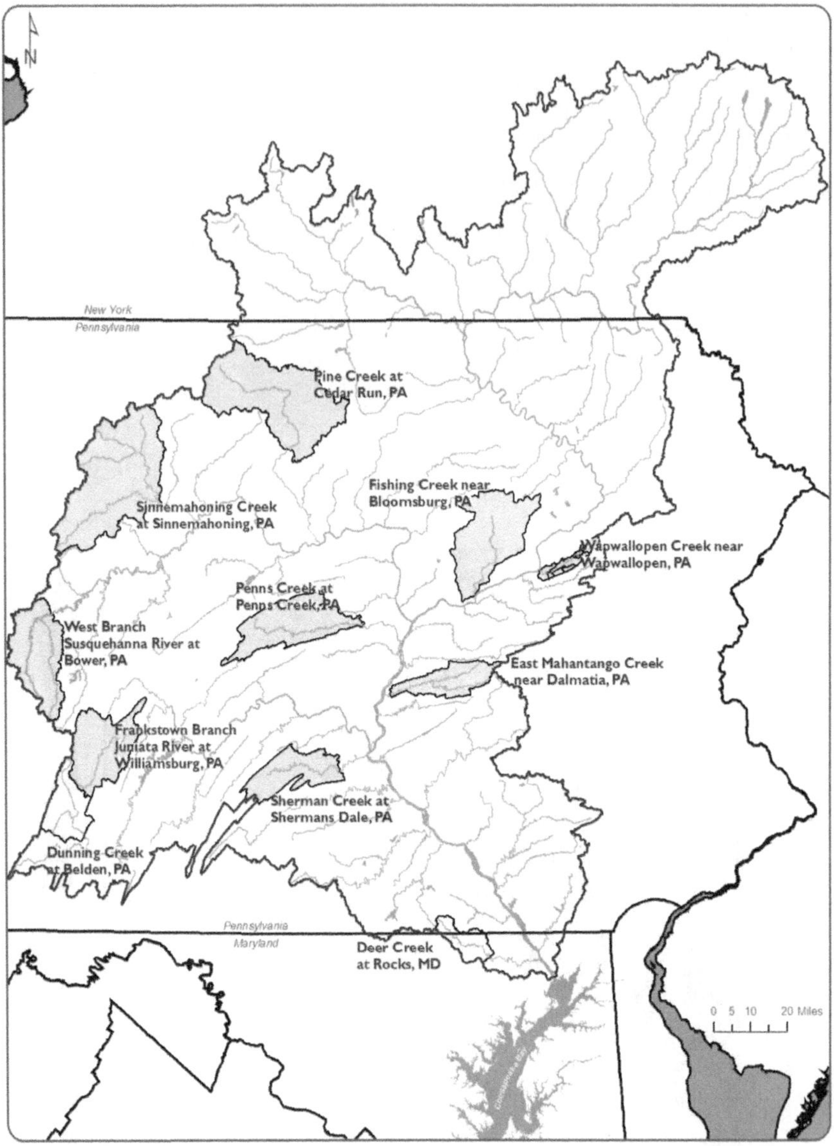

Fig. 9.6. The watersheds with abrupt increased annual minimum streamflows around 1970.

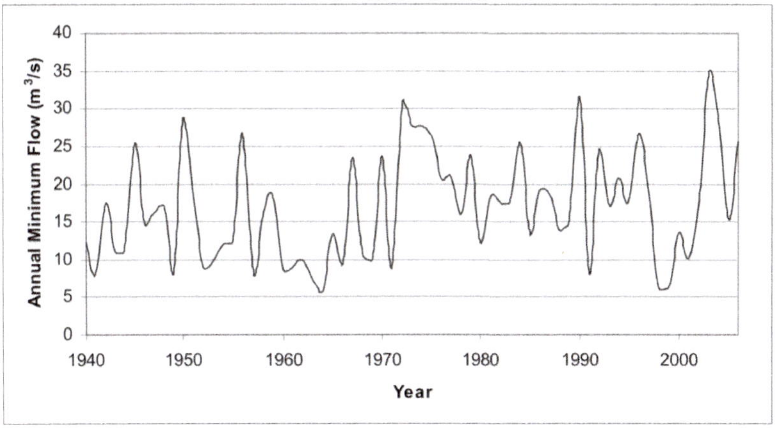

Fig. 9.7. Annual median streamflows of the Sinnemahoning Creek at Sinnemahoning, Pennsylvania.

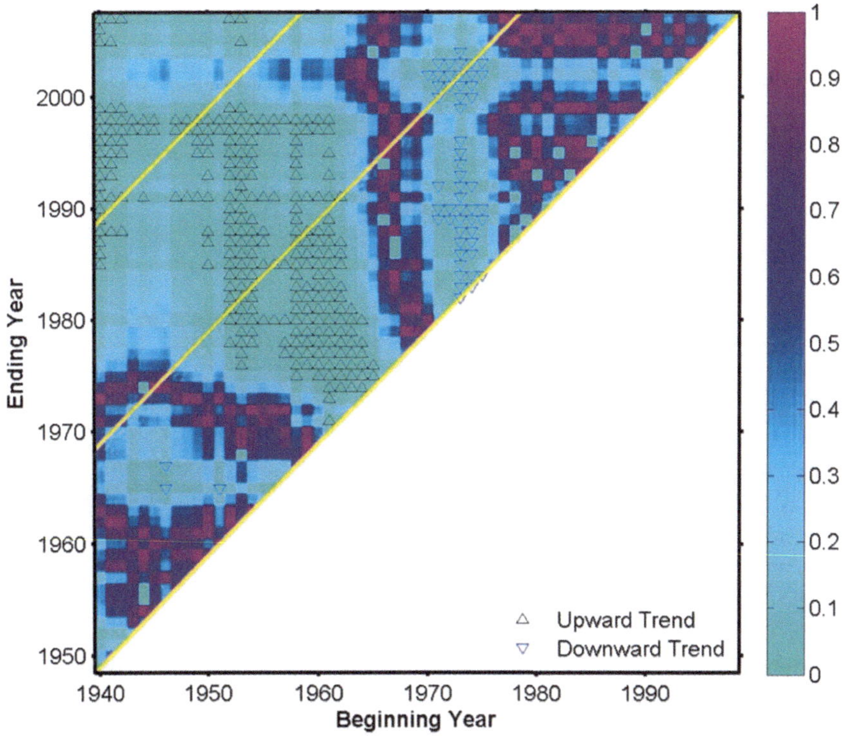

Fig. 9.8. Results of the trend pattern detection in the annual median streamflows of the Sinnemahoning Creek at Sinnemahoning, Pennsylvania. Upward triangles indicate significant increasing trend at the significance level of 0.05 and downward triangles indicate significant decreasing trend at the significance level of 0.05.

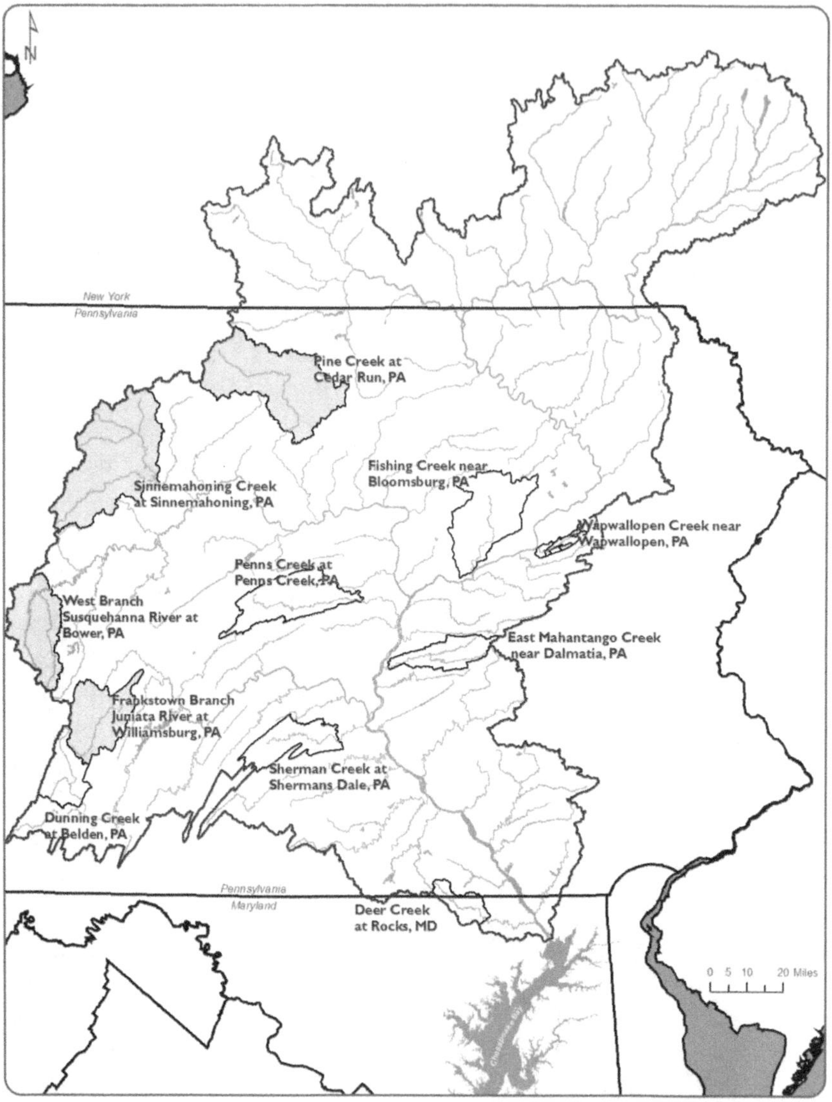

Fig. 9.9. The watersheds with abrupt increased annual median streamflows around 1970.

For annual maximum streamflows, none of the 11 watersheds of the study area show an abrupt increase around 1970 as seen in the annual minimum and median streamflows in Sinnemahoning Creek. For example, Fig. 9.10 shows the annual maximum streamflows at Sinnemahoning Creek and Fig. 9.11 shows the results of the multiple tests. The majority of the multiple monotonic trend tests are insignificant with limited scattered significant trends over relatively short-term periods.

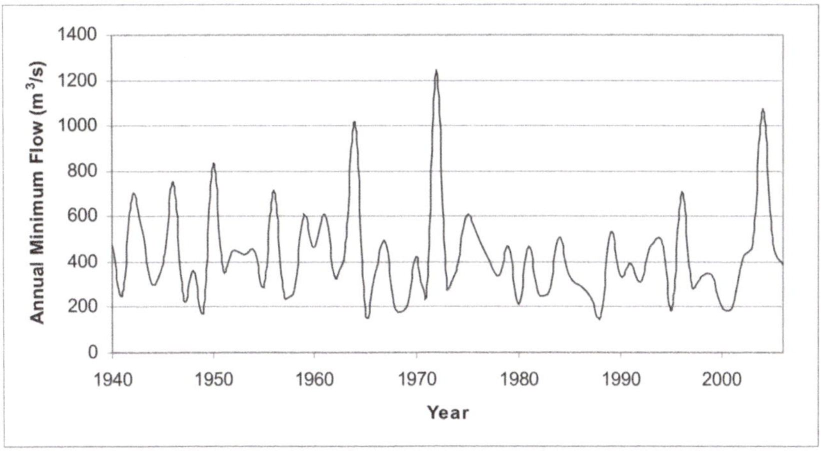

Fig. 9.10. Annual maximum streamflows of the Sinnemahoning Creek at Sinnemahoning, Pennsylvania.

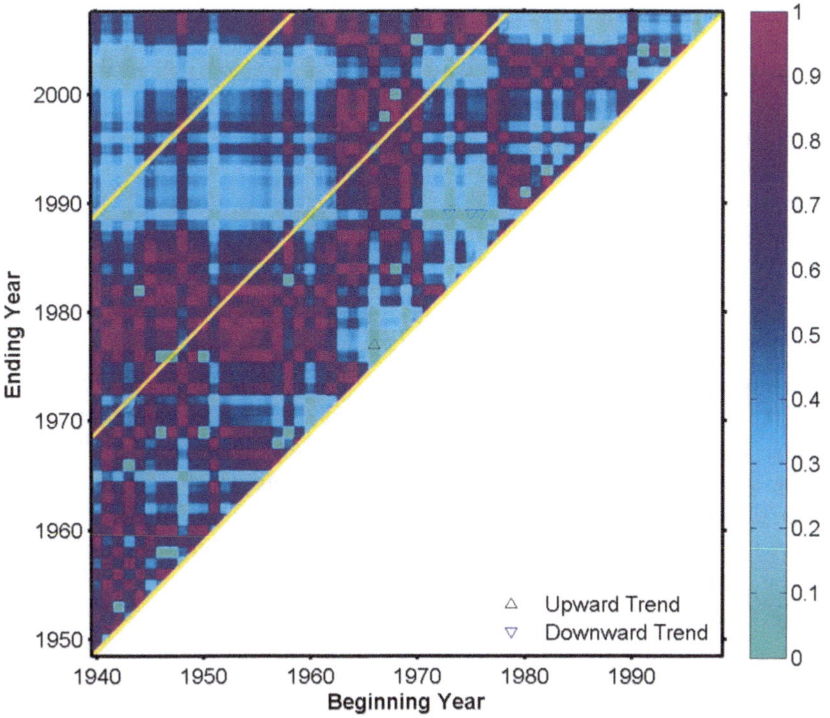

Fig. 9.11. Results of the trend pattern detection in the annual maximum streamflows of the Sinnemahoning Creek at Sinnemahoning, Pennsylvania. Upward triangles indicate significant increasing trend at the significance level of 0.05 and downward triangles indicate significant decreasing trend at the significance level of 0.05.

9.4.2 Monthly Streamflow Time Series

The trend pattern detection is conducted for monthly average baseflow, average storm runoff, average total streamflow, and monthly minimum total streamflow with the visualization of the multiple trend tests. As the abrupt step increase over a short period around 1970 is the most common change for the monthly streamflow statistics, the results of monthly streamflow time series are summarized based on whether the abrupt increase has occurred over the short period around 1970. Table 9.2 shows the results of monthly streamflow at Sinnemahoning Creek at Sinnemahoning, Pennsylvania. For Sinnemahoning Creek, the step increase around 1970 occurred in monthly average baseflow and minimum total flow in September through December. In September and November, average storm runoff and average total flow have seen the step increase around 1970.

Table 9.2. The results of monthly average baseflow, average storm runoff, average total flow, and minimum total flow of the Sinnemahoning Creek at Sinnemahoning, Pennsylvania

Hydrologic time series	*Whether step increase occurred around 1970?*			
	September	*October*	*November*	*December*
Average baseflow	Yes	Yes	Yes	Yes
Average storm runoff	Yes	No	Yes	No
Average total flow	Yes	No	Yes	No
Minimum total flow	Yes	Yes	Yes	Yes

The results of monthly streamflow across all watersheds are shown in Table 9.3. The value in individual cells represents the number of watersheds that show the step increase around 1970. There is no step increase around 1970 for any monthly streamflow time series examined for January through May. The step increase of the monthly streamflow time series happened primarily in the summer and fall seasons, which are usually characterized by low flows in the SRB. This is consistent with the observation of the step increase of annual minimum flow at most of the watersheds investigated.

9.5 Conclusions

Individual monotonic trend tests are capable of detecting the increasing or decreasing trends. However, trend pattern and change points cannot be identified by monotonic trend tests. When prior hypotheses of the change points are known, the step change before and after the hypothesized change points can be explored by the rank-sum test, two-sample t-test, and analysis of covariance. However, more often the prior hypothesized change points are not known. The use of multiple monotonic trend tests with varying beginning and ending

Table 9.3. Number of watersheds with an abrupt increase around 1970 in monthly flow statistics

Hydrologic time series	*Number of watersheds with step increase around 1970*											
	Jan	*Feb*	*Mar*	*Apr*	*May*	*Jun*	*Jul*	*Aug*	*Sep*	*Oct*	*Nov*	*Dec*
Average baseflow	0	0	0	0	0	0	2	2	5	4	2	4
Average storm runoff	0	0	0	0	0	0	0	0	3	1	2	0
Average total flow	0	0	0	0	0	0	0	0	4	0	2	1
Minimum total flow	0	0	0	0	0	1	3	4	7	8	2	6

times can be used to detect the trend pattern. In this study, the graphical representation of the multiple trend tests was used to aid in the detection of changes in the streamflow time series. This methodology (computational and graphical) was applied to 11 selected watersheds distributed across the Susquehanna River Basin of the United States.

The analysis of the results of this study indicated that the four watersheds located in the Appalachian Plateau experienced a step increase around 1970 in the annual median flow and annual minimum flow, while the four watersheds located in the Valley and Ridge experienced a step increase around 1970 in the annual minimum flow. However, none of the watersheds experienced a step increase in the annual maximum flow. The trend changes are seen to be geographically arranged with the greatest change at inland locations, and with the amount of change decreasing toward the Atlantic coast. The abrupt step increases around 1970 occur in the summer and fall seasons, which are usually characterized by low flows in the Susquehanna River Basin (SRB). This is consistent with the step increase in annual minimum flow in most of the watersheds under study. This implies that there has been an increase in the amount of recharge received by aquifers in the SRB during low flow months in the Appalachian Plateau and valley and ridge.

Overall, it can be concluded that the methodology of multiple Mann-Kendall tests with varying beginning and ending times, and the graphical interpretation of the results, as demonstrated in this study, are valuable screening tools for the detection of trend patterns in hydrologic time series. Although the SRB is not a geographically large basin, the watersheds within this basin still show varying trend patterns which indicate that fine spatial resolution trend analyses may be warranted.

References

Bhutiyani, M.R., Vishwas, S.K. and Pawar, N.J. (2008). Changing streamflow patterns in the rivers of northwestern Himalaya: Implications of global warming in the 20th century. *Current Science*, **95(5):** 618-626.

Changnon, S.A. and Demissie, M. (1996). Detection of changes in streamflow and floods resulting from climate fluctuations and land use-drainage changes. *Climatic Change*, **32(4):** 411-421.

Douglas, E.M., Vogel, R.M. and Kroll, C.N. (2000). Trends in floods and low flow in the United States: Impact of spatial correlation. *Journal of Hydrology*, **240:** 90-105.

Groisman, P.Y., Knight, R.W. and Karl, T.R. (2001). Heavy precipitation and high streamflow in the contiguous United States: Trends in the 20th century. *Bulletin of American Meteorological Society*, **82:** 219-246.

Helsel, D.R. and Hirsch, R.M. (1992). Statistical Methods in Water Resources. Elsevier Science Publishing Company Inc., New York, 522 pp.

Hirsch, R.M. and Slack, J.R. (1984). A non-parametric trend test for seasonal data with serial dependence. *Water Resources Research*, **20:** 727-732.

Kalra, A., Piechota, T.C., Davies, R. and Tootle, G.A. (2008). Changes in U.S. streamflow and Western U.S. snowpack. *Journal of Hydrologic Engineering, ASCE*, **13(3):** 156-163.

Kendall, M.G. (1962). Rank Correlation Methods. 3rd edition. Hafner Publishing Co., New York.

Koutsoyiannis, D. and Montanari, A. (2007). Statistical analysis of hydroclimatic time series: Uncertainty and insights. *Water Resources Research*, **43**, W05429, DOI: 10.1029/2006WR005592.

Koutsoyiannis, D., Montanari, A., Lins, H.F. and Cohn, T.A. (2009). Climate, hydrology and freshwater: Towards an interactive incorporation of hydrological experience into climate research – discussion of "The implications of projected climate change for freshwater resources and their management". *Hydrological Sciences Journal*, **54(2):** 394-405.

Kroll, C.N., Luz, J.G., Allen, T.B. and Vogel, R.M. (2004). Developing a watershed characteristics database to improve low streamflow prediction. *Journal of Hydrologic Engineering, ASCE*, **9(2):** 116-125.

Kulkarni, A. and von Storch, H. (1995). Monte Carlo experiments on the effect of serial correlation on the Mann-Kendall test of trend. *Meteorologische Zeitschrift*, **4(2):** 82-85.

Lettenmaier, D.P., Wood, E.F. and Wallis, J.R. (1994). Hydro-climatological trends in the continental United States, 1948-88. *Journal of Climate*, **7:** 586-607.

Lins, H.F. and Slack, J.R. (1999). Streamflow trends in the United States. *Geophysical Research Letters*, **26(2):** 227-230.

Lins, H.F. and Stakhiv, E.Z. (1998). Managing the nation's water in a changing climate. *Journal of the American Water Resources Association*, **34(6):** 1255-1264.

Machiwal, D. and Jha, M.K. (2006). Time series analysis of hydrologic data for water resources planning and management: A review. *Journal of Hydrology and Hydromechanics*, **54(3):** 237-257.

Mann, H.B. (1945). Nonparametric test against trend. *Econometrica*, **13:** 245-259.

Marsh, B. and Lewis, P.E. (1995). Landforms and human habitat. *In:* E.W. Miller (editor), A Geography of Pennsylvania. Pennsylvania State University Press, University Park, PA, pp. 17-43.

McCabe, G.J. and Wolock, D.M. (2002). A step increase in streamflow in the conterminous United States. *Geophysical Research Letters*, **29(24):** 2185-2188.

Miller, W.P. and Piechota, T.C. (2008). Regional analysis of trend and step change observed in hydroclimatic variables around the Colorado River basin. *Journal of Hydrometeorology*, **9(5):** 1020-1034.

Milly, P.C.D., Betancourt, J., Falkenmark, M., Hirsch, R.M., Kundzewicz, Z.W., Lettenmaier, D.P. and Stouffer, R.J. (2008). Stationarity is dead: Whither water management. *Science*, **319:** 573-574.

Milly, P.C.D., Dunne, K.A. and Vecchia, A.V. (2005). Global pattern of trends in streamflow and water availability in a changing climate. *Nature*, **438:** 347-350.

Pennsylvania Department of Environmental Protection (2009). Pennsylvania Water Atlas of the State Water Plan. Pennsylvania Department of Environmental Protection, Harrisburg, PA, 332 pp.

Shahin M., Van Oorschot, H.J.L. and De Lange, S.J. (1993). Statistical Analysis in Water Resources Engineering. A.A. Balkema, Rotterdam, The Netherlands, 394 pp.

Slack, J.R. and Landwehr, J.M. (1992). Hydro-Climatic Data Network: A U.S. Geological Survey Streamflow Data Set for the United States for the Study of Climate Variations, 1874-1988. U.S. Geological Survey Open File Report 92-129.

SRBC (2006). Conowingo Pond Management Plan. SRBC Publication No. 242, Susquehanna River Basin Commission (SRBC), Harrisburg, PA, 153 pp.

Vogel, R.M., Bell, C.J. and Fennessey, N.M. (1997). Climate streamflow and water supply in the northeastern United States. *Journal of Hydrology*, **198(1-4):** 42-68.

von Storch, H. (1995). Misuses of statistical analysis in climate research. *In:* H. von Storch and A. Navarra (editors), Analysis of Climate Variability: Applications of Statistical Techniques. Springer-Verlag, Berlin, Germany, pp. 11-26.

Wu, H., Soh, L.-K., Smal, A. and Chen, X.-H. (2008). Trend analysis of streamflow drought events in Nebraska. *Water Resources Management*, **22:** 145-164.

Yue, S., Pilon, P. and Cavadias, G. (2002a). Power of the Mann-Kendall and Spearman's rho tests for detecting monotonic trends in hydrological series. *Journal of Hydrology*, **259:** 254-271.

Yue, S., Pilon, P., Phinney, B. and Cavadias, G. (2002b). The influence of autocorrelation on the ability to detect trend in hydrological series. *Hydrological Processes*, **16:** 1807-1829.

Zhang, Z. and Kroll, C. (2007a). Closer look at the baseflow correlation method. *Journal of Hydrologic Engineering, ASCE*, **12(2):** 190-196.

Zhang, Z. and Kroll, C. (2007b). The baseflow correlation method with multiple gauged sites. *Journal of Hydrology*, **347(3-4):** 371-380.

Zhang, Z., Dehoff, A.A. and Pody, R.D. (2010a). A new approach to identify trend pattern of streamflows. *Journal of Hydrologic Engineering, ASCE*, **15(3):** 244-248.

Zhang, Z., Dehoff, A.A., Pody, R.D. and Balay, J.W. (2010b). Detection of streamflow change in the Susquehanna River Basin. *Water Resources Management*, **24(10):** 1947-1964.

Zhu, Y. and Day, R.L. (2005). Analysis of streamflow trends and the effects of climate in Pennsylvania, 1971 to 2001. *Journal of the American Water Resources Association*, **41(6):** 1393-1405.

10

Analysis of Trends in Low-Flow Time Series of Canadian Rivers

10.1 Introduction

The main objective of studies on analysis of trends is to ascertain how the statistical characteristics (e.g., mean and variance) of hydrological variables change over time at a given location or at a number of locations in a watershed/region. From the historical perspective, much of the earlier studies on temporal trends in time series of hydrological variables were focussed on water quality related parameters. Most of the earlier studies, reported during 1970s and 1980s, have been reviewed and documented in the work of Helsel and Hirsch (1992) and Hipel and McLeod (1994). Quite recently, interest in the investigation of trends in time series of hydrological variables has increased enormously and numerous studies have been undertaken in different parts of the world. It is difficult to present an exhaustive account of these studies in this chapter and therefore only some of these studies are listed here: Chiew and McMahon (1993), Yulianti and Burn (1998), Lins and Slack (1999), Douglas et al. (2000), Yue et al. (2002b), Robson (2002), Xiong and Shenglian (2004), Hannaford and Marsh (2006), Dixon et al. (2006), Fu et al. (2007), Khaliq et al. (2008, 2009a, 2009b) and Khaliq and Gachon (2010) for trends in streamflows (e.g. mean annual, low and high flows); Hisdal et al. (2001) for trends in hydrological droughts; Suppiah and Hennessy (1998), Haylock and Nicholls (2000), Kunkel et al. (2003), Krishnamurthy et al. (2009) and Kumar et al. (2010) for trends in precipitation related variables (e.g., annual or seasonal total precipitation, frequency and magnitude of extreme events

---

Invited contribution by M.N. Khaliq – Adaptation and Impacts Research Section, Environment Canada, Place Bonaventure, Northeast Tower, 800 Gauchetière Street West, Office 7810, Montreal, Quebec H5A 1L9, Canada and L. Sushama – Centre ESCER, University of Quebec at Montreal, Montreal, Quebec H3C 3P8, Canada.

and dry days). Scientific research on the identification of trends in time series of hydrological variables is still continuing, however using improved approaches and with an enhanced focus on the interpretation of trends. It is important to note that the majority of the studies on trends over the last two decades were driven mainly by concerns of climate change and less due to the influence of other factors like agricultural and industrial developments that could also influence time evolution of hydrological variables.

Many different trend analysis methods and their modifications particularly to address the influence of serial correlations, when assessing local/site significance of trends, and cross correlations, when assessing field/regional significance of trends, have been proposed in the literature. These methods are presented and discussed in Chapter 4 of this book (Section 4.3). The trend analysis methods are further addressed in Section 10.2 of this chapter under four topics: (i) assumptions about the data distribution (parametric and nonparametric), (ii) the type of trend model (linear and nonlinear), (iii) assumptions about the serial structure of the hydrological time series (i.e., serial independence versus dependence), and (iv) assessment of field significance. These items play a fundamental role for a sound and comprehensive analysis of trends in a particular watershed or in a region. A later section of this chapter contains a case study on trend analysis in time series of annual and seasonal low flows observed at selected gauging stations included in the Canadian Reference Hydrometric Basin Network (RHBN). The river basins of RHBN are minimally affected by human activities and therefore provide an excellent dataset for investigation of trends in time series of hydrological variables. Concluding remarks are provided in the last section of the chapter. It should be noted that most of the contents of this chapter are based on the review of various trend analysis methods presented in Khaliq et al. (2009a) and analyses of Monte Carlo simulated and observational data reported in Khaliq et al. (2008, 2009a, b) and Khaliq and Gachon (2010) due to focus of the case study on Canadian basins. Also, appropriate figures from the published literature have been included and new insights about the analysis of trends are presented and discussed.

10.2 Components of Trend Analysis Framework

In order to perform a meaningful trend analysis for a given problem, one has to address a number of issues that affect the overall outcome of such an analysis. Some of these topics are discussed below. Undoubtedly, good quality and longer observational records are equally important topics that also deserve adequate attention.

10.2.1 Assumptions about Data Distribution

In general, trend analyses are performed using parametric and nonparametric approaches. An example of the parametric approach could be a non-stationary generalized extreme value (GEV) distribution (Khaliq et al., 2006 and references therein) and that of the nonparametric approach could be the Mann-Kendall (MK) trend test (Mann, 1945; Kendall, 1975). Though less widely used for the analysis of trends, the advantage of the parametric approach is that it allows one to investigate trend as well as modelling of hydrological observations. Compared to these aspects of the parametric approach, the nonparametric approach allows investigation of trends only. Based on the extensive literature survey, Machiwal and Jha (2008) and Khaliq et al. (2009a) found that the nonparametric approach, particularly the MK test, had been widely used for analyzing trends in hydrologic time series. The reason for this inclination towards the nonparametric approach could be fewer assumptions that are required for the application of this approach compared to the parametric approach, wherein, among other factors, one has to assume underlying distribution of data and type of the trend model. Wide use of the MK test is just a coincidence because the equivalent Spearman rank order correlation (SROC) test (Dahmen and Hall, 1990) is as powerful as the MK test. To compare the ability of the SROC and MK tests for identifying trends, Monte Carlo simulation method was used. Hundred thousand samples were generated from the Normal and GEV (with shape parameter equal to ± 0.15) distributions, with mean and coefficient of variation equal to unity, and each sample was superimposed by linear trends with values ranging from -0.03 to 0.03, with an interval of 0.0025, before applying the MK and SROC tests. Rejection rates of the null hypothesis of no trend are shown in Fig. 10.1 that strongly support the above assertion, i.e. the power of the MK and SROC tests in identifying trends, represented in terms of rejection rates of the null hypothesis, is indistinguishable. It has been shown in the literature (e.g., Yue et al., 2002a) that the trend identification ability of the MK and SROC tests depends on the type of the underlying parent distribution of the data. The results shown in Fig. 10.1 for samples of eight different sizes generated from the Normal and GEV distributions also support this observation.

It is important to mention here that least squares linear regression (Haan, 1977) that requires the data to be normally distributed is another parametric approach that is also commonly used for the analysis of trends. Application of the parametric approaches for investigation of trends in hydrological time series is not addressed here. For that the reader is referred to Kundzewicz and Robson (2000) and Khaliq et al. (2006), among others. Because of its popularity and wide use, the MK trend test in its original and modified forms (to be discussed later) is used for detailed analysis of trends in simulated and observed data in the remainder of this chapter.

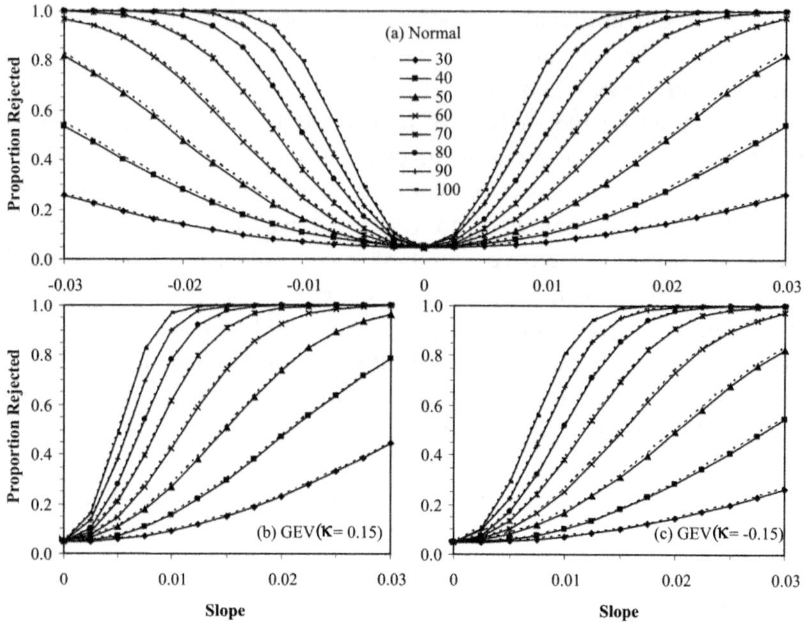

Fig. 10.1. A Monte Carlo simulation based comparison of the ability of the MK and SROC tests for identifying trends in samples from (a) Normal and (b–c) GEV (with shape parameter $\kappa = \pm 0.15$) distributions. Rejection rates (shown on the y-axis) correspond to 5% significance level and solid (dotted) lines correspond to MK (SROC) test. Because of approximate symmetry, only right halves of the plots corresponding to positive (increasing) trends are shown in Figs 10.1(b, c).

10.2.2 Type of Trend Model

The widely used MK test identifies only the presence or absence of a trend through a test of significance and it does not specify whether the trend is linear or nonlinear. Therefore, in most of the studies on trends, the MK test was combined with a nonparametric estimator of monotonic trend which is commonly known as Sen's slope estimation technique (Sen, 1968); details about the Sen's slope estimation are provided in Chapter 4 of this book. Though a monotonic trend represents rather a general case (that could include a step change at some time point over the period of observations or an exponential or logarithmic trend, etc.), it was often assumed as linear for practical applications. In addition to this, a least squares linear regression-type trend was also estimated in numerous studies. Thus, in most of the studies on trends, a monotonic or a linear trend was assumed. On the other hand, it is straightforward to investigate presence of linear as well as any type of nonlinear trends using parametric approaches that involve modelling of hydrological variables. For example, both linear as well as nonlinear trends in characteristics of flood flow series were considered by Strupczewski et al. (2001) for modelling non-stationary frequency-magnitude relationships for Polish rivers.

10.2.3 Assumptions about Serial Structure: Independence vs Short- and Long-term Persistence

It is documented in the literature that the performance of the MK test is seriously affected by serial structure of the time series being tested, i.e. if a time series is positively (negatively) correlated then the MK test will suggest a significant trend more (less) often than it will for an independent series (von Storch, 1995). The serial structure of a hydrological time series could exhibit short-term persistence (STP) or long-term persistence (LTP) or no persistence at all. In the STP case, it may resemble to serial structures of autoregressive moving average (ARMA) type stochastic process (Box and Jenkins, 1970) and in the LTP case, it may resemble to serial structures of fractional Gaussian noise or fractional autoregressive integrated moving average (FARIMA) type stochastic process (Hosking, 1984). To address the influence of STP on trend significance, Hamed and Rao (1998) and Yue and Wang (2004) proposed two different modified MK tests (respectively referred to as MMK1 and MMK2 hereafter) based on the effective sample size approach, originally introduced by Bayley and Hammersley (1946) and implemented by Lettenmaier (1976) to develop the modified SROC test of trend. Since the influence of LTP on trend significance is considerably different from that of STP, Cohn and Lins (2005) proposed adjusted likelihood ratio test (ALRT) based on FARIMA-type time series modelling and simulation approach that was further elaborated in Khaliq et al. (2009b). In addition, Hamed (2008) proposed another modified test, MK scaling (MKS) test, to address the influence of LTP on trend significance.

The influence of STP on the performance of the MK test is demonstrated in Fig. 10.2, using hundred thousand samples generated from an AR(1) process

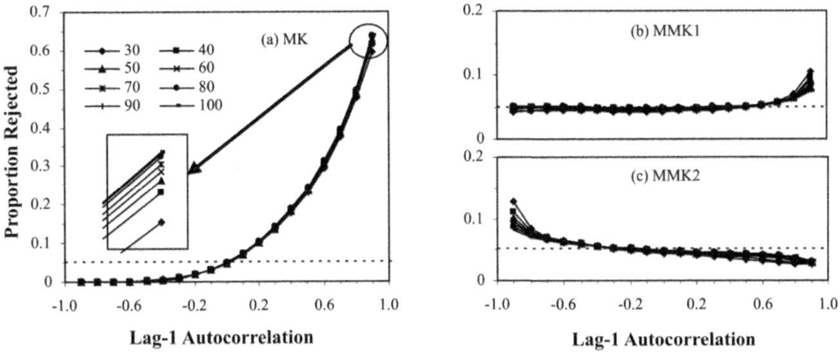

Fig. 10.2. Rejection rates of (a) MK, (b) MMK1, and (c) MMK2 tests for eight sample sizes ranging from 30 to 100, with an interval of 10. The inset in Fig. 10.2(a) is an enlarged, off scale, view of the rejection rates of the MK test. The dotted line corresponds to the nominal significance level (0.05). Autocorrelations of data were used in Fig. 10.2(c) and those of their ranks in Fig. 10.2(b). Figure adopted and modified from Khaliq et al. (2009b).

assuming mean and variance equal to unity and innovation term taken from Normal (0, 1) distribution whose variance appropriately scaled for each selected lag-1 autocorrelation considered. Fig. 10.2(a) suggests that the MK test correctly converges to the nominal significance level for independent cases (i.e. when lag-1 autocorrelation is nearly zero) and it considerably deviates from the nominal significance level as the degree of serial association departs from zero resulting in biased tests. For example, when the autocorrelation is of the order of 0.6, the rejection rate is about 30% compared to the nominal rate of 5%. In addition, the rejection rate appears to be independent of the sample size but it shows a slight increasing tendency as the sample size increases [see the inset in Fig. 10.2(a)]. The MMK1 and MMK2 tests, applied with the assumption of an autoregressive process of order-1 [i.e., AR(1)], are able to address the influence of serial dependence on trend significance for a larger part of the range of autocorrelations [Figs 10.2(b, c)]. On overall basis, the MMK1 test performs relatively better than the MMK2 test. Thus, the results of simulation experiments suggest that when the assumption of an AR(1) process holds true, the MMK1 test would provide better estimates of trend significance than the MK and MMK2 tests.

Two other approaches namely the pre-whitening (PW) (von Storch, 1995) and trend-free pre-whitening (TFPW) (Yue et al., 2002b) were also suggested to address the influence of serial dependence of type AR(1) on the significance of trends. Yue et al. (2002b) and Fleming and Clark (2002) found that if both trend and autocorrelation are present in a time series then the PW approach renders a positively (negatively) autocorrelated time series appearing less (more) trendy. To address this issue, Wang and Swail (2001) introduced a modified iterative PW approach. Though the procedure of the TFPW approach appears to be plausible, it has serious difficulties in preserving the nominal rejection rates of the null hypothesis at a given significance level. This could be due to the influence of autocorrelation on trend and vice versa. The PW and TFPW approaches are not considered further in this chapter. In addition, a resampling based approach, block bootstrap (BBS) (Kundzewicz and Robson,

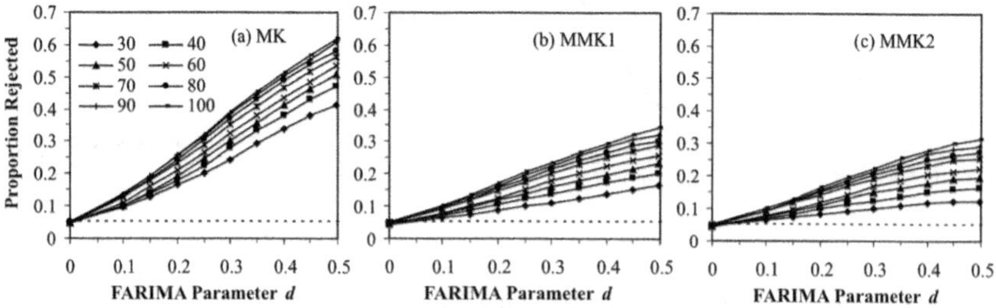

Fig. 10.3. Proportion of time series with significant trends identified with the (a) MK, (b) MMK1 and (c) MMK2 tests for eight sample sizes ranging from 30 to 100, with an interval of 10. For the MMK tests, only the first autocorrelation, whether found significant or non-significant, is considered. Figure adopted from Khaliq et al. (2009b).

2000, 2004; Khaliq et al., 2008), can also be used to address the influence of serial dependence on trend significance provided data over longer time periods are available. The BBS approach is more flexible and hence it can address the influence of autocorrelations of higher lags also and not just that of the lag-1 autocorrelation. The BBS approach is considered for the analysis of trends in the case study presented in Section 10.3.

In a similar manner as presented above for the STP case, hundred thousand samples were simulated from the FARIMA (0, d, 0) model with values of the fractional differencing parameter d ranging from 0.001 to 0.498. In its one parametric form, FARIMA (0, d, 0) is the most simplistic LTP model. Other complicated structures of FARIMA model with autoregressive and moving average components were not explored because of simplicity reasons. With the increase in the value of the parameter d, the strength/intensity of LTP increases. For an independent time series, the value of the parameter d is zero. Thus, $d = 0.001$ would approximately result in an independent time series and $d = 0.498$ would result in a highly long-term persistent time series. It is important to note that the FARIMA is a stationary model and therefore it would generate stationary time series meaning that the statistical characteristics of such time series do not change over time. The results of this investigation are shown in Fig. 10.3. The MK, with the independence assumption, and the MMK1 and MMK2 tests, with the AR(1) assumption, were used to identify trends in each simulated time series. It is clear from Fig. 10.3 that the rejection rate of the null hypothesis increases as the sample size increases meaning that the MK, MMK1 and MMK2 tests would result in significant trends more often for the longer samples than for the smaller samples. The MK test converges to the nominal significance level nearly for all sample sizes when there is no/weak LTP suggesting that the MK test is unbiased. However, it considerably deviates from the nominal significance level in the presence of strong LTP. For example, for a sample size of 50 and $d = 0.498$, there are nearly 50% chances that the MK test would suggest a significant trend, given that there is no trend under the assumption of LTP. Such trends are not real because they are merely due to fluctuations of the behaviour of the LTP model. Figures 10.3(b, c) also suggest that STP based MMK1 and MMK2 tests are not adequate to address the influence of LTP on trend significance. These observations as well as the results of the STP based investigation presented above suggest that proper verification of the serial structure of the time series being tested for temporal changes is very important before applying the MK test for identifying trends.

10.2.4 Field Significance Analysis

As the presence of positive (negative) serial correlation in a hydrological time series inflates (deflates) the rate of rejecting the null hypothesis of no trend, the presence of positive cross correlation among a gauging network will inflate the rate of rejecting the null hypothesis of no field significance of trends, while it is true (Douglas et al., 2000). The term 'field significance', in contrast to 'point or local significance', has been brought into hydrology from climatology/meteorology. This approach was introduced in the work of Livezey and Chen (1983). When the cross correlation in a gauging network is negligible, the theory of binomial distribution can be used to identify field significance. However, when the cross correlation cannot be ignored (e.g. observed and climate model simulated rainfall fields), the methods based on the Monte Carlo simulations and vector block bootstrap resampling approach can be used to determine field significance of identified trends. Details of all these methods along with their advantages and disadvantages can be found in Livezey and Chen (1983), Yue et al. (2003), Elmore et al. (2006) and Khaliq et al. (2009a).

In the above mentioned techniques of field significance analysis, the number of sites with significant trends at a given level of local significance (i.e., the significance level used to identify trends at each of the selected sites) are counted and assessed if this number has arisen due to purely coincidence. Because of the involved counting procedure, these methods of field significance testing are generally categorized as counting techniques. These techniques have been criticized because of the integer valued nature of the result of the counting procedure and because of the binary view of the results of local testing. Local null hypotheses that are very strongly rejected (i.e. local *p*-values that are very much smaller than the local significance level) carry no greater weight in the field significance test than do local tests for which the *p*-values are only slightly smaller than the local significance level (Wilks, 2006). In addition to this issue, the above tests only indicate whether the overall results are field significant or not but they do not specify where and how the results are field significant. These shortcomings of the counting procedures to field significance assessment can in general be improved upon through the use of test statistics that depend on the magnitudes of individual *p*-values of the local tests. One of such kind of tests is the false discovery rate (FDR) test proposed by Benjamini and Hochberg (1995), which is a relatively new statistical procedure for simultaneous evaluation of multiple tests by recognizing that a certain number of false rejections of the null hypothesis are to be expected. Ventura et al. (2004) and Wilks (2006) demonstrated through extensive simulation experiments that the FDR test is robust to spatial correlations. This procedure works with any statistical test for which one can generate a *p*-value. Thus, as long as the effects of serial structure of time series is taken care of appropriately for evaluating at-site *p*-values in a hydrological network/region, the FDR test could be applied for field significance analysis. A step-by-step procedure for applying the FDR test can be found in Wilks (2006) and an application of this method to hydrological time series in Khaliq et al. (2009a).

10.3 A Case Study of Trend Analysis in Time Series of Annual and Seasonal Low Flows

10.3.1 Study Area and Data

Canada is a vast country with various distinct climatic regions. Hare and Thomas (1979) delimited Canadian territory into eleven major homogenous ecoclimatic regions, based on similarity in physiography, i.e., land formation, climate, currents and distribution of flora and fauna within a region. These ecoclimatic regions include the luxuriant Pacific rainforest region with warm, humid Mediterranean climate on the west coast, the Maritimes with Maritime climate of the Atlantic on the east coast, the Arctic consisting of frozen, windswept, treeless polar deserts in the north and the Prairies (sun-drenched grain fields and grasslands) and boreal forests on the south (Yue and Pilon, 2005). Canada has more lakes than any other country and some of the largest freshwater reserves in the world. Typically, spring melt alone or spring melt with rain generates spring floods that are several orders of magnitude larger than the winter and summer low flows. Spring high flows are followed by a decline in flow which is revived occasionally by summer/fall rainstorms. In

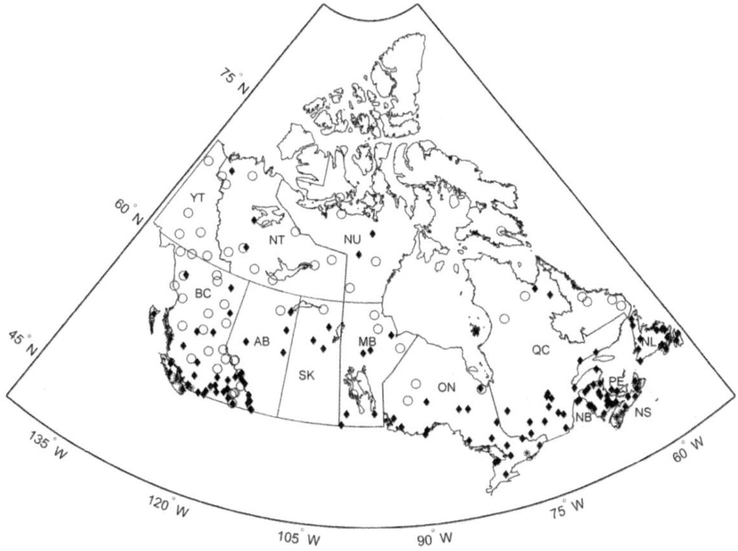

Fig. 10.4. Location and seasonal classification of 201 RHBN gauging stations. Empty circles, asterisks and filled diamonds correspond to stations where annual 7-day minimum flows were observed during the winter season only, during the summer season only and during both winter and summer seasons, respectively. The two letter abbreviations are: YT–Yukon Territory, NT–Northwest Territories, NU–Nunavut, BC–British Columbia, AB–Alberta, SK–Saskatchewan, MB–Manitoba, ON–Ontario, QC–Quebec, NB–New Brunswick, NS–Nova Scotia, PE–Prince Edward Island and NL–Newfoundland. Source: Khaliq et al. (2008).

some parts of the country, both annual high and low flows could come from more than one generating mechanism (Waylen and Woo, 1982, 1987).

Daily streamflows from the Canadian RHBN were used in the case study. The RHBN, a subset of the national hydrometric network, shown in Fig. 10.4 was identified for use in the detection, monitoring, and assessment of climate change across the country (Brimley et al., 1999; Harvey et al., 1999). The RHBN covers most of Canada's major hydrologic regions, although there are gaps in some regions of the country and there are no RHBN stations north of 70 degrees latitude. The river basins of RHBN are characterized by either pristine or stable hydrological conditions. Originally, the RHBN consisted of 249 hydrometric stations, including continuous and seasonal streamflow and continuous lake level stations. According to Khaliq et al. (2008, 2009b), this network has evolved over the years and it consisted of 229 hydrometric stations based on the best available information at that point in time. Of these 229 stations, 201 stations have continuous year-round streamflow observations; the remaining 28 stations are either lake level stations or stations which do not have year-round continuous records. Of the total 201 stations, the stations with relatively longer records (i.e. ≥ 50 years) and with not more than three missing years were considered for the present study. There are 49 hydrometric stations that satisfy this criterion; their description is provided in Fig. 10.5. Continuous daily streamflow data of the selected 49 hydrometric stations were obtained from the Water Survey of Canada's HYDAT data archive (http://www.wsc.ec.gc.ca/hydat/H20, accessed on 15 January 2008). For most of these stations (45), data until 2003 were analyzed except for stations located in the province of Quebec. For four hydrometric stations in the province of Quebec, data up to the end of 2000 were analyzed.

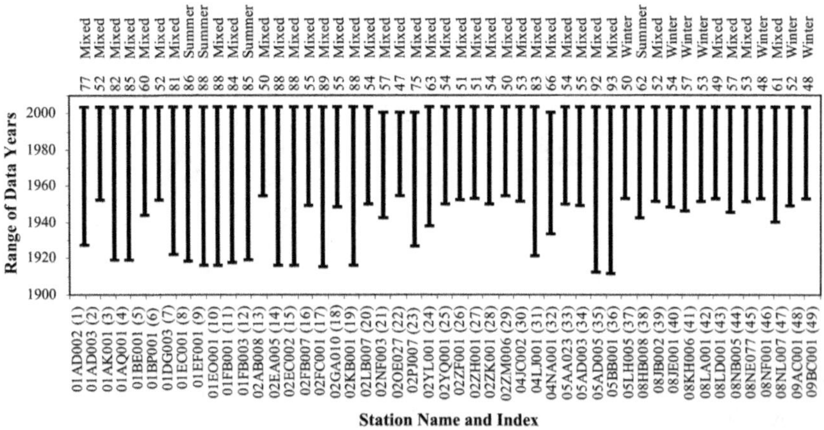

Fig. 10.5. List of 49 Canadian RHBN stations with record length ≥ 50 years (allowing three missing years). Number of years of record used in the analysis and the seasonal (i.e. winter, summer and mixed) classification of low flows for a given station are shown along the top x-axis. Station names and indices are shown along the bottom x-axis. Figure modified from Khaliq et al. (2009b).

10.3.2 Seasonality of Low Flows

Low flows in Canadian streams exhibit a seasonal behaviour (Waylen and Woo, 1987; Sushama et al., 2006; Khaliq et al., 2008) and they occur because of two different seasonal mechanisms. Firstly, the low flows occur as a result of storage depletion following below freezing temperatures during the winter season. Secondly, the low flows occur as a result of lack of precipitation and increased evaporation due to higher temperatures during the summer season. Seasonality of annual 1-, 7-, 15- and 30-day low flows is shown in Fig. 10.6. This figure suggests that it is important to study the temporal behaviour of low flows on seasonal scales because the annual scale alone would not be able to capture the seasonality of low flow regimes of river basins included in the RHBN. Also, for many of the stations, samples of low flows derived on the basis of annual time scale may appear physically inconsistent. Based on the seasonality of annual 7-day low flows, Khaliq et al. (2008) classified the RHBN stations into three categories: (a) stations with low flows occurring in winter only, (b) stations with low flows occurring in summer only, and (c) stations with low flows occurring in both winter and summer seasons (i.e., mixed low flows). This seasonal classification of 201 RHBN stations is shown in Fig. 10.4 and that of the selected 49 RHBN stations, with longer records, is shown in Fig. 10.5. For this seasonal categorization, six-monthly winter (December to May) and summer (June to November) seasons were used. For the selected 49 stations, time series of summer low flows would be the same as those of annual low flows for four stations and time series of winter low flows would be the same as those of annual low flows for seven stations and they would be different from each other for the remaining 38 stations due to

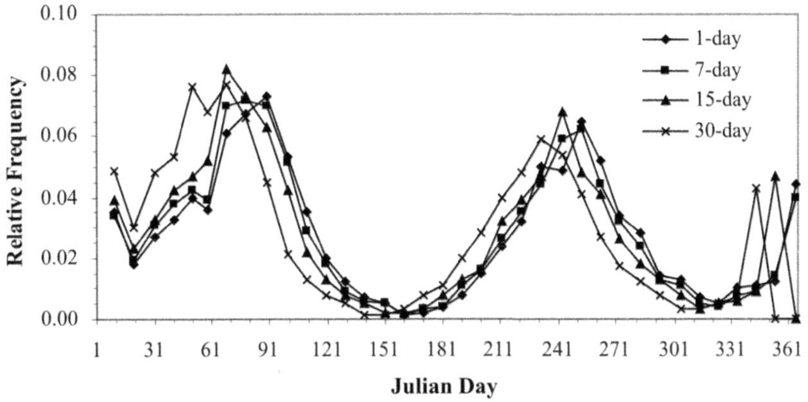

Fig. 10.6. Relative frequencies of starting dates of occurrences of 1-, 7-, 15- and 30-day annual low flows observed at 201 RHBN stations during 8590 station years. For developing these frequency plots, each month was divided into three non-overlapping time windows: first window: 1 to 10 days; second window: 11 to 20 days; and third window: remaining days of the month. *Source:* Khaliq et al. (2008).

mixed low flows. In the analyses presented in this chapter, characteristics of low flows and temporal changes in their magnitudes were studied separately for each of the three (annual, winter and summer) time scales.

10.3.3 STP- and LTP-like Serial Structures

The low flow time series could exhibit STP or LTP or no persistence at all. For the investigation of STP, an AR(1) is assumed and therefore the statistical significance of the first autocorrelation alone was assessed. Lag-1 autocorrelation values of annual, winter and summer low flow time series are shown in Fig. 10.7. About 78 (22), 67 (33) and 59 (41) percent of annual, winter and summer low flow time series were found positively (negatively) autocorrelated. Thus, positive autocorrelations appear to dominate the low flow regimes of RHBN stations. Out of the 49 stations, the number of stations where annual, winter and summer low flow time series were found significantly autocorrelated at 5% level is 8, 8 and 4, respectively. Though the number of stations with significant autocorrelations at 5% level is not very large, autocorrelation for many of the remaining low flow time series was found marginally significant at 10% level (Fig. 10.7), suggesting that it is important to consider the effect of serial dependence on trend significance.

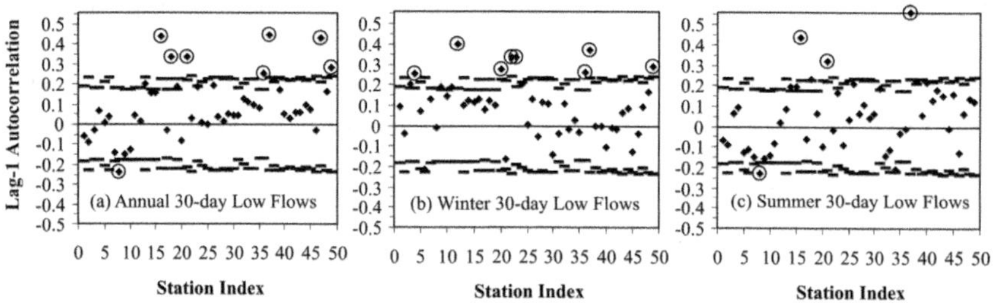

Fig. 10.7. Lag-1 autocorrelations of (a) annual, (b) winter and (c) summer 30-day low flows. For each case considered, upper and lower values of the 90% confidence interval are shown using horizontal dashes. The autocorrelations that were found significant at the 5% level are circled for clarity. Station indices (1 to 49) are the same as shown in Fig. 10.5.

The presence of LTP was investigated by estimating Hurst exponent H (Hurst, 1951). The $0.5 < H < 1$ range corresponds to a persistent process, $0 < H < 0.5$ range corresponds to an antipersistent process and $H = 0.5$ corresponds to a purely independent process in an asymptotic sense. Several methods have been developed to estimate the Hurst exponent (Taqqu et al., 1995; Doukhan et al., 2002). However, five selected techniques were applied in this study: (1) rescaled adjusted range statistic (RARS) (Mielniczuk and Wojdyllo, 2007), (2) aggregated standard deviation (ASD) (Koutsoyiannis, 2003, 2006), (3) FARIMA (p, d, q) modelling approach (Hosking, 1984), where p and q respectively stand for the number of autoregressive and moving average parameters assumed here not greater than one and $d = H - 0.5$ is the fractional differencing parameter, (4) detrended fluctuation analysis (DFA) (Peng et al., 1994; Kantelhardt et al., 2001) assuming local segmental trends of linear type and (5) maximum likelihood based estimator (MLE) (Hipel and McLeod, 1994; Hamed, 2008). In order to determine whether the estimated H value by using the first four selected techniques for a given low flow time series significantly deviates from 0.5, a simulated distribution of H was developed by using a large number of randomly generated samples. The random samples with sample size being equal to observed sample were generated from a white noise process (i.e. normally distributed values with zero mean and unit variance). From the simulated distribution of H, 2.5th and 97.5th percentile values of H were obtained to define a 95% confidence interval for each of the selected techniques. For the MLE method, 95% confidence interval was obtained using the parametric approximation described in Hipel and McLeod (1994) and Hamed (2008).

Observed values of H along with 95% confidence intervals for selected 10 stations (out of 49) with longer records (ranging from 75 to 93 years) are shown in Fig. 10.8. According to the results obtained for the RARS method, none of the estimated H values fall at or outside the upper limit at 95% confidence interval. However, the results for the ASD method suggest the possibility of LTP in annual and winter low flow time series at a single station (05BB001) while the results of the FARIMA method suggest the possibility of

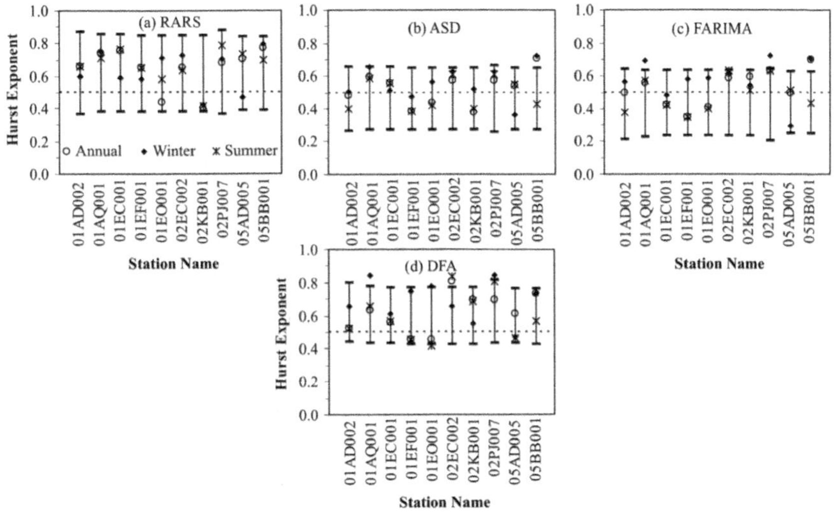

Fig. 10.8. The Hurst exponent estimated using the (a) RARS, (b) ASD, (c) FARIMA modelling and (d) DFA methods for time series of annual, winter and summer low flows observed at 10 selected stations. The 95% confidence intervals for each of the four methods and selected stations were obtained from the simulated distribution of the Hurst exponent developed by generating 10,000 random samples, of size equal to the observed samples, from Normal (0, 1) distribution. Dotted horizontal line corresponds to an independent time series.

LTP in annual low flows at a single station (05BB001) and in winter low flows at three stations (01AQ001, 02PJ007 and 05BB001). The DFA method suggests the possibility of LTP in annual low flows at a single station (02EC002), in winter low flows at three stations (01AQ001, 01EO001 and 02PJ007) and in summer low flows at stations 02EC002 and 02PJ007 (a marginal case). A similar investigation using the MLE method for all the 49 stations suggests significant values of H for 12, 13 and 10 annual, winter and summer low flow time series, respectively. The number of time series which could possibly be assumed to exhibit LTP would increase in the case of smaller than 90% confidence intervals. Differences between the results of various methods for estimating Hurst exponent have also been noticed in some earlier studies, e.g. Montanari et al. (1997), who suggested using FARIMA modelling technique to estimate Hurst exponent.

10.3.4 Results of Trend Analysis

The analyses of STP and LTP diagnostics presented above suggest that the independence assumption or the STP and LTP assumptions do not hold for all of the low flow time series collectively. Therefore, to realize the influence of each of these assumptions on trend significance, estimates of trend significance for all of the low flow time series were obtained separately assuming independence, STP and LTP. For the independence case, the original MK test was applied without considering the effect of serial dependence. For the MMK1 test, first autocorrelation of ranks of data was considered, irrespective of it being significant or non-significant. By doing so, even small departures from independence would contribute in modifying the trend significance. Because of the influence of trend on autocorrelations and vice versa, this test was applied after removing an estimate of the linear trend obtained using the Sen's slope estimation technique (Sen, 1968). However, a more reasonable alternative would be a joint estimate of both first autocorrelation and linear trend following the iterative procedure described in Khaliq and Gachon (2010) that is consistent with the trend analysis procedure using time series modelling and simulation approach introduced in the work of Cohn and Lins (2005). For the MK-BBS test (i.e., when the MK test was combined with the BBS approach), the number of contiguous significant autocorrelations, starting from the first one, was determined and their effect was considered for estimating trend significance. Thus, for those time series for which none of the first few contiguous autocorrelations were found significant, the p-values for the MK and MK-BBS tests should exactly match but they would differ slightly because of the involvement of bootstrap resampling procedure for estimating trend significance. For the case of LTP, both MKS and ALRT tests were used. Step by step instructions for applying the MKS test are available in Hamed (2008) and those of the ALRT in Cohn and Lins (2005). The MKS test was developed on the basis of scaling approach and the ALRT on the basis of time series modeling and simulation approach. Exactly the same procedure as described in Cohn and Lins (2005) was used except that the non-normality of low flow time series was taken care of by adopting a three parameter gamma marginal distribution. For these two tests, estimates of trend significance were obtained with the assumption of LTP only if the estimated value of H was greater than 0.5, otherwise STP was assumed (i.e., MMK1 for the case of MKS test and AR(1) process in the case of ALRT). Collective results of trend significance were considered under the name of the LTP based test for the sake of discussion and convenience of presentation. Similar to the MMK1 test, an estimate of linear trend using the Sen's slope estimation technique was removed from observations before applying the MKS test. It must be noted that for the case of ALRT, the magnitude of linear trend as well as model parameters were estimated by optimizing the likelihood function of the FARIMA model.

The number of stations where the trends were found significant (at 5% significance level) with the above five tests is shown in Fig. 10.9. From the results of Fig. 10.9, the effect of LTP on trend significance is obvious, i.e., some of the significant trends noted with the assumption of independence and STP simply disappeared.

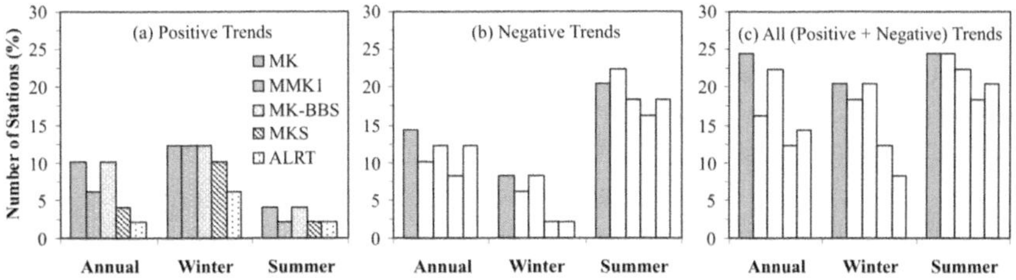

Fig. 10.9. Number of stations with significant (at 5% significance level) (a) positive (or upward) and (b) negative (or downward) trends observed in time series of annual, winter and summer 30-day low flows. Number of all stations with significant trends is shown in panel (c). The positive and negative type of trend was decided on the basis of the sign of the MK test statistic.

It is difficult to clearly appreciate the influence of STP and LTP on trend significance from the results shown in Fig. 10.9, where only one significance level was used. The effect of these serial structures on trend significance was explored further using selected tests p-values, since it is the p-value of the trend test which is affected by any of these three assumptions. The differences between the p-values obtained with the MK test from those obtained with the MK-BBS and MKS tests are shown in Fig. 10.10. The results for the STP case shown in Figs 10.10(a-c) demonstrate that the p-values increased for positively autocorrelated time series and decreased for negatively autocorrelated time series. This suggests that it is very likely that the MK test with the independence assumption would produce significant trends more (less) frequently for positively (negatively) autocorrelated time series. This observation is in agreement with the results presented in Fig. 10.2 using simulated data. In Fig. 10.10, the higher range of p-value differences shown in Figs 10.10(d-e) for the LTP case compared to those shown in Figs 10.10(a-c) for the STP case (note the scale difference) suggests that it is very less likely to find trends with the LTP based tests. In other words, it means that many of the significant trends with the independence assumption can be explained merely on the basis of the serial structure of the time series.

Overall, without reference to positive and negative type of trends, the trends in annual, winter and summer low flows identified with the MK test and those with the STP based tests were found field significant at the 5% significance level using the FDR test; for the case of LTP based tests, trends in summer low flows were found field significant at the 5% significance level and those in winter low flows were found field significant only at 10% significance level. A similar analysis performed separately for positive and negative trends led to the following conclusions. Negative trends in annual and summer low flows were found field significant at the 5% significance level for the case of MK and both STP based tests and those in winter flows were not found field significant even at the 10% significance level. Positive trends in annual and summer low flows were found field significant at 5% significance level only for the case of MK and MMK1 tests and not for the case of MK-BBS test, while those in winter low flows were found field significant for the case of MK and both STP based tests. For the LTP based tests, only negative (positive) trends in summer (winter) low flows were found field significant at 5% (10%) significance level. The Canadian RHBN spans over many different climatic regions and hence it would be more sensible to perform field significance analysis on the basis of suitably defined

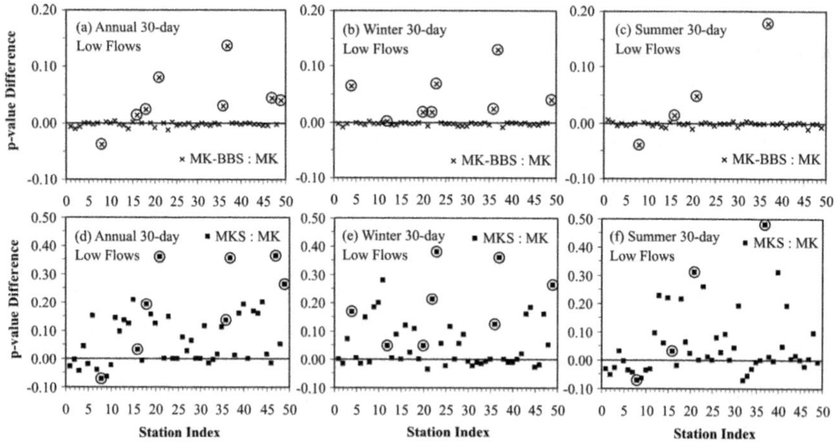

Fig. 10.10. The effect of assuming (a-c) only STP and (d–f) LTP and/or STP, where applicable, on the significance of trends shown as differences in *p*-values (e.g. *p*-value for the MK-BBS test minus *p*-value for the MK test). The time series for which the first (or the first few) autocorrelation(s) was (were) found significant are circled. Station indices are the same as shown in Fig. 10.5. The six figures are adopted from Khaliq et al. (2009b) and modified to improve their clarity.

hydrological homogeneous regions or on the basis of Canadian climatic regions and their plausible subdivisions such as those studied in Plummer et al. (2006) and Mladjic et al. (2011).

10.4 Concluding Remarks

In this chapter, a framework of trend analysis is outlined and importance of various fundamental items (like the distributional assumptions and type of trend model, assumptions about the serial structure of the hydrological time series and the influence of cross correlations) is discussed for performing a sound and comprehensive analysis of trends in a given watershed/region. The influence of serial dependence on the performance of the MK test is studied through Monte Carlo simulations by generating and testing time series of known serial structures of type AR(1) and FARIMA in order to establish general benchmarks. This is followed by a case study on analysis of observed annual, winter and summer 30-day low flows at selected stations, with longer records, included in the Canadian RHBN in order to explore sensitivity of trend significance to STP- and LTP-like serial structures. The choice of 30-day low flows is made, assuming that longer duration low flow indicators are more likely to reflect the influence of basin storage in terms of persistence compared to those of short duration and high flow indicators. For the analysis of observed low flows, the MMK1 and MK-BBS tests to address the influence of STP on trend significance and the MKS and ALRT tests to address the influence of LTP on trend significance were used. For comparison purposes, the original MK test was applied assuming no serial dependence within observations. The results of simulated and observed data suggest considerable influence of serial dependence on trend significance; it means it is very less likely to find significant trends in the presence of STP- and LTP-like serial structures. The implication of this finding is that the MK test, if applied with the independence assumption, will suggest trends more (less) frequently if positive (negative) autocorrelations prevail in a hydrological gauging network. For example, the results of simulations suggest that in the presence of a strong LTP, there are more than 50% (60%) chances that the original MK test would suggest a significant trend for a sample of size 50 (100), given that there is no trend with the LTP assumption.

The above discussion and results of simulated and observed low flow data suggest that it is important to systematically investigate and take into consideration the influence of serial dependence on trend significance. However, having recognized the role of STP- and LTP-like serial structures on trend significance, the risk is that it is very likely that an investigator would end up misdiagnosing a weak to moderate LTP as STP or no persistence at all. This is due to the large uncertainty associated with the Hurst exponent estimated from small size samples. In addition, this exponent also appears to be sensitive to the method of estimation. Thus, longer observational records as well as more than one estimation methods for the Hurst exponent are required to reliably diagnose LTP-like behaviour.

Lastly, for a reliable estimate of trend significance, large samples of high quality data are very important and results of trend analyses from small samples, typically of 20 to 30 years long, should be viewed with caution because the apparent short-term regime changes could be an artifact of the fluctuating behaviour of the underlying observation generating hydrological mechanism.

Acknowledgements

The authors are thankful to Prof. Madan Kumar Jha of Indian Institute of Technology Kharagpur and Dr. Deepesh Machiwal of Central Arid Zone Research Institute, India for their constructive comments and suggestions that led to an improved chapter.

References

Bayley, G.V. and Hammersley, J.M. (1946). The effective number of independent observations in an autocorrelated time series. *Journal of Royal Statistical Society*, **8(1B):** 184-197.

Benjamini, Y. and Hochberg, Y. (1995). Controlling the false discovery rate: A practical and powerful approach to multiple testing. *Journal of Royal Statistical Society Series B*, **57:** 289-300.

Box, G.E.P. and Jenkins, G.M. (1970). Time Series Analysis: Forecasting and Control. Holden-Day, San Francisco, 592 pp.

Brimley, B., Cantin, J.F., Harvey, D., Kowalchuk, M., Marsh, P., Ouarda, T.B.M.J., Phinney, B., Pilon, P., Renouf, M., Tassone, B., Wedel, R. and Yuzyk, T. (1999). Establishment of the Reference Hydrometric Basin Network (RHBN), Environment Canada Research Report, Ontario, Canada, 41 pp.

Chiew, F.H. and McMahon, T.A. (1993). Detection of trend or change in annual flow of Australian rivers. *International Journal of Climatology*, **13:** 643-653.

Cohn, T.A. and Lins, H.F. (2005). Nature's style: Naturally trendy. *Geophysical Research Letters*, **32**, L23402, doi: 10.1029/2005GL024476.

Dahmen, E.R. and Hall, M.J. (1990). Screening of Hydrological Data. Publication No. 49, International Institute for Land Reclamation and Improvement (ILRI), Netherlands, 58 pp.

Dixon, H., Lawler, D.M. and Shamseldin, A.Y. (2006). Streamflow trends in western Britain. *Geophysical Research Letters*, **32,** L19406, doi: 10.1029/2006GL027325.

Douglas, E.M., Vogel, R.M. and Knoll, C.N. (2000). Trends in flood and low flows in the United States: Impact of spatial correlation. *Journal of Hydrology*, **240:** 90-105.

Doukhan, P., Oppenheim, G. and Taqqu, M.S. (2002). Theory and Applications of Long-Range Dependence. Birkhauser, Boston.

Elmore, K.L., Baldwin, M.E. and Schultz, D.M. (2006). Field significance revisited: Spatial bias errors in forecasts as applied to the Eta model. *Monthly Weather Review*, **134:** 519-531.

Fleming, S.W. and Clark, G.K.C. (2002). Autoregressive noise, deserialization, and trend detection and quantification in annual river discharge time series. *Canadian Water Resources Journal*, **27(3):** 335-354.

Fu, G.B., Charles, S.P., Viney, N.R., Chen, S.L. and Wu, J.Q. (2007). Impacts of climate variability on streamflow in the Yellow River. *Hydrological Processes*, **21:** 3431-3439.

Haan, C.T. (1977). Statistical Methods in Hydrology. The Iowa State University Press, Ames, Iowa, 378 pp.

Hamed, K.H. (2008). Trend detection in hydrologic data: The Mann-Kendall trend test under the scaling hypothesis. *Journal of Hydrology*, **349:** 350-363.

Hamed, K.H. and Rao, A.R. (1998). A modified Mann-Kendall trend test for autocorrelated data. *Journal of Hydrology*, **204:** 219-246.

Hannaford, J. and Marsh, T. (2006). An assessment of trends in UK runoff and low flows using a network of undisturbed catchments. *International Journal of Climatology*, **26:** 1237-1253.

Hare, F.K and Thomas, M.K. (1979). Climate Canada. John Wiley and Sons Canada Limited, Toronto.

Harvey, K.D., Pilon, P.J. and Yuzyk, T.R. (1999). Canada's reference hydrometric basin network (RHBN). Proceedings of the Annual Conference of Canadian Water Resources Association: Partnerships in Water Resources Management, June 1999, Halifax, Nova Scotia.

Haylock, M. and Nicholls, N. (2000). Trends in extreme rainfall indices for an updated high quality data set for Australia, 1910-1998. *International Journal of Climatology*, **20:** 1533-1541.

Helsel, D.R. and Hirsch, R.M. (1992). Statistical Methods in Water Resources. Elsevier, Amsterdam, 522 pp.

Hipel, K.W. and McLeod, A.I. (1994). Time Series Modeling of Water Resources and Environmental Systems. Elsevier, Amsterdam, 1013 pp.

Hisdal, H., Stahl, K., Tallaksen, L.M. and Demuth, S. (2001). Have streamflow droughts in Europe become more severe or frequent? *International Journal of Climatology*, **21:** 317-333.

Hosking, J.R.M. (1984). Modeling persistence in hydrological time series using fractional differencing. *Water Resources Research*, **20(12):** 1898-1908.

Hurst, H.E. (1951). Long term storage capacity of reservoirs. *Transactions of ASCE*, **116:** 776-808.

Kantelhardt, J.W., Bunde, E.K., Rego, H.H.A. and Halvin, S. (2001). Detecting long-range correlations with detrended fluctuation analysis. *Physica A: Statistical Mechanics and its Applications*, **295(3-4):** 441-454.

Kendall, M.G. (1975). Rank Correlation Methods. Charles Griffin and Co. Ltd., London, U.K.

Khaliq, M.N., Ouarda, T.B.M.J., Ondo, J.-C., Gachon, P. and Bobée, B. (2006). Frequency analysis of a sequence of dependent and/or non-stationary hydro-meteorological observations: A review. *Journal of Hydrology*, **329(3-4):** 534-552.

Khaliq, M.N. and Gachon, P. (2010). Pacific Decadal Oscillation climate variability and temporal pattern of winter flows in Northwestern North America. *Journal of Hydrometeorology*, **11:** 917-933.

Khaliq, M.N., Ouarda, T.B.M.J. and Gachon, P. (2009b). Identification of temporal trends in annual and seasonal low flows occurring in Canadian rivers: The effect of short- and long-term persistence. *Journal of Hydrology*, **369(1-2):** 183-197.

Khaliq, M.N., Ouarda, T.B.M.J., Gachon, P. and Sushama, L. (2008). Temporal evolution of low flow regimes in Canadian rivers. *Water Resources Research*, **44**, W08436, doi: 10.1029/2007WR006132.

Khaliq, M.N., Ouarda, T.B.M.J., Gachon, P., Sushama, L. and St-Hilaire, A. (2009a). Identification of hydrological trends in the presence of serial and cross correlations: Review of selected methods and their application to annual flow regimes of Canadian rivers. *Journal of Hydrology*, **368(1-4):** 117-130.

Koutsoyiannis, D. (2003). Climate change, the Hurst phenomenon, and hydrological statistics. *Hydrological Sciences Journal*, **48(1):** 3-24.

Koutsoyiannis, D. (2006). Nonstationary versus scaling in hydrology. *Journal of Hydrology*, **324:** 239-254.

Krishnamurthy, C.K.B., Lall, U. and Kwon, H.-U. (2009). Changing frequency and intensity of rainfall extremes over India from 1951 to 2003. *Journal of Climate*, **22:** 4737-4746.

Kumar, V., Jain, S.K. and Singh, Y. (2010). Analysis of long-term rainfall trends in India. *Hydrological Sciences Journal*, **55(4):** 484-496.

Kundzewics, Z.W. and Robson, A.J. (2000). Detecting Trend and other Changes in Hydrological Data. World Climate Program–Data and Monitoring, WMO/TD-No. 1013, World Meteorological Organization, Geneva, 158 pp.

Kundzewicz, Z.W. and Robson, A.J. (2004). Change detection in hydrological records: A review of the methodology. *Hydrological Sciences Journal*, **49(1):** 7-19.

Kunkel, K.E., Easterling, D.R., Redmond, K. and Hubbard, K. (2003). Temporal variations of extreme precipitation events in the United States: 1895-2000. *Geophysical Research Letters*, **30**, 1900, doi: 10.1029/2003GL018052.

Lettenmaier, D.P. (1976). Detection of trend in water quality data from record with dependent observations. *Water Resources Research*, **12(5):** 1037-1046.

Lins, H.F. and Slack, J.R. (1999). Streamflow trends in the United States. *Geophysical Research Letters*, **26(2):** 227-230.

Livezey, R.E. and Chen, W.Y. (1983). Statistical field significance and its determination by Monte Carlo techniques. *Monthly Weather Review*, **111:** 46-59.

Machiwal, D. and Jha, M.K. (2008). Comparative evaluation of statistical tests for time series analysis: An application to hydrologic time series. *Hydrological Sciences Journal*, **53(2):** 353-366.

Mann, H.B. (1945). Nonparametric tests against trend. *Econometrica*, **13:** 245-259.

Mielniczuk, J. and Wojdyllo, P. (2007). Estimation of Hurst exponent revisited. *Computational Statistics and Data Analysis*, **51:** 4510-4525.

Mladjic, B., Sushama, L., Khaliq, M.N., Laprise, R., Caya, D. and Roy, R. (2011). Canadian RCM projected changes to extreme precipitation characteristics over Canada. *Journal of Climate*, doi: 10.1175/2010JCLI3937.1.

Montanari, A., Rosso, R. and Taqqu, M. (1997). Fractionally differenced ARIMA models applied to hydrologic time series: Identification, estimation and simulation. *Water Resources Research*, **33(5):** 1035-1044.

Peng, C.-K., Buldyrev, S.V., Halvin, S., Simons, M., Stanley, H.E. and Goldberger, A.L. (1994). Mosaic organization of DNA nulcleotides. *Physical Review*, **49:** 1685-1689.

Plummer, D.A., Caya, D., Frigon, A., Côté, H., Giguère, M., Paquin, D., Biner, S., Harvey, R. and de Elia, R. (2006). Climate and climate change over North America as simulated by the Canadian RCM. *Journal of Climate*, **19:** 3112-3132.

Robson, A.J. (2002). Evidence for trends in UK flooding. *Philosophical Transactions of Royal Society, Series A*, **360:** 1327-1343.

Sen, P.K. (1968). Estimates of the regression coefficient based on Kendall's tau. *Journal of American Statistical Association*, **63:** 1379-1389.

Strupczewski, W.G., Singh, V.P. and Mitosek, H.T. (2001). Non-stationary approach to at-site flood frequency modeling. III. Flood analysis of Polish rivers. *Journal of Hydrology*, **248:** 152-167.

Suppiah, R. and Hennessy, K.J. (1998). Trends in total rainfall, heavy rain events and number of dry days in Australia, 1910-1990. *International Journal of Climatology*, **10:** 1141-1164.

Sushama, L., Laprise, R., Caya, D., Frigon, A. and Slivitzky, M. (2006). Canadian RCM projected climate change signal and its sensitivity to model errors. *International Journal of Climatology*, **26(15):** 2141-2159.

Taqqu, M.S., Teverovsky, V. and Willinger, W. (1995). Estimators for long-range dependence: An empirical study. *Fractals*, **3:** 785-798.

Ventura, V., Paciorek, C.J. and Risbey, J.S. (2004). Controlling the proportion of falsely rejected hypotheses when conducting multiple tests with climatological data. *Journal of Climate*, **17:** 4343-4356.

von Storch, H. (1995). Misuses of statistical analysis in climate research. *In:* H. von Storch and A. Navarra (editors), Analysis of Climate Variability: Applications of Statistical Techniques, Springer-Verlag, Berlin, pp. 11-26.

Wang, X.L. and Swail, V.R. (2001). Changes of extreme wave heights in northern hemisphere oceans and related atmospheric circulation regimes. *Journal of Climate*, **14:** 2204-2221.

Waylen, P.R. and Woo, M.-K. (1982). Prediction of annual floods generated by mixed processes. *Water Resources Research*, **18:** 1283-1286.

Waylen, P.R. and Woo, M.-K. (1987). Annual low flows generated by mixed processes. Hydrological *Sciences Journal*, **32(3):** 371-383.

Wilks, D.S. (2006). On "field significance" and false discovery rate. *Journal of Applied Meteorology and Climatology*, **45:** 1181-1189.

Xiong, L. and Shenglian, G. (2004). Trend test and change-point detection for the annual discharge series of the Yangtze River at the Yichang hydrological station. *Hydrological Sciences Journal*, **49(1):** 99-112.

Yue, S. and Pilon, P. (2005). Probability distribution type of Canadian annual minimum streamflow. *Hydrological Sciences Journal*, **50(3):** 427-438.

Yue, S. and Wang, C.Y. (2004). The Mann-Kendall test modified by effective sample size to detect trend in serially correlated hydrological series. *Water Resources Management*, **18:** 201-218.

Yue, S., Pilon, P. and Cavadias, G. (2002a). Power of the Mann-Kendall and Spearman's rho tests for detecting monotonic trends in hydrological series. *Journal of Hydrology*, **259:** 254-271.

Yue, S., Pilon, P. and Phinney, B. (2003). Canadian streamflow trend detection: Impacts of serial and cross-correlation. *Hydrological Sciences Journal*, **48(1):** 51-63.

Yue, S., Pilon, P., Phinney, B. and Cavadias, G. (2002b). The influence of autocorrelation on the ability to detect trend in hydrological series. *Hydrological Processes*, **16:** 1807-1829.

Yulianti, J.S. and Burn, D.H. (1998). Investigating links between climatic warming and low streamflow in Prairies region of Canada. *Canadian Water Resources Journal*, **23(1):** 45-60.

11

Exploring Trends in Climatological Time Series of Orissa, India Using Nonparametric Trend Tests

11.1 Introduction

Scientific literature and successive assessment reports of the Intergovernmental Panel on Climate Change (IPCC, 2001; IPCC, 2007; Min et al., 2011) show that the net anthropogenic radiative forcing causes the global warming and intensification of hydrological cycle with consequent increase in the occurrence of extreme weather events. To trace the future of water resources under climate change, climate research uses simulation models well known as general circulation models (GCMs) for forecasting (Koutsoyiannis and Montanari, 2007). Trend analysis of paleoclimatic observation has been an important tool to test the presence of a systematic component (i.e., signal) against the background of natural variability and randomness (i.e. noise) of the instrumental record of hydroclimatic time series (e.g., Zhang et al., 2001; Bhutiyani et al., 2007; Wilson et al., 2010). Huntington (2006) reported that trends in hydrologic variables are consistent with an intensification of the water cycle. However, substantial uncertainty in trends exists due to regional differences of response variables and unavailability of datasets.

Frequent extreme weather events of recent years are increasingly more pronounced in the Indian sub-continent due to large dependence of the population (~68% of above one billion population) on the climate-sensitive agriculture and allied sectors (O'Brien et al., 2004; De et al., 2005). For example, the impact of the deficit of 19% in the Indian summer monsoon rainfall in 2002 is estimated to be of billions of dollars i.e. well over 1% of the gross domestic production (Gadgil et al., 2004). Using both GCMs and regional

---

Invited contribution by Dileep K. Panda and A. Kumar – Directorate of Water Management (ICAR), Chandrasekharpur, Bhubaneswar - 751 023, Orissa, India.

simulation models, several efforts have been made to link the climate change with the hydrology of India, and to project the future climate scenarios (e.g., Singh and Kumar, 1997; Kumar et al., 2006; Tripathy et al., 2006). Although intensification of the hydrological cycle is a unanimous projection for India, there still remains considerable uncertainty about the regional variations and projections of the response of hydroclimatic variables. The uncertainty may be due to the complex physiography and large spatial and temporal variations of the climate, which include the subfreezing Himalayan range, the tropical coastal climate, the rainy climate of the northeastern states, and the arid Great Indian Desert. Furthermore, the GCMs do not resolve the mesoscale processes that play key role in the climate feedback processes (Pan et al., 2004). Additionally, the Indian summer monsoon being the prime modulator of the hydroclimatic variability, is influenced by the action and interaction among the factors such as El Nino/southern oscillation (ENSO) (Kumar et al., 1999; Ashrit et al., 2001), sea surface temperature (SST) (Krishnan et al., 2003); deforestation (Gupta et al., 2005); Eurasian snow cover (Bamzai and Shukla, 1999); and the aerosols (Menon et al., 2002; Ramanathan et al., 2005; Ramanathan et al., 2007).

References to trend analysis of hydroclimatic variables of India are mostly confined to the study of temperature patterns. A general warming temperature trend was observed for India given substantial spatial and temporal differences in trend magnitude (e.g. Arora et al., 2005; Kothawale and Kumar, 2005; Fowler and Archer, 2006). However, bringing consensus about the rainfall trends of India possess difficulty, may be due to the high spatial and temporal variability of rainfall. Different parts of the country exhibit significant increasing and decreasing trends in monsoon rainfall during different timescales (Dash et al., 2007; Ramesh and Goswami, 2007). Recently several studies show a significant increase in the occurrence of extreme rainfall events (Sen Roy and Balling Jr, 2004; Goswami et al., 2006; Rajeevan et al., 2008). However, the rainfall amount and the number of rainy days in both the early and late monsoon exhibit a decreasing trend implying a shorter monsoon over India (Ramesh and Goswami, 2007).

The Orissa state of India is the most climate change-affected region due to frequent occurrence of hydrologic extremes in the recent past (Swiss Re, 2002; Mirza, 2003). The Centre of Environmental Studies (CES, 2007) reported that the erratic behaviour of climate of Orissa is primarily due to the combination of anthropogenic factors such as deforestation, extensive construction activities, uncontrolled mining, elimination of water bodies and extensive carbon consumption over a period of time. Further, a minor change in the pressure anomaly of the Bay of Bengal can have profound hydrological impact on the land mass of Orissa due to its geographical location (Fig. 11.1) at the head of the Bay where the weather forms (CES, 2007). Applying nonparametric methods to the GCM output, Ghosh and Majumdar (2007) predicted a severe drought condition for Orissa. They attributed this future drying scenario to the global warming due to greenhouse effect, sensitivity of rainfall to ENSO, and

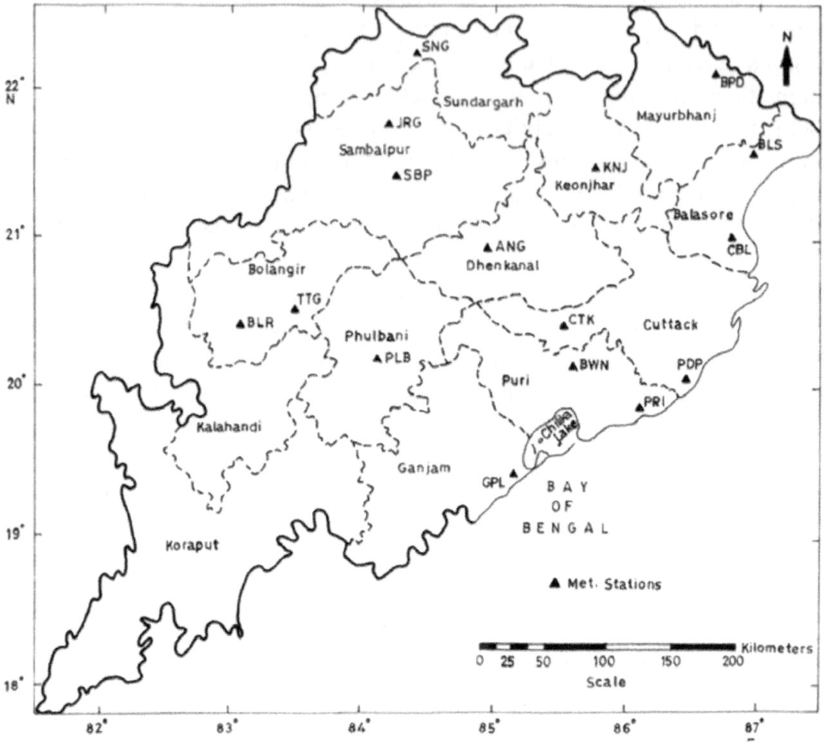

Fig. 11.1. The geographical location of the study area. Four predominant physiographic zones include the coastal zone (Balasore, Cuttack, Puri, Ganjam), the eastern ghat zone (Koraput, Phulbani, Kalahandi), the central table land (Bolangir, Dhenkanal, Sambalpur), and the northern plateau (Keonjhar, Mayurbhanj, Sundargarh).

the coastal setting of Orissa. Also, Panda et al. (2007) found a significant declining trend of groundwater levels in Orissa, which is due to the systematic forcing mechanism of drought in conjunction with human stress and high temperatures.

The Bay of Bengal plays an important role in the global climate system: it is one of the intense heat source regions in the global tropics (Ding, 1994; Chu and Wang, 1997), and also possesses the largest global maximum of summer monsoon rainfall (Hoyos and Webster, 2007). Further, the northwest Bay of Bengal and the adjoining landmass is the region of higher aerosol optical depth (e.g., Menon et al., 2002; Chylek et al., 2006; Prasad et al., 2006). It is, therefore, of scientific interest to study the long-term response of hydroclimatic variables of Orissa due to its geographical location at the head of the Bay of Bengal. The aim of this chapter is to identify and quantify the trends of the hydroclimatic variables such as annual rainfall, monthly minimum and maximum air temperatures, and monthly average relative humidity of the eastern Indian state of Orissa, using the nonparametric statistical tests.

11.2 The Study Area and Climate

The Orissa state lies on the east coast of India, adjacent to the north Bay of Bengal and close to the south of the normal position of monsoon trough with 17°47′-22°33′ N latitude and 81°31′-87°30′ E longitude (Fig. 11.1). The geographical area of Orissa is 1,55,707 km^2 (cultivated area 62,000 km^2; forest area 58,130 km^2). The state has a coastline of 384 km. The long-term average annual rainfall is about 1482 mm, and the mean air temperature ranges from a minimum of 12 °C to a maximum of 39 °C. The area is featured by the presence of mountains, hills, hillocks, rivers and rivulets. The predominant physiographic zones are: the coastal zone having undivided districts (Balasore, Cuttack, Puri, Ganjam), the eastern ghat zone (Koraput, Phulbani, Kalahandi), the central table land (Bolangir, Dhenkanal, Sambalpur), and the northern plateau (Keonjhar, Mayurbhanj, Sundargarh). Overall slopes of the area are from northern and northeast direction to west and southwest, and from southern and southwest direction to east and then to the coastal plains. The river system and their tributaries of the state carry an annual surface runoff of $132 \times 10^9\ m^3$ ($95 \times 10^9\ m^3$ from Orissa) (Lenka, 2001). The Mahanadi is the largest river of the area, which drains 42% geographical area with a length of 494 km in Orissa. The Mahanadi River branches off into several streams at the Naraj gauging station, and carries an annual runoff of $66.88 \times 10^9\ m^3$. The Brahmani River drains 14% geographical area with a length of 541 km. The annual runoff of the Brahmani River is $28.48 \times 10^9\ m^3$. The major hydrogeologic settings, covering 80% of the geographical area, are pre-Cambrian hard-rock formation, which includes granites, gneisses, schist, khondalites and charnockites.

Variability of both rainfall and the total cyclonic disturbances has been above normal since the 1960s, leading to occurrence of more droughts and floods in the study area. However, the extreme events of the late 1990s have been more severe. A record-breaking heat wave in 1998 claimed 2200 human lives, which has been linked to the El Nino. The super cyclone of 1999 was the strongest and deadliest of the region with a recorded wind speed of about 356 km h^{-1} and sea water surge of 8-10 m high (Mirza, 2003). Over 10,000 people were killed, and the coastal ecosystem in particular was seriously affected due to the cyclone. Further, the droughts in 2000 and 2002, and floods in 2001 and 2003 have also affected the economy and environment of the state.

11.2.1 Anthropogenic Activities

The Orissa state is rich in minerals and ores having deposit of iron ore, coal, bauxite, chromites, nickel, lead, copper and limestone. The state accounts for

18% of the explored mineral and ore of country (Sene-Johansen, 1995). The coal belts of Orissa are situated on the bank of the rivers Mahanadi and Brahmani. Availability of water and minerals have led to setting up of thermal power plants and coal-based industries like aluminum, steel, fertilizer, cement plants, etc. In Orissa, ~22 × 10^6 tonnes of coal is consumed annually for power generation to meet the domestic and industrial energy requirements. The National Thermal Power Corporation (NTPC), National Aluminum Company (NALCO), and Talcher Thermal Power Station (TTPS) are the largest consumers of coal in the state with requirements of 36,000, 14,000 and 7000 tonnes of coal per day, respectively. A direct correlation between coal consumption and the regions having high greenhouse gas built-up has been established (Garg et al., 2001). Further, biomass burning, and construction activities are also the sources of atmospheric perturbation. In Orissa, more than 80% of the population (36 million in 2001) live in rural area, where the biofuels meet the major chunk of domestic energy requirement. The Cuttack district is the hotspot district for biofuel use in India (Garg et al., 2001). Large-scale deforestation for industrialization, reservoirs construction and agriculture are also the sources of CH_4 emission in the study region (Rao, 1993).

Industrial and mining operations like drilling, blasting, loading, hauling, dumping, crushing, transportation, and processing of ore and mine products emit radiatively active air pollutants such as carbon monoxide (CO), sulphur dioxide (SO_2), oxides of nitrogen (NO_x), suspended particulate material (SPM) and unburned hydrocarbons (SER, 2006; OSPCB, 2007). Around 10 $\times 10^6$ tonnes (NTPC 4.62 $\times 10^6$ tonnes; NALCO 2.14 $\times 10^6$ tonnes) of fly ash and 1.4 $\times 10^6$ tonnes of blast furnace slag are generated every year in the state. Total solid waste from the major industrial sectors has been estimated to be 25×10^6 tonnes. The aluminium production process (production capacity 0.58×10^6 tonnes yr^{-1}) emits SO_2, per flouro carbon, and hexa flouro ethane, which contribute to global warming having high residence period (CSE, 2006). The chromium mines in the Sukinda Valley is the largest open cast ore mines in the world, and the mining is associated with the pollution due to the hexavalent chromium, overburden dump, and mine water discharge. One tonne of chro-mium mining generates around 10 tonnes of overburden, which contains hexavalent chromium in the overburden with a concentration range 12-311 mg kg^{-1} (OSPCB, 2004). Recently, the Blacksmith Institute (2007) has designated the Sukinda Valley region as one of the highly polluted places of the world. All these anthropogenic activities are likely to perturb the atmosphere, resulting in global warming with consequent changes in the climate. Therefore, it is imperative to explore the long-term climatological trends of the study area.

11.3 Methodology

In this study, monthly rainfall data of 13 districts of Orissa (Fig. 11.1) for 44-year period (1960-2003) were collected from the Directorate of Economics and Statistics, Government of Orissa. The district rainfall represents the average of rainfall of all the administrative blocks of that district. The state revenue department maintains one raingauge for each of the 324 blocks. In addition, monthly maximum temperature (T_{max}), monthly minimum temperature (T_{min}), monthly average relative humidity at 8:30 hours (RH_{fn}) and at 17:30 hours (RH_{an}) for 16 meteorological stations of the study area for 15-year period (1987-2001) were obtained from the report 'Climatological Data of Orissa 1987-2001' published by Directorate of Economics and Statistics, Government of Orissa. The above-mentioned data were also available for two newly established stations (ANG, TTG) at an industrial region for 8-year period (1994-2001). For trend analysis, the two new stations having short records were included for the spatial coverage and importance of the stations at the industrial region. Trends in three climatic time series, i.e., annual rainfall, maximum and minimum air temperature and relative humidity, were explored by applying Mann-Kendall test and Spearman's rho test. Furthermore, the trends were quantified by applying Sen's slope estimation technique and homogeneity of the trends was tested by applying χ^2-based homogeneity test.

The locally weighted scatter-plot smooth (LOWESS) procedure, a robust nonparametric method for estimating the regression surfaces (Burn and Elnur, 2002; Broers and Grift, 2004), was used to draw the smoothed plots for the annual rainfalls of the districts of Orissa. The Yule-Kendall skewness (S_k), a resistance measure of the shape of the distribution (Ferro et al., 2005), was calculated using the 25th percentile (Q_1), 50th percentile (Q_2), and 75th percentile (Q_3) as $(Q_1 - 2Q_2 + Q_3)(Q_3 - Q_1)^{-1}$. The probability plot correlation coefficient (PPCC) (Helsel and Hirsch, 1995) was computed for the test of normality of the rainfall time series. Furthermore, the change in the linear relationship was tested using the Chow's F-test (Wilby et al., 2004) with rainfall as the response variable and year as independent variable between the sub-periods.

11.4 Results and Discussion

11.4.1 Trend and Variability in Annual Rainfall Time Series

Figure 11.2 shows the cases of differential annual rainfall pattern in different physiographic regions of Orissa during 44-year period (1960-2003). The anomalous events in terms of both drought and flood are increasingly more conspicuous in recent years. The LOWESS plots and visual inspection of the annual rainfall indicated a subtle shift in the rainfall pattern of most of the

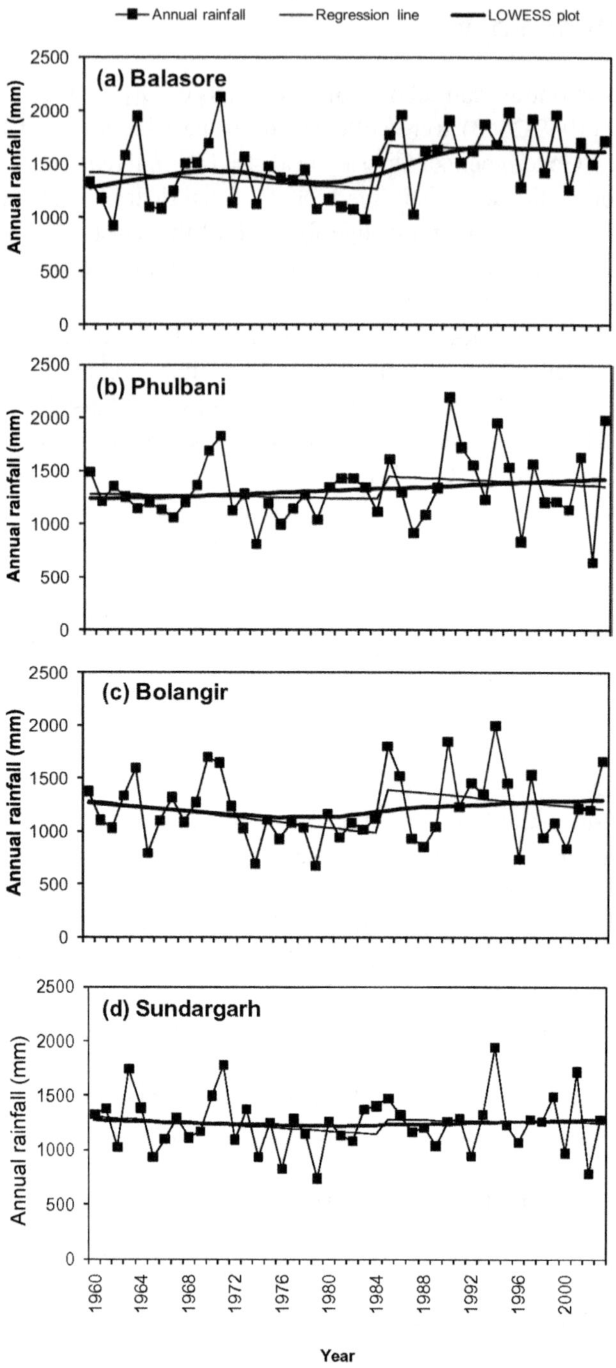

Fig. 11.2. Annual rainfalls for (a) Balasore in coastal physiography, (b) Phulbani in the eastern ghat zone, (c) Bolangir in the central table land, and (d) Sundargarh in the northern plateau.

Table 11.1. Basic statistics of annual rainfall in different districts of Orissa for the first and second sub-series

District	*Sub-series*	*Mean*	*SD*	*Min*	Q_1	Q_2	Q_3	*Max*	S_k	*Observed t-statistic*	*Observed F-statistic*
Balasore	First	1347	297	922	1102	1331	1512	2133	-0.12	-3.57	3.05
	Second	1652	267	1033	1496	1690	1910	1985	0.07		
Cuttack	First	1317	240	946	1146	1323	1422	1859	-0.28	-2.62	1.69
	Second	1555	336	897	1396	1653	1747	2085	-0.46		
Puri	First	1230	201	895	1103	1225	1345	1669	0.00	-2.07	2.38
	Second	1401	315	795	1175	1479	1635	1981	-0.32		
Ganjam	First	1187	221	776	1012	1206	1377	1545	-0.06	-1.96	1.85
	Second	1336	271	891	1148	1302	1412	1963	-0.16		
Koraput	First	1392	263	924	1198	1339	1491	2126	0.03	0.32	0.28
	Second	1369	204	1154	1222	1310	1408	1944	0.06		
Phulbani	First	1258	214	813	1135	1208	1355	1830	0.34	-1.40	0.63
	Second	1402	405	638	1133	1340	1631	2194	0.17		
Kalahandi	First	1295	293	671	1123	1250	1411	1964	0.12	-0.34	0.58
	Second	1332	406	496	1073	1269	1567	2175	0.21		
Balangir	First	1138	260	669	1029	1103	1272	1696	0.39	-1.63	2.73
	Second	1300	370	732	940	1231	1534	2000	0.02		
Dhenkanal	First	1200	258	653	1021	1207	1315	1854	-0.27	-2.38	0.54
	Second	1384	251	954	1201	1451	1554	1725	-0.42		
Sambalpur	First	1209	254	742	1084	1230	1366	1757	-0.04	-1.30	0.36
	Second	1315	279	789	1128	1343	1450	1981	-0.34		
Keonjhar	First	1322	278	855	1138	1363	1449	2010	-0.45	-0.66	1.48
	Second	1377	274	928	1152	1376	1573	1869	-0.06		
Mayurbhanj	First	1428	252	927	1277	1392	1559	2010	0.19	-0.63	0.48
	Second	1479	275	1071	1287	1442	1591	2171	-0.02		
Sundargarh	First	1228	250	737	1095	1256	1375	1781	-0.14	-0.51	0.57
	Second	1269	268	787	1072	1266	1332	1945	-0.49		

Note: SD = Standard Deviation; Min = Minimum; Max = Maximum; Q_1 = 25th Percentile; Q_2 = 50th Percentile (median); Q_3 = 75th Percentile; S_k = Yule-Kendall Skewness.

districts during the mid 1980s. To compare the changes in the location and shape parameters of the rainfall distribution, the complete time series was divided into two sub-series, i.e., Sub-series 1 (1960-1984) and Sub-series 2 (1985-2003). The basic statistics of annual rainfall presented in Table 11.1 show that an increase in average for the period 1985-2003 is associated with an increase in the variability (standard deviation). This indicates that the wet period is more prone to uncertainty compared to the corresponding dry period. Studying the changes in percentiles and quantiles, however, is of interest to understand the nature of extreme climatic events, and also has greater societal impacts (Beniston et al., 2004; Ferro et al., 2005). The minimum and maximum annual rainfall have moved towards opposite direction in the recent period in comparison to the earlier period suggesting that both drought and flood have become more severe. The negative value of the Yule-Kendall skewness (S_k) was observed largely in the coastal zone and the northern plateau of Orissa (Table 11.1), which indicates that the wet years have outnumbered the dry years and a few unusually low rainfall years have influenced the shape of the distribution. However, the Eastern Ghat region and part of the central table land of the state have experienced more dry years and a few unusually high rainfall years as evident from the positive S_k.

The PPCC test of normality indicated (not shown) that the annual rainfall for the whole time series is nearly normally distributed for most of the districts. Hence, change in origin of the annual rainfall for different sub-periods was examined using the t-test. Furthermore, the change in the linear relationship between the sub-periods was tested using the Chow's F-test. Results of t-test and F-test show that neither the shift in origin (average) nor the structural changes have taken place for the sub-series of annual rainfall as both the observed t- and F-statistics are less than their respective critical values. In most of the districts, annual rainfall exhibited negative linear relationship with the year for the sub-series indicating a decreasing trend. However, the positive linear relationship between annual rainfall and year was observed for most of the districts when the complete time series was considered. This indicates that the severe anomalous years during second sub-series have influenced the overall trend. In India, an increasing trend of extreme rainfall events has been reported (Sen Roy and Balling Jr, 2004; Goswami et al., 2006). The number of raining days, however, has decreased. Similarly, Liu et al. (2005) observed that the increasing proportion of rainfall in China has been contributed by the heavy rainfall events, and also noted a decreasing trend of light rainfall events.

11.4.2 Trends in Seasonal Rainfall Time Series

The monsoon season comprising June, July, August and September months, contributes about 80% of the mean annual rainfall of Orissa. Rainfall from the non-monsoon months has been clubbed into pre-monsoon season (February,

March, April and May) and post-monsoon season (October, November, December and January), as the non-monsoon months receive no rainfall or very negligible rainfall. The descriptive statistics of the seasonal rainfall for the study period includes: pre-monsoon season (mean 129 mm; standard deviation 99 mm), June (mean 199 mm; standard deviation 96 mm), July (mean 307 mm; standard deviation 110 mm), August (mean 334 mm; standard deviation 116 mm), September (mean 220 mm; standard deviation 92 mm), and post-monsoon season (mean 136 mm; standard deviation 119 mm), respectively. The long-term normal rainfall for the monsoon season are 213 mm, 351 mm, 335 mm, 236 mm in June, July, August and September, respectively. This indicates a reduction in monsoon season rainfall for the study period. The seasonal hydroclimatic variables are not normally distributed in this study, and therefore, the nonparametric statistical tests were employed for trend identification and quantification. Both the Mann-Kendall (MK) and the Spearman rho (SR) tests were applied to the seasonal rainfall in two time frames, i.e. 1985-2003 and 1960-2003. Before application of the Mann-Kendall test, the presence of significant serial correlation was tested and the pre-whitening was done.

Figure 11.3 shows the cumulative frequency distribution of MK and SR trend test results of 78 season-stations rainfall time series (six seasons for each of the 13 stations). The positive trend indicates wetting trend and the negative trend indicates drying trend. Under the null hypothesis of no trend, the frequency curve should have been around the axis resulting zero mean trend. Further, under the assumption that both the drying and the wetting trends have equal distribution, the frequency curve should divide the axis into two halves with equal proportions of area above and below the axis. For the time frame 1985-2003, 60% and 62% of the season-stations have experienced the drying trends based on the MK and SR tests, respectively [Fig. 11.3(a)]. However, no significant trend was observed except one for the SR test. The mean observed test-statistic value of MK test is –0.2 (standard deviation 0.88), and the mean observed test-statistic value of SR test is –0.20 (standard deviation 0.79). However, for the time frame 1960-2003 [Fig. 11.3(b)], 44% of the season-stations have experienced the drying trends based on the MK and SR tests. The mean observed test-statistic value of MK test is 0.26 (standard deviation 1.14), and the mean observed test-statistic value of SR test is 0.26 (standard deviation 1.12). To illustrate the seasonal distribution of the rainfall trend, the MK test results were used. Figure 11.3(a) shows that the negative drying trends are more conspicuous in the monsoon and the post-monsoon seasons for the time frame 1985-2003. However, for the time frame 1960-2003, positive (wetting) rainfall trends have been more in the pre-monsoon, June, August seasons. The results of this study are in agreement with findings of several other studies conducted in the study area or other parts of India (e.g., Mohapatra and Mohanty, 2004; Ramanathan et al., 2005; Chung and Ramanathan, 2006; Goswami et al., 2006; Dash et al., 2007; Ramesh and Goswami, 2007).

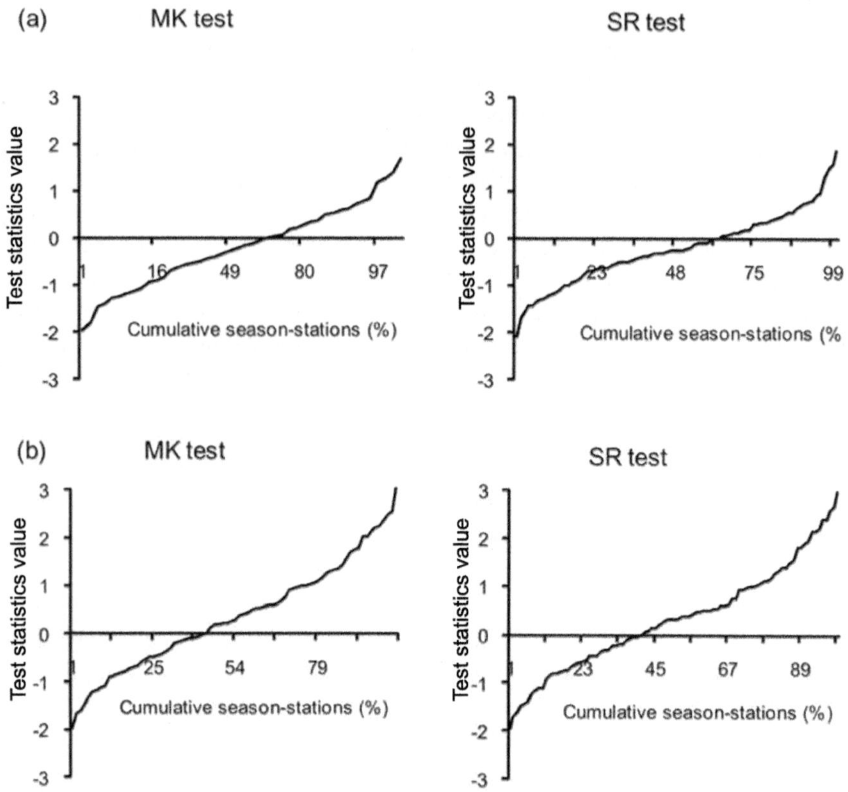

Fig. 11.3. Cumulative frequency distribution of MK and SR trend test results of 78 rainfall time series (6 seasons × 13 stations) for the time frame (a) 1985-2003, and (b) 1960-2003, respectively.

In both the time frames considered in this study, the non-monsoon seasons experienced a wetting trend. The annual rainfall does not show the traces of drying monsoon trend. Increase in the pre- and post-monsoon seasons rainfall promoted by strengthened low-pressure systems in the Bay of Bengal, and anomalous raining events may have influenced the annual rainfall amount. This finding is in agreement with results of Dash et al. (2007). The box plots of the Kendall slope (β) (not shown) for quantifying the seasonal rainfall trends for both the time frames indicated a decreasing and more fluctuating seasonal rainfall pattern in recent years. The monsoon season rainfall in June, July, August and September months decreased by 1.67, 1.80, 1.10 and 1.64 mm per year, respectively. However, for the second time frame, only two monsoon season months (July and September) exhibited drying trend at the rate 0.19 and 0.74 mm per year, respectively.

The result of the trend tests showed high spatial and seasonal variability of monsoon season rainfall, and thus, homogeneity of the trend was examined. The homogeneity test was used to test homogeneity in only monsoon rainfall trends and the results are presented in Table 11.2. For the time frame 1985-

Table 11.2. Homogeneity tests of monsoon rainfall trends for different time frames

Sources of	*Data Series 1: 1985-2003*			*Data Series 2: 1960-2003*		
variation	χ^2-*cal*	*d.f.*	*Significance*	χ^2-*cal*	*d.f.*	*Significance*
Total variation	30.76	52	–	68.24	52	–
Homogeneity	23.17	51	–	67.41	51	–
Season	0.67	3	NS	28.06	3	S ($p \leq 0.05$)
Station	5.38	12	NS	8.91	12	NS
Season station	17.13	36	NS	30.43	36	NS
Trend	7.59	1	S ($p \leq 0.05$)	0.84	1	Not required

Note: d.f. = Degrees of Freedom; NS = Not Significant; S = Significant.

2003, the overall trend ($\chi^2_{Trend,\ 1}$) is significant at $\alpha = 0.05$ level (critical $\chi^2_{0.995,\ 1} = 3.84$). This result indicates an overall drying trend of monsoon rainfall ($Z = -0.38$) without any significant spatial and seasonal heterogeneity. For the second time frame (1960-2003), however, the seasonal heterogeneity ($\chi^2_{Season,\ p-1}$) is found to be significant at $\alpha = 0.05$ level (critical $\chi^2_{0.995,\ 3} = 7.81$). Based on the results of tests for heterogeneity among the seasons, the June and September months of the monsoon season are found to be significantly heterogeneous with χ^2-values of 20.22 and 7.85, respectively. It can be inferred that the June month exhibited a significant wetting trend ($Z_s = 1.24$), and the September month exhibited a significant drying trend ($Z_s = -0.78$) for the second time frame. Higher frequency of drying trend in monsoon rainfall time series of Orissa in the present study, and the future occurrence of extreme weather events with probable severe droughts reported by Ghosh and Majumdar (2007) necessitates a more detailed future research.

11.4.3 Trends in Temperature Time Series

The basic statistics of mean monthly maximum temperature (T_{max}) and mean monthly minimum temperature (T_{min}) for 16 stations of the study area are presented in Table 11.3. The standard deviation values of the temperatures are relatively lesser for coastal belt (Balasore, Chandbali, Paradeep, Cuttack, Puri, Bhubaneswar, Gopalpur) compared to the non-coastal region, which indicates a stable T_{max} and T_{min} in coastal belt. Although the coastal region is densely populated, the mean T_{max} of the non-coastal region is more than that of the coastal region. However, the average T_{min} of the non-coastal region is less than that of the coastal region. Most of the stations exhibit positive skewness for T_{max} and negative skewness (S_k) for T_{min}. This indicates that unusually high and low temperatures influenced the distribution of T_{max} and T_{min}, respectively. The topographic features, location of the station from the coast, and the anthropogenic interventions are the plausible factors for the spatial difference of the temperature pattern. The coal-based industrial regions such as ANG, JRG, TTG and SNG show higher temperatures as obvious from the location parameters.

Figure 11.4 shows the cumulative frequency distribution of MK and SR trend test results of 192 temperature time series (12 seasons for each of the 16 stations) for the T_{max} and T_{min}, respectively. The cumulative frequency distribution of T_{max} shown in Fig. 11.4(a) exhibits a pronounced warming trend as 99% of the season-stations show positive trends for both the MK and SR trend test. However, the cumulative frequency distribution of T_{min} experiences a cooling trend as 93% and 94% of the season-stations show negative trends for the MK and SR trend test, respectively [Fig. 11.4(b)]. Higher cases of positive trends of the MTR [Fig. 11.4(c)] are also observed,

Table 11.3. Basic statistics of monthly maximum temperature (T_{max}) and monthly minimum temperature (T_{min}) in °C for the period 1987-2001

Station	*Altitude (m MSL)*	*Variable*	*N*	*Mean*	*SD*	*Min*	Q_1	Q_2	Q_3	*Max*	S_k
Balasore (BLS)	18.8	T_{max}	180	33.37	3.79	25.1	30.85	33.20	35.35	44.0	-0.04
		T_{min}		20.55	4.81	9.6	16.55	22.20	24.50	27.1	-0.42
Chandbali (CBL)	4.8	T_{max}	180	33.96	3.80	23.9	31.40	33.65	36.20	44.6	0.06
		T_{min}		20.46	4.54	10.3	16.95	22.20	24.00	29.9	-0.49
Paradeep (PDP)	7.6	T_{max}	180	31.91	2.78	25.2	30.55	31.80	33.75	41.9	0.22
		T_{min}		21.32	4.22	7.8	18.60	22.65	24.50	29.6	-0.37
Cuttack (CTK)	25.7	T_{max}	180	34.25	3.59	24.8	31.65	33.90	36.55	44.1	0.08
		T_{min}		19.10	4.68	7.5	16.00	20.00	22.20	36.9	-0.29
Puri (PRI)	4.8	T_{max}	180	32.02	2.39	20.8	30.50	32.25	33.65	40.8	-0.11
		T_{min}		22.92	4.00	13.3	19.80	23.95	26.40	37.6	-0.26
Bhubaneswar (BWN)	45	T_{max}	180	34.25	3.60	27.6	31.75	34.10	36.10	44.6	-0.08
		T_{min}		20.82	4.66	9.5	17.65	22.40	24.60	29.9	-0.37
Gopalpur (GPL)	16	T_{max}	180	32.28	2.65	25.0	30.55	32.25	34.15	40.3	0.06
		T_{min}		21.92	4.06	12.0	19.05	22.85	25.40	28.4	-0.20
Phulbani (PLB)	462.6	T_{max}	180	32.65	4.59	24.4	29.35	31.65	35.30	43.5	0.23
		T_{min}		13.00	6.51	-0.6	8.30	12.30	17.95	24.6	0.17
Angul (ANG)	138	T_{max}	96	36.60	4.74	28.5	33.23	35.40	40.05	47.0	0.36
		T_{min}		17.37	4.73	4.0	13.50	18.80	21.03	26.5	-0.41
Bolangir (BLR)	189.2	T_{max}	180	34.10	5.82	23.1	29.50	33.63	37.90	47.7	0.02
		T_{min}		16.33	5.58	3.1	13.30	16.40	20.10	27.1	0.09

(Contd.)

Table 11.3 (*Contd.*)

Station	*Altitude (m MSL)*	*Variable*	*N*	*Mean*	*SD*	*Min*	Q_1	Q_2	Q_3	*Max*	S_k
Jharsuguda (JRG)	228	T_{max}	180	34.71	5.16	26.0	31.00	33.55	38.10	48.0	0.28
		T_{min}		19.33	5.91	5.5	13.85	20.20	24.15	39.5	-0.23
Sambalpur (SBP)	148	T_{max}	180	34.89	4.96	25.2	31.45	34.10	38.45	47.5	0.24
		T_{min}		18.82	5.87	3.6	14.35	20.45	23.55	31.6	-0.33
Titilgarh (TTG)	210	T_{max}	132	36.61	5.90	27.0	31.40	36.10	40.90	49.8	0.01
		T_{min}		20.15	5.43	4.5	16.30	22.00	23.95	32.0	-0.49
Baripada (BPD)	53.5	T_{max}	180	33.90	4.41	24.9	31.05	33.40	36.60	44.6	0.15
		T_{min}		19.75	5.45	8.4	15.55	21.55	24.45	33.7	-0.35
Keonjhar (KNJ)	485	T_{max}	180	32.21	4.37	20.1	29.30	31.40	34.90	43.6	0.25
		T_{min}		17.79	5.61	5.2	13.27	19.10	22.22	32.3	-0.30
Sundargarh (SNG)	240	T_{max}	144	34.13	5.39	22.3	30.40	33.00	37.80	46.0	0.30
		T_{min}		15.05	5.50	1.9	10.52	16.76	19.25	28.2	-0.43

Note: N = Number of observations; SD = Standard deviation; Min = Minimum; Max = Maximum; Q_1 = 25th Percentile; Q_2 = 50th Percentile (median); Q_3 = 75th Percentile; S_k = Yule-Kendall Skewness.

which indicate that both T_{max} and T_{min} are moving in opposite direction. Around 46% and 60% of the season-stations experienced significant warming trend of T_{max} at $\chi = 0.05$ and 0.10 levels, respectively. The significant warming trends of T_{max} were more concentrated in the non-monsoon seasons. For T_{min}, around 39 and 48% of the season-stations experienced significant cooling trend at $\alpha = 0.05$ and 0.10 levels, respectively. Increasing number of cooling trends was observed in the pre-monsoon summer season. The Kendall slopes for trend quantification indicated that the T_{max} exhibited an overall warming trend at the rate 0.37 °C yr^{-1} (standard deviation 0.18). The seasonal distribution of the Kendall slope indicated that the pre-monsoon summer months experienced more warming trends. The T_{min} exhibited a cooling trend at the rate 0.32 °C yr^{-1} (standard deviation 0.23). Comparatively higher rate of cooling was experienced in the pre-monsoon summer months. The test of homogeneity (Table 11.4) indicates that the overall trend ($\chi^2_{\text{Trend, 1}}$) for both T_{max} and T_{min} are significant at $\chi = 0.05$ level (critical $\chi^2_{0.995,\ 1} = 3.84$). This result indicates an overall warming trend for T_{max} ($\bar{Z} = 1.85$) and cooling trend for T_{min} ($\bar{Z} = -1.62$) without any significant spatial and seasonal heterogeneity.

The opposite trends of T_{max} and T_{min} indicate that the governing factors are also different. Greenhouse gases in the atmosphere accelerate the warming process. The observed pronounced warming trend of T_{max} in the study area may be attributed to the build-up of greenhouse gases due to coal combustion, land-use change, and other anthropogenic factors (Garg et al., 2001; CSE, 2006). Earlier, Rao (1993) attributed the warming trend in the Mahanadi river basin of Orissa to the increase in the greenhouse gases, specially CO_2 and CH_4, and the changes in the land-use pattern. Several other researchers have also observed the warming trends of seasonal and annual temperatures for different parts of India (e.g., Hingane et al., 1985; Arora et al., 2005; Kothawale and Kumar, 2005; Flower and Archer, 2006). More interesting result is the simultaneous cooling trend of the night temperatures i.e. T_{min}, and consequent widening trend of the MTR. The aerosol loads may have influenced the observed cooling trends by reducing the surface receipt of solar radiation and rainfall efficiency as discussed in the previous section. Cooling trends have also been observed at other parts of India (Hingane et al., 1985; Yadav et al., 2004; Flower and Archer, 2006). The deforestation and drying rainfall trend may have partially contributed to the cooling trend of Orissa.

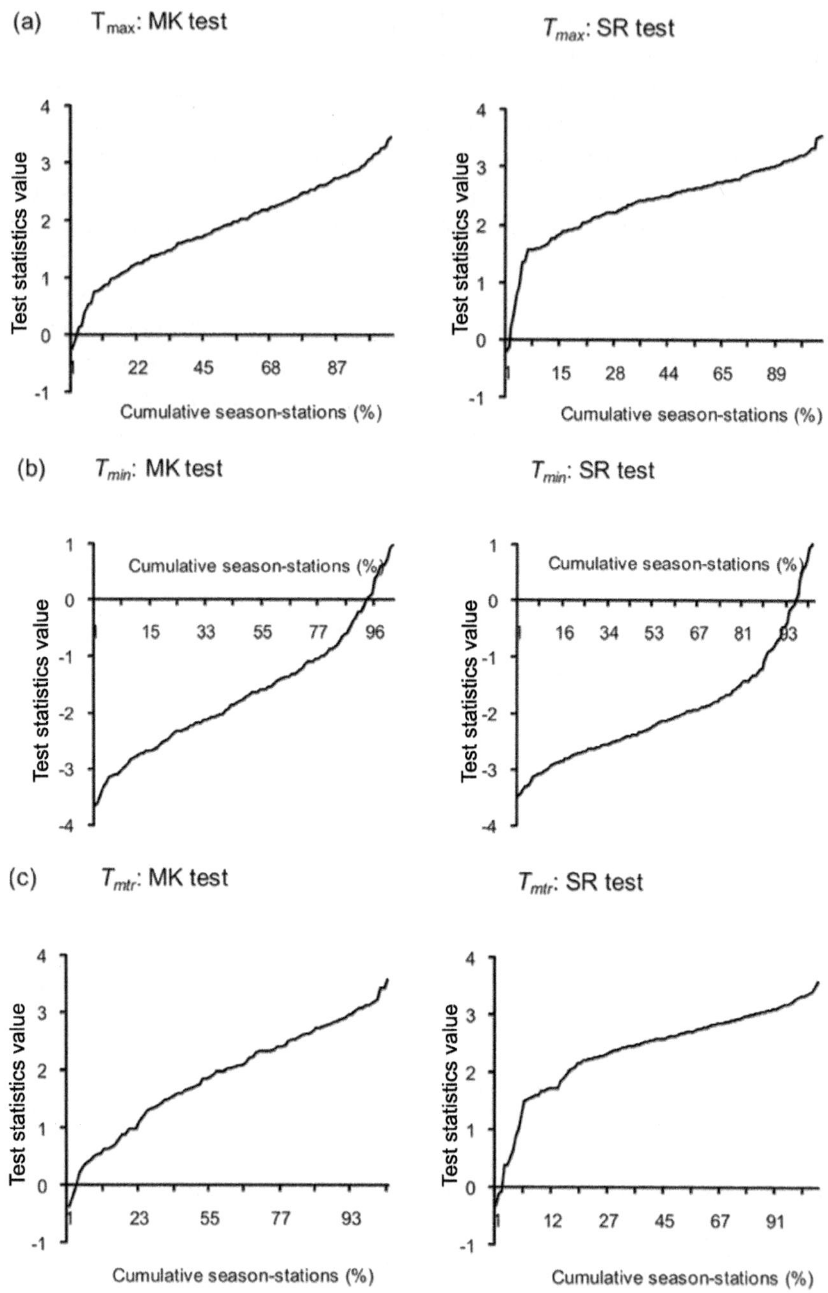

Fig. 11.4. Cumulative frequency distribution of MK and SR trend test results of 192 temperature time series (12 seasons × 16 stations) for (a) monthly maximum temperature (T_{max}), (b) monthly minimum temperature (T_{min}), and (c) monthly temperature range (T_{mtr}), respectively.

Table 11.4. Homogeneity tests of trends of monthly maximum temperature and monthly minimum temperature for the period 1987-2001

Sources of variation	*Monthly maximum temperature (T_{max})*			*Monthly minimum temperature (T_{min})*		
	χ^2-cal	*d.f.*	*Significance*	*χ^2-cal*	*d.f.*	*Significance*
Total variation	765.54	192	–	679.95	192	–
Homogeneity	104.12	191	–	174.43	191	–
Season	12.13	11	NS	11.27	11	NS
Station	21.42	15	NS	77.75	15	NS
Season station	70.57	165	NS	85.41	165	NS
Trend	661.42	1	S ($p \leq 0.05$)	505.52	1	S ($p \leq 0.05$)

Note: *d.f.* = Degrees of freedom; NS = Not significant; S = significant.

11.4.4 Trends in Time Series of Relative Humidity

The descriptive statistics of monthly average relative humidity for 16 stations of the study area recorded at 8.30 hours (RH_{fn}) and at 17.30 hours (RH_{an}) are presented in Table 11.5. The water vapour over the coastal-belt is more in comparison to the inland areas. However, unusually low values of relative humidity have influenced the shape of distribution as obvious from the negative skewness (S_k) in most of the stations. The spatial and temporal variability of water vapour is due to a combination of factors such as the proximity of the location to the Bay of Bengal, topographic features, and anthropogenic factor like land-use changes. Decrease in vapour fluxes has occurred due to reduction in evapotranspiration at places where forest lands were converted to agricultural lands (Mishra and Das, 1984; Douglas et al., 2006).

Figure 11.5 displays the cumulative frequency distribution for the results of MK and SR trend test for 192 season-stations relative humidity time series (12 seasons for each of the 16 stations). For RH_{fn}, 67 and 65% of 192 season-stations relative humidity time series exhibit positive trends based on the MK and SR tests, respectively [Fig. 11.5(a)]. However, for RH_{an}, 50 and 56% of 192 season-stations relative humidity time series exhibit positive trends based on the MK and SR tests, respectively [Fig. 11.5(b)]. Comparatively higher cases of positive trends indicate an increasing moisture load in the atmosphere. Significant positive trends, although less in number, were more concentrated in the monsoon season months. The Kendall slopes indicated that both RH_{fn} and RH_{an} exhibited an overall moistening trend at a rate of 0.07% yr^{-1} (standard deviation 0.48) and 0.01% yr^{-1} (standard deviation 0.53), respectively. The location parameters of the box plot showed that the trends of RH_{fn}, and RH_{an} were more stable in the monsoon and post-monsoon seasons. The test of homogeneity of the RH_{fn} trend results (Table 11.6) indicates that the overall trend ($\chi^2_{Trend,\,1}$) is significant at $\chi = 0.05$ level (critical $\chi^2_{0.995,\,1} = 3.84$). This positive overall RH_{fn} trend ($\bar{Z} = 0.37$) indicates an overall moistening trend of

Table 11.5. Basic statistics of monthly average relative humidity (%) in forenoon (RH_{fn}) and afternoon (RH_{an}) for the period 1987-2001

Station	*Altitude (m MSL)*	*Variable*	*N*	*Mean*	*SD*	*Min*	Q_1	Q_2	Q_3	*Max*	S_k
Balasore (BLS)	18.8	RH_{fn}	180	72.24	9.43	18	67	72	80	87	0.23
		RH_{an}		71.39	7.96	29	66	72	78	85	0.00
Chandbali (CBL)	4.8	RH_{fn}	180	77.37	9.66	28	73	78	84	93	-0.25
		RH_{an}		67.69	14.38	13	57	68	80	90	0.00
Paradeep (PDP)	7.6	RH_{fn}	180	78.95	5.52	61	75	80	83	89	0.04
		RH_{an}		72.85	8.34	50	66	73	80	86	-0.44
Cuttack (CTK)	25.7	RH_{fn}	180	75.97	10.24	28	73	78	82	90	-0.11
		RH_{an}		63.78	14.63	12	55	64	76	89	0.14
Puri (PRI)	4.8	RH_{fn}	180	81.73	4.79	67	78	82	86	92	0.00
		RH_{an}		79.91	7.53	60	75	82	86	94	-0.29
Bhubaneswar (BWN)	45	RH_{fn}	180	80.64	4.36	67	78	81	84	90	0.09
		RH_{an}		78.32	7.18	33	74	81	84	89	0.04
Gopalpur (GPL)	16	RH_{fn}	180	80.62	9.06	28	78	82	86	93	-0.07
		RH_{an}		77.98	11.27	17	72	81	86	91	-0.27
Phulbani (PLB)	462.6	RH_{fn}	180	76.37	12.09	34	70	80	86	92	-0.29
		RH_{an}		66.96	16.84	5	55	70	81	96	-0.18
Angul (ANG)	138	RH_{fn}	96	76.16	7.13	59	71	78	81	88	-0.40
		RH_{an}		61.52	14.51	31	49	63	75	84	-0.12
Bolangir (BLR)	189.2	RH_{fn}	180	69.68	13.47	30	59	73	81	93	-0.27
		RH_{an}		59.86	16.99	23	44	61	76	88	-0.03

Jharsuguda (JRG)	228	RH_{fn}	180	68.03	14.52	36	58	71	80	90	-0.18
		RH_{an}		52.44	20.86	15	35	51	73	85	0.16
Sambalpur (SBP)	148	RH_{fn}	180	71.07	14.65	30	63	75	82	99	-0.26
		RH_{an}		59.79	19.62	18	46	62	78	96	0.01
Titilgarh (TTG)	210	RH_{fn}	132	66.86	13.40	39	57	69	78	89	-0.14
		RH_{an}		59.76	17.40	19	47	62	75	92	-0.09
Baripada (BPD)	53.5	RH_{fn}	180	74.45	16.56	24	70	79	86	94	-0.13
		RH_{an}		64.79	21.46	11	51	69	84	93	-0.10
Keonjhar (KNJ)	485	RH_{fn}	180	69.70	13.50	30	60	73	81	90	-0.24
		RH_{an}		62.79	18.91	17	47	66	81	92	-0.13
Sundargarh (SNG)	240	RH_{fn}	144	69.06	14.50	31	59	72	80	97	-0.22
		RH_{an}		60.82	16.89	26	49	62	75	94	-0.05

Note: N = Number of observations; SD = Standard deviation; Min = Minimum; Max = Maximum; Q_1 = 25th Percentile; Q_2 = 50th Percentile (Median); Q_3 = 75th Percentile; S_k = Yule-Kendall Skewness.

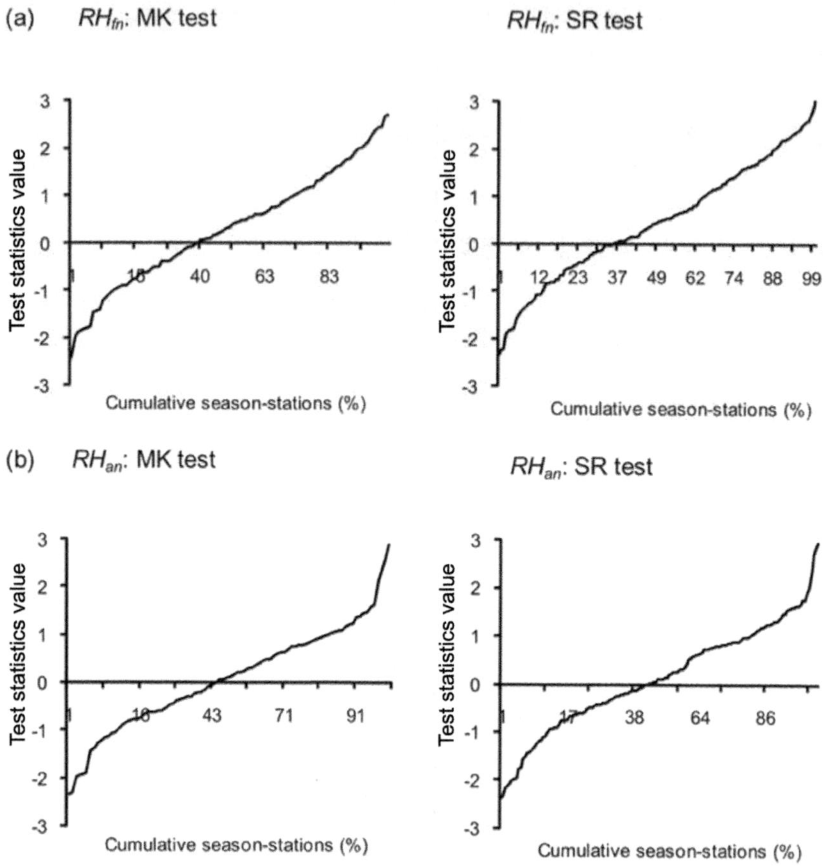

Fig. 11.5. Cumulative frequency distribution of MK and SR trend test results of 192 relative humidity time series (12 seasons × 16 stations) for (a) monthly average relative humidity recorded at 8.30 hours (RH_{fn}), and (b) monthly average relative humidity recorded at 17.30 hours (RH_{an}), respectively.

Table 11.6. Results of trends homogeneity tests of monthly average relative humidity in forenoon (RH_{fn}) and afternoon (RH_{an}) for the period 1987-2001

Sources of variation	*Monthly average relative humidity in forenoon (RH_{fn})*			*Monthly average relative humidity in afternoon (RH_{an})*		
	χ^2-cal	*d.f.*	*Significance*	*χ^2-cal*	*d.f.*	*Significance*
Total variation	261.05	192	-	174.58	192	-
Homogeneity	234.01	191	-	171.98	191	-
Season	21.36	11	NS	26.97	11	NS
Station	109.28	15	NS	39.64	15	NS
Season station	103.37	165	NS	105.37	165	NS
Trend	27.04	1	S ($p \leq 0.05$)	2.60	1	NS

Note: *d.f.*= Degrees of freedom; NS = Not significant; S = Significant; χ^2–cal = Calculated chi-square test-statistics.

the forenoon relative humidity without any significant spatial and seasonal heterogeneity. However, results of the test of homogeneity of trend for the RH_{an} indicates no heterogeneity with respect to seasons and stations, and also a non-significant overall positive ($\bar{Z}$= 0.12) trend.

The result of increasing trend of relative humidity is consistent with that evidenced by Wang and Gaffen (2001). The increasing trends of the relative humidity may be attributed to the abundant moisture supply from the Bay of Bengal. The minimum temperature also contributes to the increase in relative humidity trends in winter. This explains the increasing trend of relative humidity in Orissa as the night temperature shows a cooling trend. However, the pronounced warming trend of daytime temperature could have increased the moisture load in the atmosphere. Combustion of coal and other fossil fuel, although in small quantity, produces water vapour (Gaffen and Ross, 1999). Further, Gorden et al. (2005) reported that the increase in water vapour is correlated with intensive food production in the Indian subcontinent, and thus expanding irrigation increases the risk for changes in the monsoon system. However, the uncertainty in both the temporal and spatial distribution of water vapour remains to be a hindrance while attributing the trends and variability of relative humidity (Gaffen and Ross, 1999). Even the day-night difference in relative humidity is larger than the amplitude of the seasonal cycle (Wang and Gaffen, 2001).

The radiative effect of water vapour feedback mechanism in climate change context is comparable to the radiative forcing from CO_2 enrichment (Fasullo and Sun, 2001). Surface relative humidity regulates the evaporation and transpiration process, and consequently has connections with both hydrological and surface energy budgets. Any increase in atmospheric moisture enhances the moisture convergence into storm, and thus amplifies the intensity of rainfall. However, the frequency and duration of rainfall decreases with consequent prevalence of drought because the total precipitation is controlled by the available surface energy (Trenberth, 1998; Trenberth et al., 2007). Therefore, the drought occurrence has increased over tropics and sub-tropics partly due to the reduced rainfall in landmass, and also due to the warming and increased atmospheric demand for moisture. Increasing trend of relative humidity for Orissa, although not significant, may also have contributed partially to the occurrence of intense weather events.

11.5 Conclusions

This study employed two nonparametric statistical tests, i.e., Mann-Kendall (MK) test and Spearman's rho (SR) test for trend detection to understand response of three hydroclimatic variables (rainfall, temperature and relative humidity) of Orissa. The annual rainfall shows that the anomalous events in terms of both drought and flood are increasingly more conspicuous in recent years. However, the mean annual rainfall and the linear relationship did not change significantly. The homogeneity test indicates that the monsoon rainfall of recent years experiences an overall drying trend. Furthermore, the mean monthly maximum temperature and mean monthly minimum temperature exhibited pronounced warming and cooling trends, respectively. The monthly temperature range showed an increasing trend. The monthly mean relative humidity showed higher percentage of increasing trends indicating an enhanced moisture load in the atmosphere. The pronounced warming with consequent increase in rainfall intensity and decrease in rainfall frequency may have led to a drying monsoon trend of the study area. Although the present study has not explored the trend attribution of the hydroclimatic variables, we speculate that the observed trends and variability are primarily due to the local anthropogenic activities. Future research is needed to estimate the local greenhouse gas and aerosol forcings for precise characterization of trends of hydroclimatic variables. The economic development and ecological security need to go hand in hand as the local anthropogenic activities can even cause global climate change. Therefore, policy should be formulated to establish the tradeoff between the two. There is a need to reduce aerosol and greenhouse gas emission by exploring the alternate energy sources.

Acknowledgements

We are very thankful to Prof. Madan Kumar Jha and Dr. Deepesh Machiwal for their helpful comments and suggestions to improve this chapter as well as for painstaking editing of this chapter.

References

Arora, M., Goel, N.K. and Singh, P. (2005). Evaluation of temperature trends over India. *Hydrological Sciences Journal*, **50(1):** 81-93.

Ashrit, R.G., Kumar, R. and Kumar, K.K. (2001). ENSO-monsoon relationship in greenhouse warming scenario. *Geophysical Research Letters*, **28:** 1727-1730.

Bamzai, A.S. and Shukla, J. (1999). Relationship between Eurasian snow cover, snow depth, and the Indian summer monsoon: An observational study. *Journal of Climate*, **12:** 3117-3132.

Beniston, M. and Stephenson, D.B. (2004). Extreme climatic events and their evolution under changing climatic conditions. *Global and Planetary Change*, **44:** 1-9.

Bhutiyani, M.R., Kale, V.S. and Pawar, N.J. (2007). Long-term trends in maximum, minimum and mean annual air temperatures across the northwestern Himalaya during the twentieth century. *Climatic Change*, **85:** 159-177.

Blacksmith Institute (2007). The world's worst polluted places. The top ten (of The Dirty Thirty). New York, NY 10035, http://www.blacksmithinstitute.org (accessed on January 22, 2010).

Broers, H.P. and Grift, B. (2004). Regional monitoring of temporal changes in groundwater quality. *Journal of Hydrology*, **296:** 192-220.

Burn, D.H. and Elnur, M.A.H. (2002). Detection of hydrologic trends and variability. *Journal of Hydrology*, **255:** 107-122.

CES (2007). *ENVIS Newsletter*. **8(1)**. Centre of Environmental Studies (CES), Forest and Environment Department, Government of Orissa, Bhubaneswar. http://www.envisorissa.org (accessed on February 24, 2010).

Chu, P. and Wang, J. (1997). Recent change in the tropical western pacific and Indian Ocean regions as detected by outgoing longwave radiation records. *Journal of Climate*, **10:** 636-646.

Chung, C.E. and Ramanathan, V. (2006). Weakening of north Indian SST gradients and the monsoon rainfall in India and the Sahel. *Journal of Climate*, **19:** 2036-2045.

Chylek, P., Dubey, M.K., Lohmann, U., Ramanathan, V., Kaufman, Y.J., Lesins, G., Hudson, J., Altmann, G. and Olsen, S. (2006). Aerosol indirect effect over the Indian Ocean. *Geophysical Research Letters*, **33:** L06806, DOI: 10.1029/2005GL025397.

Dash, S.K., Jenamani, R.K., Kalsi, S.R. and Panda, S.K. (2007). Some evidence of climate change in twentieth century India. *Climatic Change*, **85:** 299-321.

De, U.S., Dube, R.K. and Rao, G.S. (2005). Extreme weather events over India in the last 100 years. *Journal of Indian Geophysical Union*, **9(3):** 173-187.

Ding, Y. (1994). Monsoon over China. Kluwer Academic Publishers. 250 pp.

Douglas, E.M., Niyogi, D., Frolking, S., Yeluripati, J.B., Pielke Sr., R.A., Niyogi, N., Vorosmarty, C.J. and Mohanty, U.C. (2006). Changes in moisture and energy fluxes due to agricultural land use and irrigation in the Indian monsoon belt. *Geophysical Research Letters*, **33:** L14403, DOI:10.1029/2006GL026550.

Fasullo, J. and Sun, D.Z. (2001). Radiative sensitivity to water vapor under all sky condition. *Journal of Climate*, **14:** 2798-2807.

Ferro, C., Hannachi, A. and Stephenson, D.B. (2005). Simple nonparametric techniques for exploring changing probability distribution of weather. *Journal of Climate*, **18:** 4344-4354.

Fowler, H.J. and Archer, D.R. (2006). Conflicting signals of climate change in the upper Indus basin. *Journal of Climate*, **19:** 4276-4293.

Gadgil, S., Vinayachandran, P.N., Francis, P.A. and Gadgil, S. (2004). Extremes of Indian summer monsoon rainfall, ENSO and equatorial Indian Ocean oscillation. *Geophysical Research Letters*, **31**, L12213, DOI: 10.1029/2004GL019733.

Gaffen, D.J. and Ross, R.J. (1999). Climatology and trends of U.S. surface humidity and temperature. *Journal of Climate*, **12:** 811-827.

Garg, A., Bhattacharya, S., Shukla, P.R. and Dadhwal, V.K. (2001). Regional and sectoral assessment of greenhouse gas emissions in India. *Atmospheric Environment*, **35:** 2679-2695.

Ghosh, S. and Mujumdar, P.P. (2007). Nonparametric methods for modeling GCM and scenario uncertainty in drought assessment. *Water Resources Research*, **43**, W07405, DOI:10.1029/2006WR005351.

Gorden, L.J., Steffen, W., Jonson, B.F., Folke, C., Falkenmark, M. and Johannessen, A. (2005). Human modification of global water vapor flows from the land surface. *Proceedings of the National Academy of Sciences USA*, **102:** 7612-7617.

Goswami, B.N., Venugopal, V., Sengupta, D., Madhusoodanan, M.S. and Xavier, P.K. (2006). Increasing trend of extreme rain events over India in a warming environment. *Science*, **314:** 1442-1445.

Gupta, A., Thapliyal, P.K., Pal, P.K. and Joshi, P.C. (2005). Impact of deforestation on Indian monsoon: A GCM sensitive study. *Journal of Indian Geophysical Union*, **9(2):** 97-104.

Helsel, D.R. and Hirsch, R.M. (1995). Statistical Methods in Water Resources. Elsevier, Amsterdam. 522 pp.

Hingane, L.S., Kumar, K.K. and Murthy, R. (1985). Long-term trends of surface air temperature in India. *Journal of Climatology*, **5:** 521-528.

Hoyos, C.D. and Webster, P.J. (2007). The role of intraseasonal variability in the nature of Asian monsoon precipitation. *Journal of Climate*, **20:** 4402-4424.

Huntington, T.G. (2006). Evidence for intensification of the global water cycle: Review and synthesis. *Journal of Hydrology*, **319:** 83-95.

IPCC (2001). Third Assessment Report: Climate Change 2001 (TAR). Intergovernmental Panel on Climate Change, http://www.ipcc.ch (accessed on January 12, 2010).

IPCC (2007). Fourth Assessment Report: Climate Change 2007 (AR4). Intergovernmental Panel on Climate Change, http://www.ipcc.ch (accessed on January 12, 2010).

Kothawale, D.R. and Kumar, K.R. (2005). On the recent changes in surface temperature trends over India. *Geophysical Research Letters*, **32**, L18714, DOI: 10.1029/2005GL023528.

Koutsoyiannis, D. and Montanari, A. (2007). Statistical analysis of hydroclimatic time series: Uncertainty and insights. *Water Resources Research*, **43**, W05429, DOI: 10.1029/2006WR005592.

Krishnan, R., Majumdar, M., Vaidya, V., Ramesh, K.V. and Satyan, V. (2003). The abnormal Indian summer monsoon of 2000. *Journal of Climate*, **16:** 1177-1194.

Kumar, K.K., Rajagopalan, B. and Cane, M.A. (1999). On the weakening relationship between the Indian monsoon and ENSO. *Science*, **284:** 2156-2159.

Kumar, K.R., Sahai, A.K., Kumar, K.K., Patwardhan, S.K., Mishra, P.K., Revadekar, J.V., Kamala, K. and Pant, G.B. (2006). High-resolution climate change scenarios for India for the 21st century. *Current Science*, **90(3):** 334-345.

Lenka, D. (2001). Agriculture in Orissa. Kalyani Publishers, New Delhi, India, 350 pp.

Liu, B., Xu, M., Henderson, M. and Qi, Y. (2005). Observed trends of precipitation amount, frequency, and intensity in China, 1960-2000. *Geophysical Research Letters*, **110**, D08103, DOI: 10.1029/2004JD004864.

Menon, S., Hansen, J., Nazarenko, L. and Luo, Y. (2002). Climate effects of black carbon aerosols in China and India. *Science*, **297:** 2250-2253.

Min, S-K., Zhang, X., Zwiers, F.W. and Hegerl, G.C. (2011). Human contribution to more-intense precipitation extremes. *Nature*, **470:** 378-381.

Mirza, M.Q. (2003). Climate change and extreme weather events: Can developing countries adapt? *Climate Policy*, **3:** 233-248.

Mishra, A. and Das, M.C. (1984). Desertification around Hirakud reservoir. *The Environmentalist*, **4(1):** 451-458.

Mohapatra, M. and Mohanty, U.C. (2004). Some characteristics of low pressure systems and summer monsoon rainfall over Orissa. *Current Science*, **87(9):** 1245-1255.

O'Brien, K., Leichenko, R., Kelkar, U., Venema, H., Aandahl, G., Tompkins, H., Javed, A., Bhadwal, S., Barg, S., Nygaard, L. and West, J. (2004). Mapping vulnerability to multiple stressors: Climate change and globalization in India. *Global Environmental Change*, **14:** 303-313.

OSPCB (2004). Annual Report, 2004-2005. Orissa State Pollution Control Board (OSPCB), Government of Orissa, Bhubaneswar, Orissa.

OSPCB (2007). Ambient Air Quality Status of Some Selected Towns of Orissa. Orissa State Pollution Control Board (OSPCB), Government of Orissa, Bhubaneswar, Orissa.

Pan, Z., Arritt, R.W., Takle, E.S., Gutowski Jr. W.J., Anderson, C.J. and Segal, M. (2004). Altered hydrologic feedback in a warming climate introduces a "warming hole". *Geophysical Research Letters*, **31**, L17109, DOI: 10.1029/2004GL020528.

Panda, D.K., Mishra, A., Jena, S.K., James, B.K. and Kumar, A. (2007). The influence of drought and anthropogenic effects on groundwater levels in Orissa, India. *Journal of Hydrology*, **343:** 140-153.

Prasad, A.K., Singh, R.P. and Kafatos, M. (2006). Influence of coal based thermal power plants on aerosol optical properties in the Indo-Gangetic basin. *Geophysical Research Letters*, **33**, L05805, DOI: 10.1029/2005GL023801.

Rajeevan, M., Bhate, J. and Jaswal, A.K. (2008). Analysis of variability and trends of extreme rainfall events over India using 104 years of gridded daily rainfall data. *Geophysical Research Letters*, **35**, L18707, DOI: 10.1029/2008GL035143.

Ramanathan, V., Chung, C., Kim, D., Bettge, T., Buja, L., Kiehi, J.T., Washington, W.M., Fu, Q., Sikka, D.R. and Wild, M. (2005). Atmospheric brown clouds: Impacts on south Asian climate and hydrological cycle. *Proceedings of the National Academy of Sciences USA*, **102(15):** 5326-5333.

Ramanathan, V., Ramana, M.V., Roberts, G., Kim, D., Corrigan, C., Chung, C. and Winker, D. (2007). Warming trends in Asia amplified by brown cloud solar absorption. *Nature*, **448(2)**, DOI:10.1038/nature06019.

Ramesh, K.V. and Goswami, P. (2007). The shrinking Indian summer monsoon. Research Report RR CM 0709, Council of Scientific and Industrial Research, New Delhi, India. http://www.cmmacs.ernet.in (accessed on January 12, 2010).

Rao, P.G. (1993). Climatic changes and trends over a major river basin in India. *Climate Research*, **2:** 215-223.

Sen Roy, S. and Balling Jr, R.C. (2004). Trends in extreme daily precipitation indices in India. *International Journal of Climatology*, **24:** 457-466.

Sene-Johansen, S. (1995). Strengthening the State Control Board, Orissa State. A Norconsult International Consultant Report, Norway, 221 pp.

SER (2006). State of Environment Report (SER). Industry Chapter II, Government of Orissa, http://www.ospboard.org/stateenv2006 (accessed on February 25, 2010).

Singh, P. and Kumar, N. (1997). Impact assessment of climate change on the hydrological response of a snow and glacier melt runoff dominated Himalayan river. *Journal of Hydrology*, **193:** 316-350.

Swiss Re (2002). Natural catastrophes and man-made disasters in 2001: Man-made losses take on a new dimension. Swiss Reinsurance Company Publication, Zurich. Sigma No. 1/2002, 18 pp.

Trenberth, K.E. (1998). Atmospheric moisture residence times and cycling: Implications for rainfall rates with climate change. *Climatic Change*, **39:** 667-694.

Trenberth, K.E., Smith, L., Qian, T., Dai, A. and Fasullo, J. (2007). Estimate of global water budget and its annual cycle using observational and model data. *Journal of Hydrometeorology*, **8:** 758-769.

Tripathy, S., Srinivas, V.V. and Nanjundiah, R.S. (2006). Downscaling of precipitation for climate change scenarios: A support vector machine approach. *Journal of Hydrology*, **330:** 621-640.

Wang, J.X.L. and Gaffen, D.J. (2001). Late-twentieth-century climatology and trends of surface humidity and temperature in China. *Journal of Climate*, **14:** 2833-2845.

Wilby, R.L., Wedgbrow, C.S. and Fox, H.R. (2004). Seasonal predictability of the summer hydrometeorology of the river Themes, UK. *Journal of Hydrology*, **295:** 1-16.

Wilson, D., Hisdal, H. and Lawrence, D. (2010). Has streamflow changed in the Nordic countries? – Recent trends and comparisons to hydrological projection. *Journal of Hydrology*, **394:** 334-346.

Yadav, R.R., Park, W.K., Singh, J. and Dubey, B. (2004). Do the western Himalayas defy global warming? *Geophysical Research Letters*, **31**, L17201, DOI: 10.1029/2004GL020201.

Zhang, X., Harvey, K.D., Hogg, W.D. and Yuzyk, T.R. (2001). Trends in Canadian streamflow. *Water Resources Research*, **37(4):** 987-998.

12

Analysis of Trend and Periodicity in Long-Term Annual Rainfall Time Series of Nigeria

12.1 Introduction

Understanding trends and variations of current and historical hydroclimatic variables is pertinent to the future development and sustainable management of water resources of a particular region. Information regarding hydroclimatological issues is important within the context of global warming, water and energy cycles and the increasing demand for water due to population and economic growth (Sankarasubramanian and Vogel, 2003; Oguntunde et al., 2006). Changes in the climate system and land cover have been widely accepted to have important consequences for regional to global water resources management and conservation. The extent to which human alteration of earth's environment affects the global hydrologic cycle is still largely unknown (Szilagyi, 2001). Valuable historical records of hydrologic patterns over complex drainage basins help to understand anthropogenic and climatic effects on large-scale terrestrial ecosystems (Vörösmarty and Sahagian, 2000). One of the very important necessities of research into climate change is to analyze and detect historical changes in the climatic system (Houghton et al., 1996). Rainfall is a principal element of the hydrological cycle, hence understanding its behaviour may be of profound social and economic significance. Detection of trends and oscillations in the rainfall time series yields important information for understanding the climate. However, rainfall changes are particularly hard to gauge, because rainfall is not uniform and varies considerably from place to place and time to time, even on small scales.

Invited contribution by Philip G. Oguntunde – Department of Agricultural Engineering, the Federal University of Technology, Akure, Nigeria and Babatunde J. Abiodun – Department of Environmental and Geographical Science, University of Cape Town, South Africa.

Several studies on analysis of rainfall time series have been carried out at different temporal scales and in different parts of the globe. Existing analyses of rainfall time series show for some areas a positive trend and a tendency towards higher frequencies of heavy and extreme rainfall in the last few decades (Houghton et al., 1996). In the Canadian Prairies, a significant increase in the amounts and number of rainfall events have been reported (Akinyemi et al., 2001). Ati et al. (2009) reported significant increase in rainfall over nine stations in northern Nigeria between 1953 and 2002. Increasing trends were also reported for Ibadan in southwest Nigeria (Oguntunde et al., 2011). Many researchers analyzed the precipitation patterns in several parts of Europe. Brazdil (1992) described fluctuations of precipitation in Europe using a time series of annual areal precipitation totals. Some of the results of precipitation analysis suggest that spatial and temporal non-uniformity in trend exists, which make generalization over large areas difficult if not impossible. Significant positive and negative trends have also been observed in the USA (Karl et al., 1995; Trenberth, 1998; Kunkel et al., 1999), Australia (Suppiah and Hennessey, 1998; Plummer et al., 1999), South Africa (Mason et al., 1999), the United Kingdom (Osborn et al., 2000) and Italy (Brunetti et al., 2000; 2001). Besides the increase in precipitation intensity, there are some indications that the overall percentage of the earth's surface affected by either drought and/or excessive wetness has increased (Dai and Trenberth, 1998). Gemmer et al. (2004) analyzed the annual rainfall series of 160 stations in China. They observed a spatial clustering of the trends in certain months, including district trend belts in east and northeast China.

Recent studies have shown that Africa has been drier over the last few decades (Nicholson et al., 2000; L'Hóte et al., 2002; Oguntunde et al., 2006). Furthermore, there are two schools of thought regarding the recent trends in Sahelian rainfall. Some researchers believe that the Sahelian drought continued till the end of the 20th century (L'Hóte et al., 2002), while others argue it may have ended in the 1990s (Ozer et al., 2003). Ojo (1987) examined the characteristics of rainfall variations between 1901 and 1985 in West Africa and discussed such characteristics as periodicities and variabilities. The study found no observable regular patterns in trends, periodicities and persistence of hydrologic consequences of rainfall variations to allow for predictability of these consequences in relation to rainfall variations. The role of rainfall trends in poor growth performance of sub-Saharan African nations relative to other developing countries, using a new cross-country panel climatic dataset in an empirical economic growth framework was examined (Barrios et al., 2010). The results showed that rainfall has been a significant determinant of poor economic growth for African nations but not for other countries.

In Nigeria, studies on rainfall analyses have been reported for different periods and locations within the country (e.g., Adefolalu, 1986; Tarhule and Woo, 1998; Bello, 1998; Olaniran, 2002; Ogolo and Adeyemi, 2009; Ati et al., 2009; Alli, 2010; Oguntunde et al., 2011). For example, Bello (1998) compared the seasonality of rainfall distribution in Nigeria for two periods, 1930-1961 and 1962-1993, and Alli (2010) studied rainfall trends and cycles for 20 rainfall stations scattered over Nigeria since 1960. Adefolalu (1986) analyzed 70-year period (1911-1980) rainfall data from 28 meteorological stations to examine trends in precipitation patterns. The results indicated a general decrease of dry season contribution to annual rainfall. In Nigeria, with more than 70% of the populace engaged in agriculture that is mostly rainfed, rainfall is the most important climatic variable owing to its manifestation as a deficient resource (droughts) or a catastrophic agent (floods). Therefore, for the purpose of sustainable water resources planning and management, trends and periodicities in long-term rainfall must be examined. The main aim of this study is to detect significant temporal and spatial trends and periodicities in the long-term (1901-2000) annual rainfall time series of Nigeria. This chapter is organized as follows: after presenting the background of this study, an overview of the study area is presented in Section 12.2 followed by methodology in Section 12.3, results and discussion in Section 12.4 and conclusions in Section 12.5.

12.2 Study Area

Nigeria is located in West Africa between latitude 4°-14°N and longitude 2°-15°E and encompasses a total area of about 925,796 km^2. The climate of Nigeria is more varied than that of any other country in West Africa. This is the result of the great length from south to the north (1100 km), which covers virtually all the climatic belts of West Africa. The climate is dominated by the influence of three main wind currents. These are the tropical maritime (MT) air mass, the tropical continental (CT) air mass and the equatorial easterlies (Ojo, 1987). The first air mass (MT) originates from the southern high-pressure belt located off the Namibian coast, and along its way picks up moisture from over the Atlantic Ocean and is thus wet. The second air mass (CT) has the high-pressure belt north of the Tropic of Cancer as its origin. This air mass is always dry as a result of little moisture it picks along its way. The first two air masses, MT and CT, meet along a starting surface called the Inter-Tropical Discontinuity (ITD). The third air mass (equatorial easterlies) is a somehow erratic cool air mass, which comes from the east and flows in the upper atmosphere along ITD. This air mass penetrates occasionally to actively undercut the MT or CT and gives rise to squall lines or dust devils (Iloeje, 2001). The entire country is grouped into three ecological zones based on latitude, which are Sahel (11°-14°N), Savannah (8°-11°N) and the Guinea (4°-8°N) zones as shown in Fig. 12.1. The climate is semi-arid in the north and humid in the south and also humid strip along the coast with an average annual rainfall of more than 2000 mm. Rainfall commences at the beginning of the rainy season around March/April from the coast (in the south), spreads through the middle belt, reaching its peak between July and September, to eventually get to the northern part very much later. The reverse of the situation also holds for the rainfall retreat period (Iloeje, 2001). About two thirds of the cropped area of Nigeria is in the north, with the rest cropped areas distributed between the middle belt and the south. According to the results of 2006 census, Nigeria is by far the most populous country in Africa with over 140 million people and population density of 138 people per km^2.

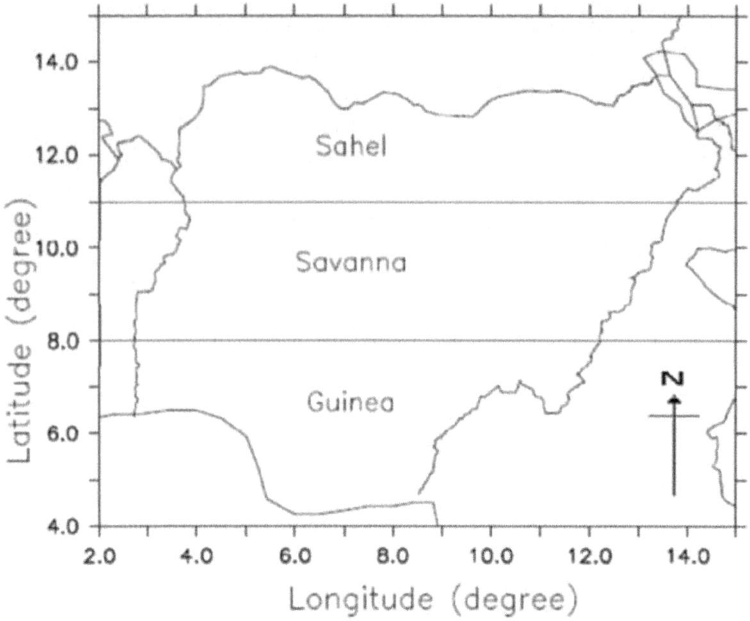

Fig. 12.1. Map of Nigeria showing the location of the delineated agro-ecological zones.

12.3 Methodology

12.3.1 Data Collection

Monthly rainfall data of Nigeria were collected from the Global Gridded Climatology of Climatic Research Unit Time Series (CRU TS 2.1) presented at a new high resolution and made available by the Climate Impacts LINK project, Climate Research Unit, University of East Anglia, Norwich, UK (Mitchell and Jones, 2005). The Climatic Research Unit (CRU) dataset is composed of monthly 0.5° latitude/longitude gridded series of precipitation; mean monthly temperature, diurnal temperature range, wet-day frequency, vapour pressure, cloud cover and ground-frost frequency (New et al., 1999). The 0.5 degree resolution dataset of monthly surface-based climate parameters cover the period 1901-2002. Amongst these parameters monthly accumulations of precipitation are generated from available gauge datasets. Although the time series extends back to 1901, it should be noted that the number of available gauges varies with time, for example, in 1901 a total of 4,957 gauges contributed to the dataset, which by 1981 has increased to 14,579 gauges. The CRU inserts synthetic zero anomaly values in regions that are "too far" from observations (i.e., farther than 450 km), while the other schemes simply interpolate over the entire distance. The annual rainfall time series data were checked for consistency and normality prior to testing for trends and periodicities.

12.3.2 Data Analysis

In this study, first of all, descriptive statistics of annual rainfall time series including minimum, maximum, range, mean, standard deviation, variance, standard error of the mean, kurtosis and skewness with their standard errors were computed by using SPSS 15.0 software. These statistics help to provide a preliminary overview of the dispersion and distribution of the data series. Details about these descriptive statistics can be found in Chapter 2 of this book.

Normality tests are used to determine whether a dataset can be described by a normal distribution or not, or to compute how likely an underlying random variable is to be normally distributed. There are many reasons for applying the normality tests to a hydrologic time series, which include data screening, outlier identification, description, assumption checking, and characterizing differences among sub-populations (groups of cases). Data screening may show that you have unusual values, extreme values, gaps in the data, or other peculiarities. Exploring the data this way may help to determine whether the statistical techniques that are intended for data analysis are appropriate. There exist different approaches to normality testing as mentioned in Chapter 3. In this study, histogram, and box and whisker plot were used as graphical methods, while the Kolmogorov-Smirnov test and the Shapiro-Wilk test were used to check the presence of normality in the rainfall datasets.

Rainfall variability index is usually computed as the standardized precipitation departure and helps to separate the available rainfall time series into different climatic regimes such as 'very dry climatic year', 'normal climatic year', 'wet climatic year', 'very wet climatic year', etc. (Lamb, 1982; L'Hóte et al., 2002). Rainfall variability index (δ) was calculated as (L'Hóte et al., 2002; Oguntunde et al., 2006):

$$\delta_i = (P_i - \mu)/\sigma \tag{1}$$

where δ_i is rainfall variability index for year i, P_i is annual rainfall for year i, and μ and σ are the mean and standard deviation of annual rainfalls for the 1901-2000 period. When δ is within ± 0.5, the year is characterized as a 'normal year'; when δ is between +0.5 and +1, it is characterized as a 'wet year'; when $\delta > +1$, it is characterized as a 'very wet year'. Similarly, when δ is between –0.5 and –1, the year is characterized as a 'dry year', when $\delta <$ –1, it is characterized as a 'very dry year' (Lamb, 1982; L'Hóte et al., 2002).

Spatial annual rainfall time series were examined for presence of trends by using one of the World Meteorological Organization (WMO) recommended nonparametric tests, i.e. Mann-Kendall test. The Mann-Kendall test is often used to explore trends in hydroclimatological time series (Salmi et al., 2002; Tosic and Unkasevic, 2005, Oguntunde et al., 2006). Details about the Mann-Kendall test can be found in Chapter 4. It should be noted that the Mann-Kendall test is non-dimensional and does not quantify the scale or the magnitude of the trend in the units of the time series under study, but is rather a measure of the correlation of variable with time and, as such, simply offers information as to the direction and a measure of the significance of the observed trends. To estimate the true slope of an existing trend, the nonparametric Sen's slope estimation method was used (Salmi et al., 2002). The Sen's slope estimation method can be used in cases where the trend can be assumed to be linear. The details about the Sen's slope estimation method can be found in Chapter 4.

Generally, before embarking on the parametric trend test or least-squares (regression) analysis, the time series data are checked for its suitability for regression analysis by checking the three assumptions of the linear regression (Montgomery et al., 2006; Kleinbaum et al., 2007): (i) the source population is normally distributed, (ii) the variance of the dependent variable in the source population is constant regardless of the value of the independent variable(s), and (iii) the residuals are independent of each other. In this study, the normality assumption for linear regression was tested using the Kolmogorov-Smirnov test (details can be found in Chapter 3). Constant variance was tested by computing the Spearman rank correlation between the absolute values of the residuals and the observed value of the dependent variable and the Durbin-Watson statistic was used to test residuals for their independence to each other. The Durbin-Watson statistic is a measure of serial correlation between the residuals. If the residuals are not correlated, the Durbin-Watson statistic will be 2 (Montgomery et al., 2006; Kleinbaum et al., 2007).

To estimate the true slope of an existing trend, the parametric method or least-squares regression analysis was used (Liu et al., 2008). This method can be used in cases where the trend can be assumed to be linear. This means that slope (Q) and intercept (B) in linear equation $f(t) = Qt + B$ are estimated by minimizing the sum of square errors between predicted and observed values. Thus, the mean values of Q and B that yield the least error of estimate for the model are selected. A t-statistic is then computed to measure the significance of the independent variable in predicting the dependent variable. The regression module of SigmaPlot 10.0 software was used in this analysis, including the verification of the assumptions.

Finally, autocorrelation and spectral methods were used to analyze periodic signals in the annual rainfall time series of three zones, namely Guinea, Savanna and Sahel. Autocorrelation is the correlation of a time series dataset signal with itself at different time lags (Phillips et al., 2008). Theoretical details about autocorrelation analysis are presented in Chapter 4. Spectral analysis, on the other hand, is a procedure for decomposing a complex time series dataset into a spectrum of cycles of different lengths. It decomposes the dataset into few underlying sinusoidal (sine and cosine) functions of particular wavelength (Jenkins and Watts, 1968; Wei, 1989). This analysis helps to uncover reoccurring cycles of different length in a time series, which at first looks like a random noise. A periodogram presents a plot of amplitude (or power) of each cycles against their frequencies (or periods).

12.4 Results and Discussion

12.4.1 Temporal Analysis

12.4.1.1 Summary of Descriptive Statistics

Salient statistical properties of the long-term (1901-2000) annual rainfall time series for different zones of Nigeria (Guinea, Savannah, Sahel and entire Nigeria) are summarized in Table 12.1. It can be seen from Table 12.1 that the annual rainfall varies mostly in the south (Guinea) with standard deviation of 158 mm, and ranges from about 1340 to 2200 mm (mean = 1830 ± 158 mm). The annual rainfall in Sahel ranges from about 430 to 970 mm and least varied with standard deviation of 114 mm among the three ecological time series examined. For the country as a whole, the annual rainfall varied between about 870 and 1500 mm (mean = 1220 ± 110 mm). All the examined series are slightly skewed to the left whereas the positive kurtosis, a measure of the relative flatness of a distribution compared with the normal distribution (zero kurtosis), indicates tapering distributions.

12.4.1.2 Rainfall Variability

Rainfall variability index for the three zones of the study area are shown in Fig. 12.2. The Rainfall variability index observed in both Guinea and Savannah seems to have overriding effect on the mean values for the entire Nigeria. There are slight differences in the distribution of rainfall variability index especially prior to the beginning of drought in 1970. Following the findings of other researchers, e.g., Nicholson et al. (2000) and L'Hote et al. (2002), four series of characteristic periods may be distinguished for the Sahel as: (i) from 1901 to 1949 (49 years), an apparently random succession of dry periods, "normal" periods and wet periods; (ii) from 1950 to 1968 (19 years), a series of 19 successive wet years; (iii) from 1969 to 1979 (11 years), an apparently succession of four dry years, five "normal" years and two wet years; and (iv) 1980 to 2000 – a series of 21 years that were dry or very dry, with five wet years. The driest decade was of the 1980s while the wettest decade was of the 1950s. Three time series of characteristic periods may be distinguished for the Guinea as: (i) from 1901 to 1953 (53 years), an apparently random succession of dry periods, "normal" periods and wet periods; (ii) from 1954 to 1969 (16 years), a series of an apparently random succession of nine wet years, four "normal" years and three very dry years; and (iii) from 1970 to 2000 (31 years), an apparently succession of 15 dry and very dry years, 11 "normal" years and five wet years. Rainfall variability characteristics exhibited by

Table 12.1. Summary of descriptive statistics for annual rainfall series of Guinea, Savannah, Sahel and entire Nigeria

Zone	*Minimum*	*Maximum*	*Range*	*Mean*	*SE*	*SD*	*Variance*	*Skewness*		*Kurtosis*	
	mm	*mm*	*mm*	*mm*	*mm*	*mm*	mm^2	-	*SE*	-	*SE*
Guinea	1339.0	2206.0	867.0	1828.5	158	157.9	24940.5	-0.06	0.24	0.31	0.48
Savannah	794.0	1406.0	612.0	1134.8	119	118.6	14075.7	-0.16	0.24	-0.28	0.48
Sahel	432.9	973.3	540.4	700.6	114	114.4	13096.9	-0.27	0.24	-0.17	0.48
Nigeria	868.1	1497.2	629.1	1221.3	110	110.1	12130.7	-0.23	0.24	0.59	0.48

Note: SE = Standard error; SD = Standard deviation.

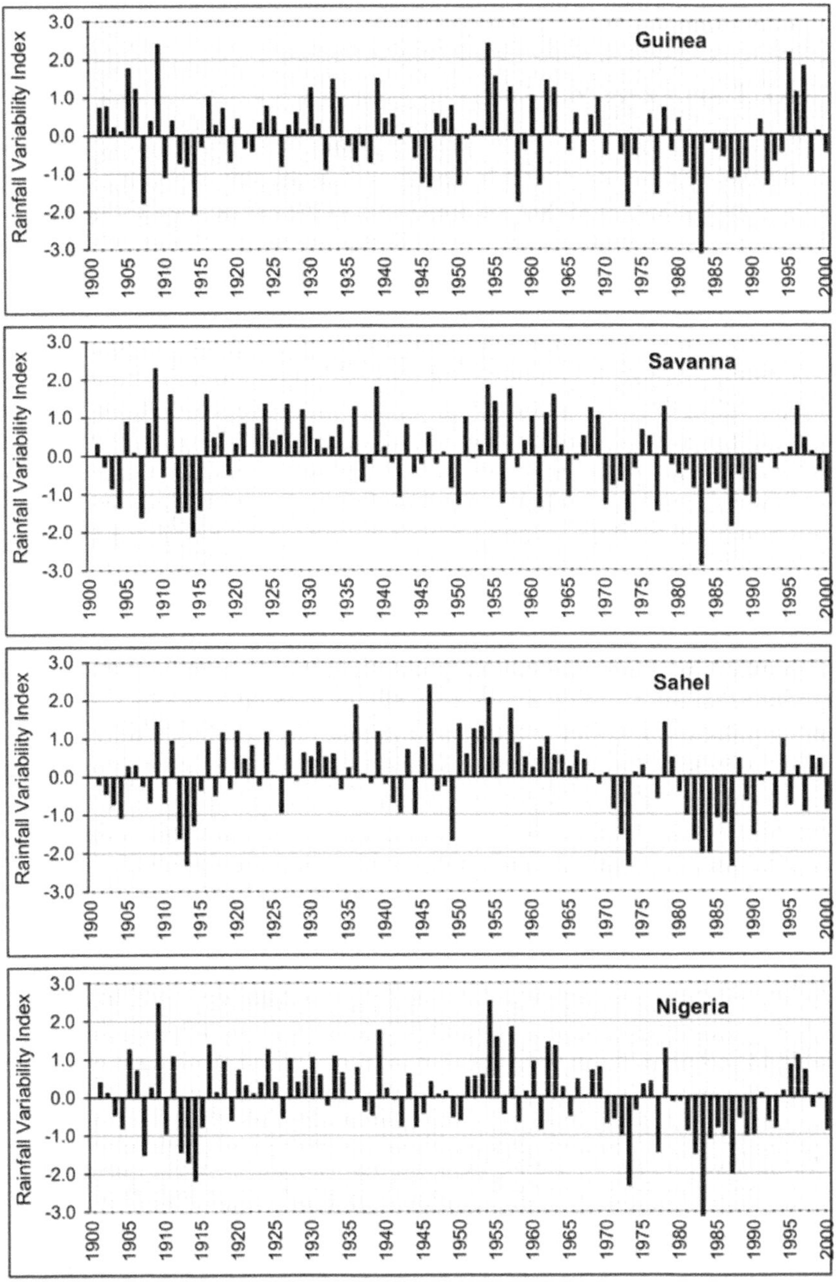

Fig. 12.2. Rainfall variability index for different zones of Nigeria.

Savannah zone and the Nigeria average are very similar. Generally, it has been wet before the 1970s, whereas in the post 1970s, there has been dramatic reduction in annual rainfall over Nigeria.

12.4.1.3 Temporal trends

1. *Normality analysis:* Prior to applying parametric trend tests, normality of the annual rainfall was tested by using both graphical and statistical tests. Using data averaged over Nigeria, histogram of the annual rainfall was plotted along with the theoretical normal distribution curve as shown in Fig. 12.3. The results showed that apart from the one outlier, 870 mm (with a distance of slightly more than three times the standard deviation from mean), the series can be approximated by a normal distribution. A similar result was obtained by plotting a box plot, presented in Fig. 12.4, where the outlier was clearly flagged better than the histogram. The results of statistical tests such as the Kolmogorov-Smirnov test and a more robust Shapiro-Wilk test are presented in Table 12.2. For these tests, *p*-value for normality determines the probability of being incorrect in concluding that the data is not normally distributed (*p*-value is the risk of falsely rejecting the null hypothesis that the data is normally distributed). If the *p*-value computed by the test is greater than the *p*-value set a priori, the test passes. To require a stricter adherence to normality then the *p*-value must be increased. The suggested value in SigmaPlot and SPSS software is 0.05. Larger *p*-values (for example, 0.10) require less evidence to conclude that the residuals are not normally distributed. One often rejects the null hypothesis when the *p*-value is less than 0.05 or 0.01, corresponding respectively to a 5% or 1% chance of rejecting the null hypothesis when it is true (Type I error). It was observed from Table 12.2 that the Kolmogorov-Smirnov test accepted the null hypothesis that our sample is normally distributed for data with 'the outlier' and a more robust Shapiro-Wilk test also barely accepted the null hypothesis. However, both the statistical tests confirmed presence of normality in annual rainfall time series after the removal of single outlier (Table 12.2). In a sample of 1000 observations, the presence of up to five observations deviating from the mean by more than three times the standard

Table 12.2. Results of two normality tests for annual rainfall time series of Nigeria

Kolmogorov-Smirnov[a] *test*			*Shapiro-Wilk test*		
Test-statistic	*df*	*Significance*	*Test-statistic*	*df*	*Significance*
(a) Annual rainfall time series with outlier					
0.058	100	0.200*	0.989	100	0.612
(b) Annual rainfall time series after removing single outlier					
0.053	99	0.200*	0.992	99	0.853

Note: [a]Lilliefors Significance Correction; *df* = degree of freedom; *Lower bound of true significance.

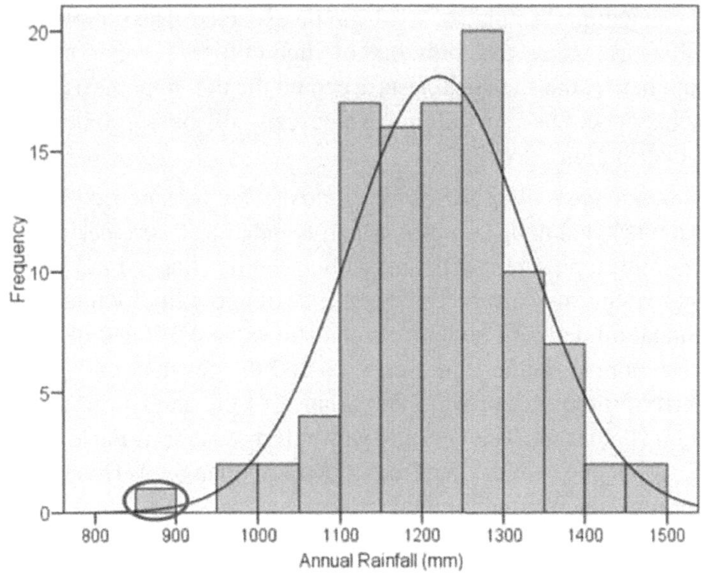

Fig. 12.3. Histogram for annual rainfall time series of Nigeria to be used as a check of normality (the outlier is enclosed in the oval shape). The histogram is overlay with the theoretical normal distribution curve.

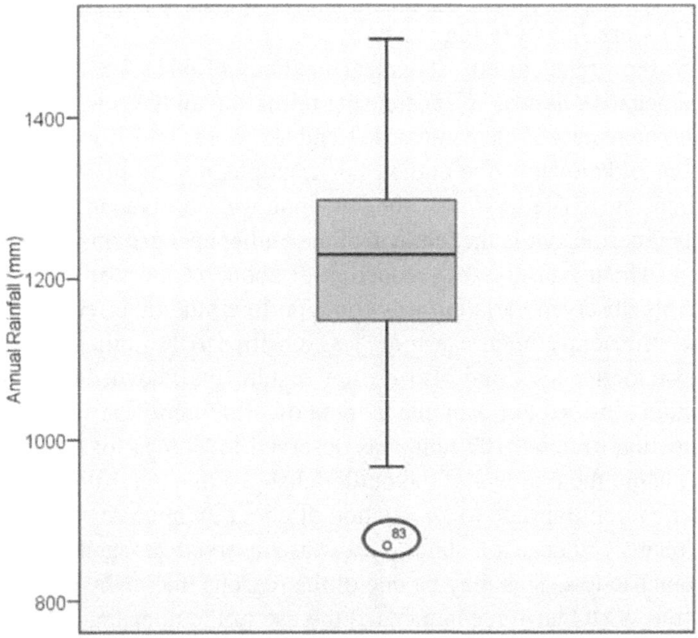

Fig. 12.4. Box plot for annual rainfall time series of Nigeria to be used as a check for normality (the outlier is encircled and corresponds to the annual rainfall of year 1983).

deviation is within the range of what can be expected. The sample size is only 100 in this case study, thus only one of such outliers seems not to be out of place, and hence, the statistical tests accepted the null hypothesis that the data with inclusion of the one extreme value can still be approximated with a normal distribution.

2. *Parametric and nonparametric trends*: The parametric Least Square Regression test and nonparametric Mann-Kendall test were performed on the annual rainfall time series with and without outlier flagged in Fig. 12.4. The Kolmogorov-Smirnov test revealed presence of normality for linear regression in the annual rainfall time series (test-statistic value 0.071 at 0.68 significance level). The annual rainfall time series passed the constant variance test (p = 0.92) verifying the assumption that variance of the annual rainfall in source population is constant. The Durbin-Watson test-statistic value is computed to be 1.56, which does not deviate from 2 by more than 0.50. This indicates that the linear regression assumption of independent residuals is true. Thus, all the three assumptions of the linear regression hold true, and therefore, the parametric trend test, i.e. least square regression test, can be applied to the annual rainfall time series in this study. The results of the parametric and nonparametric trend tests for the four annual rainfall time series are summarized in Table 12.3. The linear regression lines depicting trends in the annual rainfall time series for the Guinea, Savannah and Sahel zones, and Nigeria over the last century are shown in Fig. 12.5.

Both the graphical and statistical methods (Table 12.3 and Fig. 12.5) detected negative trends, i.e., decreasing rainfall in all the zones and in entire Nigeria. The results of the parametric trend test were significantly affected by the presence/removal of the outlier. For example, a trend of –0.90 mm/year, which translate to a reduction of about 90 mm, was observed in Guinea for the 1901-2000 period, while the removal of the outlier changed this trend to –0.71 mm/year, which translate to a reduction of about 71 mm for the same time series. This observed variation was common to all the time series examined. However, the nonparametric tests are less sensitive to the outliers as compared to the parametric tests and also do not require the knowledge of the data distribution *a priori*. For example, a trend of –1.07 mm/year, which translate to a reduction of about 107 mm, was observed in Guinea for the 1901-2000 periods, while the removal of the outlier changed this trend slightly to –0.98 mm/year, which translate to a reduction of about 98 mm. For the parametric test, a relative change of about 21% was observed as against 8% in the nonparametric test. This may be one of the reasons that many researchers as well as the WMO have recommended the use of the nonparametric methods for trend detection in hydroclimatological time series (Mitchell et al., 1966; Liu et al., 2008). It should be noted that, in the above example, both the magnitude and direction of the outlier's departure from the sample mean contribute significantly to the overall trend estimated.

Table 12.3. Results of parametric and nonparametric trend tests for annual rainfalls of three zones and entire Nigeria

Data type	*Time series*	*Least square regression test*			*Mann-Kendall test*		
		Mean slope (mm/year)	*Test-statistic*	*Significance*	*Median slope (mm/year)*	*Test-statistic*	*Significance*
Including outlier	Guinea	-0.90	-1.65	+	-1.07	-1.88	+
	Savannah	-0.75	-1.85	+	-0.85	-1.94	+
	Sahel	-0.63	-1.60	ns	-0.63	-1.51	ns
	Nigeria	-0.76	-2.02	*	-0.83	-2.16	*
Excluding outlier	Guinea	-0.71	-1.37	ns	-0.98	-1.71	+
	Savannah	-0.63	-1.59	ns	-0.80	-1.77	+
	Sahel	-0.55	-1.40	ns	-0.57	-1.35	ns
	Nigeria	-0.63	-1.75	+	-0.74	-2.00	*

Note: +Trend is significant at 1% significance level; * Trend is significant at 5% significance level; ns: Trend is not significant.

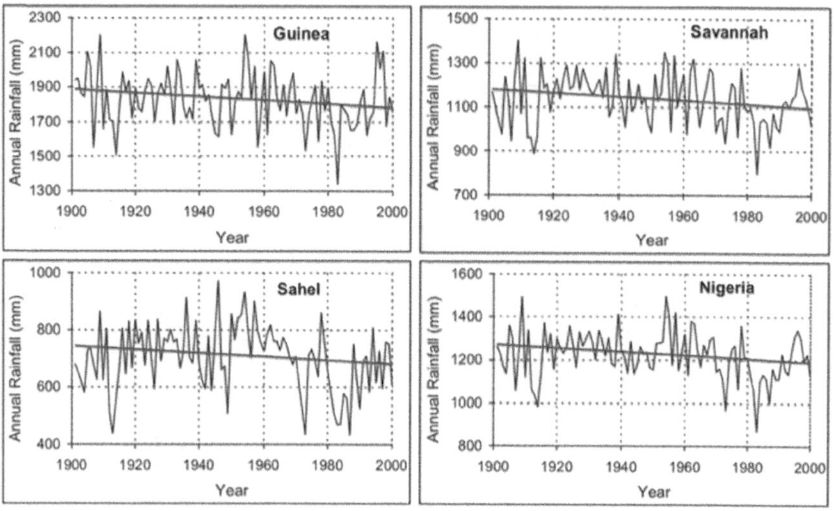

Fig. 12.5. Time series plots of annual rainfall for Guinea, Savannah, Sahel, and entire Nigeria. Linear regression lines are shown as solid thick lines.

12.4.2 Spatial Analysis

12.4.2.1 Spatial Distribution of Rainfall

The distribution of annual rainfall and the corresponding values of coefficient of variation (CV) are shown in Figs 12.6 and 12.7, respectively. The mean annual rainfall generally decreased with latitude (Fig. 12.6). Its value ranged from about 400 mm near the Lake Chad in the northeast corner to about 2500 mm at the southern part of Nigeria. Spatial pattern of the CV increased with latitude as rains become more varied northwards. The coefficient of variation values vary from 5 to 25% for the major portion of the southern and the middle-belt of the study area covering entirely the Guinea and the Savannah zones, whereas the CV values range between 15 and 35% for the Sahel zone. Cumulative distribution function (*cdf*) of the mean annual rainfall for 100-year period (1901-2000) is shown in Fig. 12.8. The *cdf* is very helpful to set threshold values below or above which certain rainfall events occurs. For example, it can be deduced from Fig. 12.8 that less than 10% of the entire Nigeria landscape experience about 500 mm of annual rainfall, 60% experience about 1300 mm while only about 10%, in the southern part, of the landscape experience very heavy storm above 2000 mm.

12.4.2.2 Spatial Trends

Pixel level trend analysis was carried out using only the nonparametric Mann-Kendall trend test. A total of 320 pixels covering the entire Nigeria were analyzed. Results obtained for different significance levels (10%, 5% and 1%) are summarised in Table 12.4. The number and percentages of pixels with negative or positive trends are reported for different significant levels. Guinea

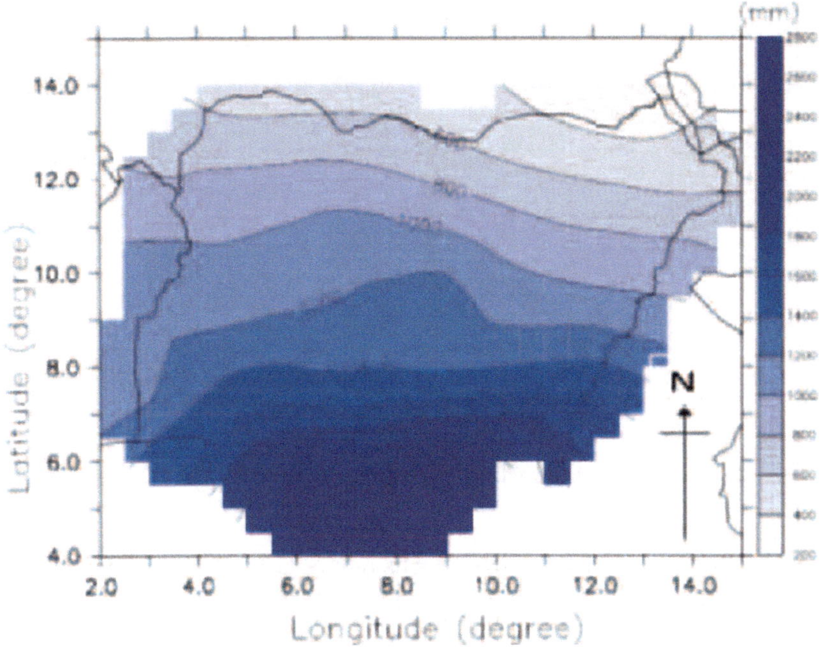

Fig. 12.6. Distribution of mean annual rainfall over Nigeria.

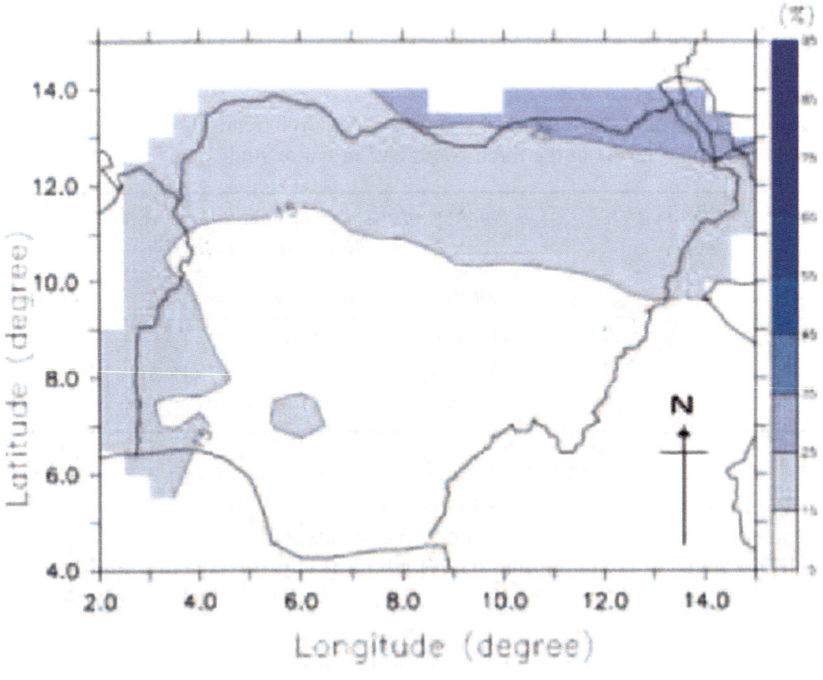

Fig. 12.7. Coefficient of variation of annual rainfall over Nigeria.

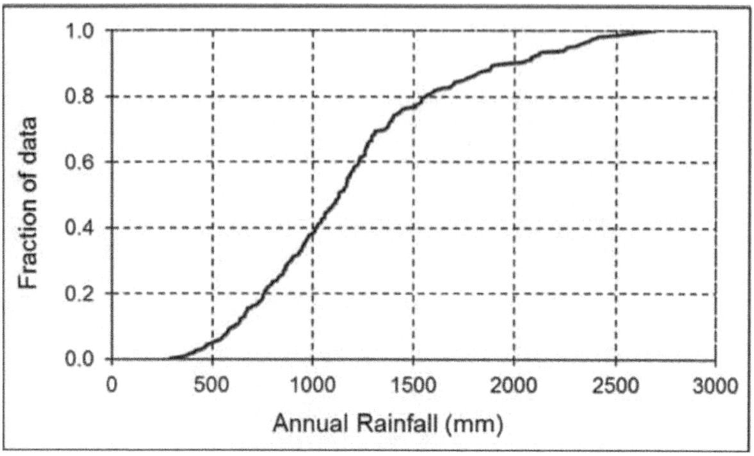

Fig. 12.8. Cumulative probability distribution curve for spatial estimates of mean annual rainfall for the 100 years (1901-2000).

zone has a total of 104 pixels, out of which only 11 or 10.6% exhibit negative significant trends at 1% significance level. However, the negative significant trends were observed for 18 pixels (17.3%) at 5% significance level and for 21 pixels or 20.2% at 10% significance level. Similarly, of the entire Nigeria, 3.8% exhibits negative significant trends at 1% significance level. Whereas, negative trends were observed for 63 pixels (19.7%) at 5% significance level, which increased to 98 pixels or 30.6% at 10% level. No pixel showed positive significant trend at any of the three significance levels.

Table 12.4. Spatial distribution of significant negative trends at three significance levels in the three zones and in entire study area

Significance level	*Number of pixels with positive or negative trends (% within parenthesis)*			
	Guinea	*Savannah*	*Sahel*	*Nigeria*
1%	11 (10.6)	0 (0.0)	1 (1.1)	12 (3.8)
5%	18 (17.3)	28 (22.6)	17 (18.5)	63 (19.7)
10%	21 (20.2)	47 (37.9)	30 (32.6)	98 (30.6)

Figure 12.9 showed the spatial distribution of the trends at 0.5 mm/year intervals. These values varied between –3.46 and +0.76 mm/year. About 90% of the entire landscape exhibited negative trends while less than 10% showed positive trends. For a better understanding of the significance of rainfall change in Nigeria, spatial distribution of the estimated values of Mann-Kendall test-statistics is presented in Fig. 12.10. The test-statistic values varied spatially from –3.33 to +0.91. The spatial pattern of the changes at 10%, 5% and 1% significance levels are vividly displayed towards the southern part of Nigeria in the Niger Delta area and in the north-central portion of the study area. The

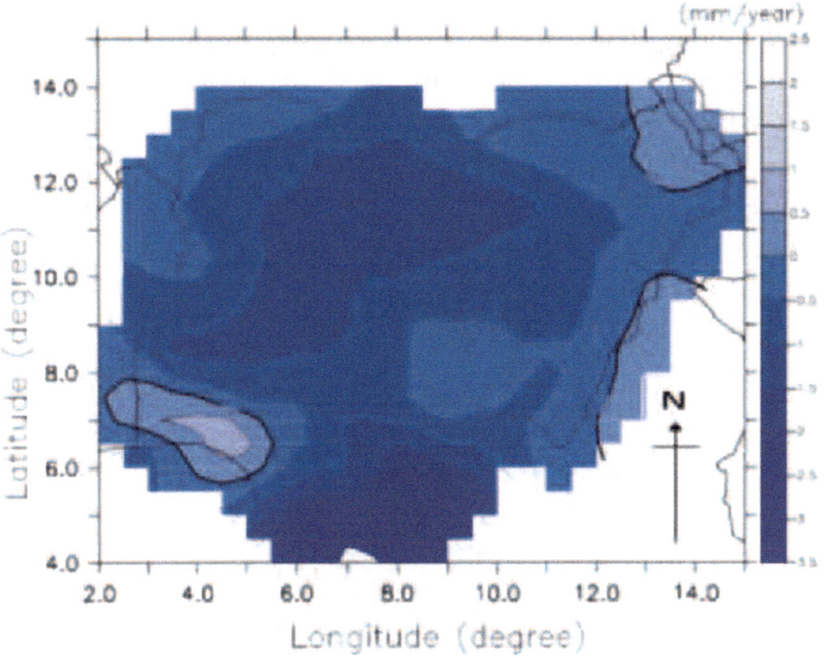

Fig. 12.9. Spatial distribution of annual rainfall trends (1901-2000).

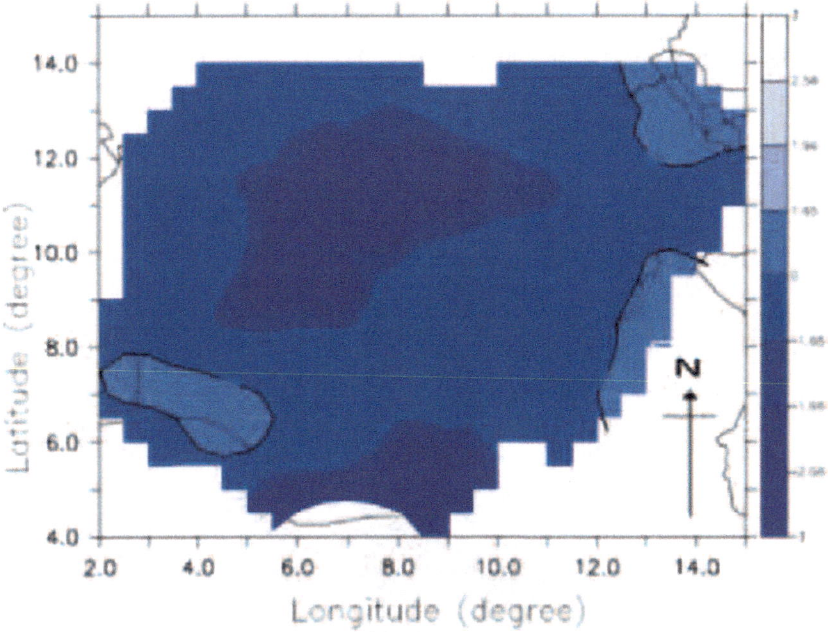

Fig. 12.10. Spatial distribution of observed values of Mann-Kendall test-statistic for the 100-year period (1901-2000) annual rainfall time series.

actual changes in rainfall in the last century over the entire Nigeria is shown in Fig. 12.11 and the area with significant change in annual rainfall values at 5% significance level is shown in Fig. 12.12. About 2.5% of the total area experienced overall rainfall change in the order of between –350 and –250

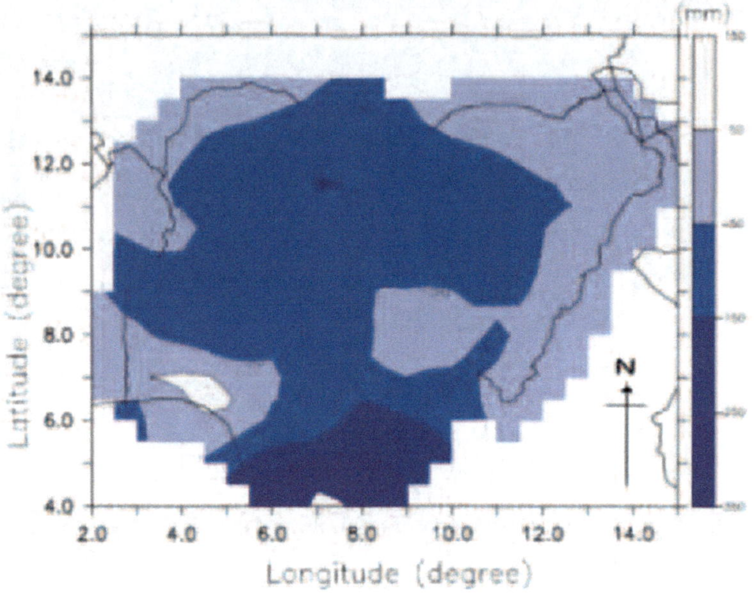

Fig. 12.11. Change in annual rainfall of Nigeria between 1901 and 2000.

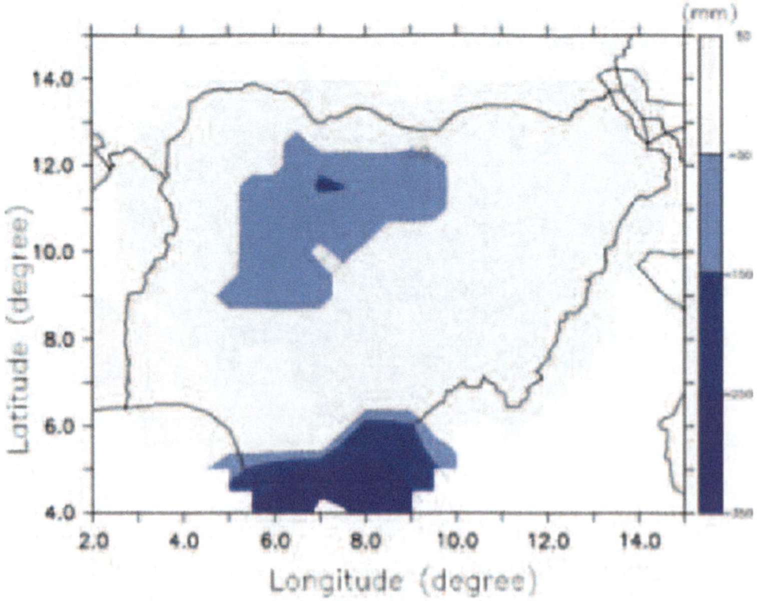

Fig. 12.12. Significant changes in annual rainfall at 5% significance level.

mm; 4.4% showed changes varying from –250 to –150 mm; 56.9% of the Nigeria landscape experienced changes between –150 and –50 mm; ± 50 mm occurred on about 34.6% while only 1.6% of Nigeria showed changes in the positive direction from +50 to +150 mm.

12.4.3 Rainfall Cycles and Periodicities

Autocorrelation plots for annual rainfall time series in Sahel, Savanna and Guinea zones are presented in Figs 12.13(a-c). The annual rainfall time series has an underlying sinusoidal pattern, because it exhibits alternating sequence of positive and negative correlation values, and the values are not decaying to zero. Such a pattern is signature of an autocorrelation of sinusoidal model. However, the signal has different strength over the zones, the strongest over Savanna and the weakest over Sahel. It is worth mentioning that this signal only emerges after applying a 5-year moving average to filter out the noise in the dataset. Without the filtering, the rainfall series over the zones is better classified as random; but the filtering enhance the performance of the autocorrelation analysis in revealing hiding periodic signal in the rainfall series. However, it is difficult to describe the characteristic of the periodic signal using the autocorrelation plots. This is better done with spectral analysis.

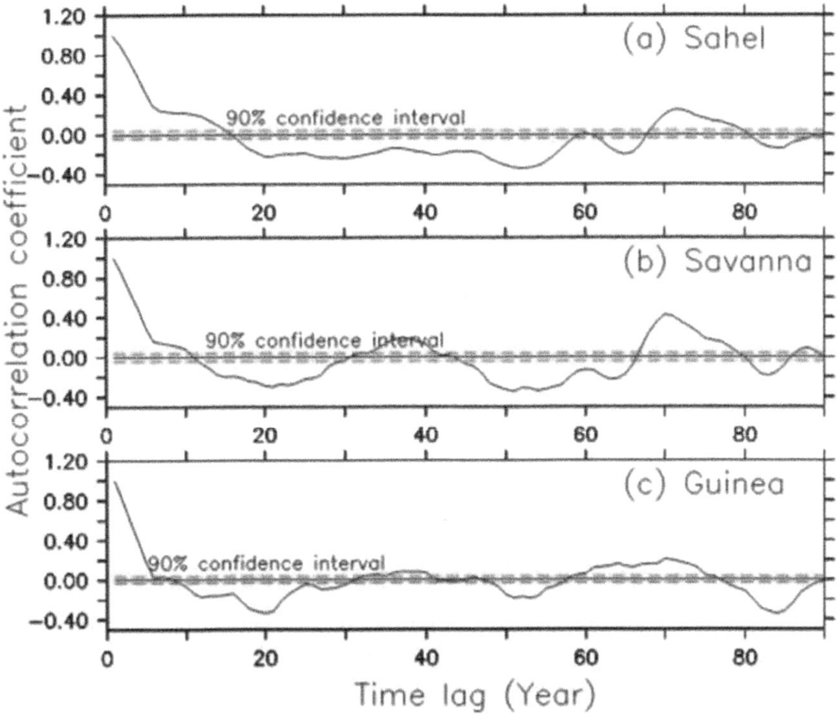

Fig. 12.13. Autocorrelation coefficients of annual rainfall over (a) Sahel, (b) Savannah and (c) Guinea zones (5-year moving average was applied to the rainfall data).

Periodograms of the annual rainfall series over three zones of the study area are presented in Figs 12.14(a-c). Three dominant peaks of rainfall cycles are evident over the zones. The cycles have periods 14 and 33 years over Sahel zone, 3 and 33 years over Savanna zone, and 3, 7 and 33 years over the Guinea zone. The 3-year oscillation period may be linked to the stratospheric Quasi-Biennial Oscillation (QBO) of equatorial zonal winds (Reed et al., 1961), the 7-year to tropical sea-surface temperatures (including El Nino-Southern Oscillation events, ENSO; Tosic and Unkasevic, 2005), the 14-year to the sunspot cycle (Makarau and Jury, 1997; Currie and O'Brien, 1988), and the 33-year to the Atlantic Multi-Decadal Oscillation (AMO) of sea surface temperature (Zhang and Delworth, 2006; Chang-Seng, 2007).

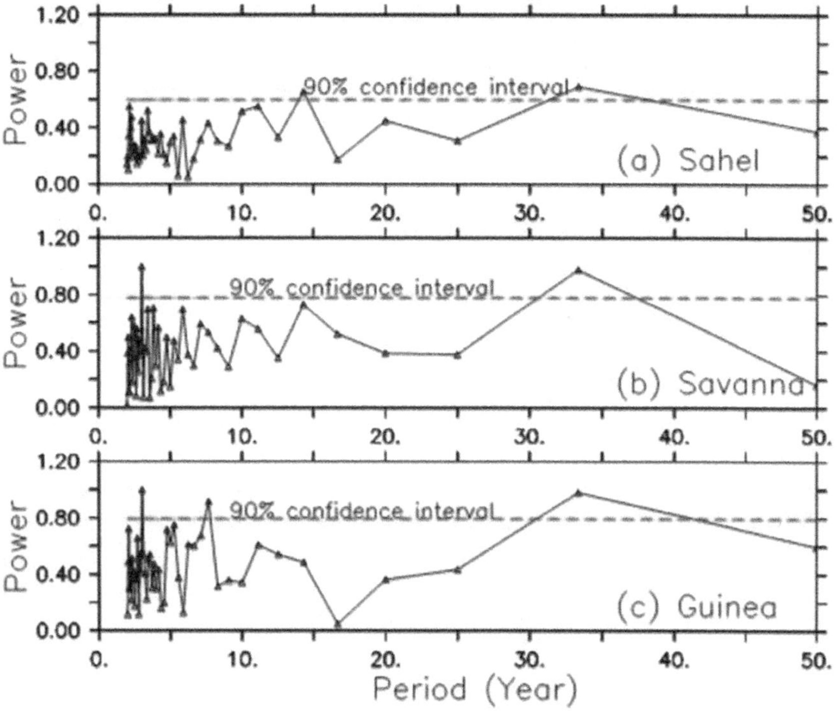

Fig. 12.14. Periodogram of annual rainfall over (a) Sahel, (b) Savannah and (c) Guinea zones.

12.5 Conclusions

Global climate change could have important effects on various environmental variables including rainfall in many countries of the world. Changes in rainfall regime directly affect agriculture, water resources management, hydrology and ecosystems. Hence, the significance of investigating the changes in the spatial and temporal rainfall patterns is imperative for suggesting suitable strategies for sustainable management of water resources, agriculture, ecosystems and environment. In this study, parametric and nonparametric statistical tests such as Least Square Regression test, Mann-Kendall test, rainfall variability index, autocorrelation and spectral analyses are used for detecting temporal and spatial trends and periodicities in 100-year period time series of annual rainfalls in Nigeria.

The observed annual rainfall variability index showed slight differences for the three zones, namely Guinea, Savannah and Sahel. Rainfall variability characteristics, exhibited by the Savannah zone and the mean rainfall over entire Nigeria, are very similar. Generally, it has been wet before the 1970s, whereas there has been dramatic reduction in annual rainfall totals over Nigeria in the post-1970s. The analysis of rainfall trends showed that more than 90% of the entire landscape was generally drying, while only 10% of the total area was getting wetter during the 20th century. The annual rainfall has been reduced significantly over 20% of the landscape. The amount of annual rainfall has been declined by 50 to 350 mm in 63.7% (590,000 km^2) portion of the entire country during the 20th century. The 100-year period Nigerian annual rainfall time series has an underlying sinusoidal signal embedded in random noise. The signatures of QBO, ENSO and sunspot cycles are evident in the annual rainfall series.

There is a need to extend the present study to include monthly and seasonal analyses so as to capture intra-annual trends and distribution of rainfall over Nigeria. Analysis of each calendar month and season will allow the identification of time characteristics peculiar to each month and season, which may be masked in annual total rainfall analysis. Furthermore, future related studies should investigate country-wide changes in other rainfall characteristics such as frequency of events, onset of rains, cessation and length of growing season.

References

Adefolalu, D.O. (1986). Rainfall trends in Nigeria. *Theoretical and Applied Climatology*, **37:** 205-219.

Akinyemi, O.O., McGinn, S.M. and Cutforth, H.W. (2001). Seasonal and spatial patterns of rainfall trends on the Canadian Prairies. *Journal of Climate*, **14:** 2177-2182.

Alli, A.A. (2010). Trends and cycles of rainfall and temperature for water resources development in Nigeria. Unpublished M.Sc. Thesis, the Federal University of Technology, Akure, Nigeria.

Ati, O.F., Stigter, C.J., Iguisi, O.E. and Afolayan, J.O. (2009). Profile of rainfall change and variability in the northern Nigeria, 1953-2002. *Research Journal of Environmental and Earth Sciences*, **1(2):** 58-63.

Barrios, S., Bertinelli, L. and Strobl, E. (2010). Trends in rainfall and economic growth in Africa: A neglected cause of the African growth tragedy. *The Review of Economics and Statistics*, MIT Press, **92(2):** 350-366.

Bello, N.J. (1998). A study of evidence of climate change based on rainfall seasonality and the reliability of rainfall regime in Nigeria. *Sustained Africa*, **4:** 30-32.

Brazdil, R. (1992). Fluctuation of atmospheric precipitation in Europe. *GeoJournal*, **27:** 275-291.

Brunnetti, M., Buffoni, L., Maugeri, M. and Nanni, T. (2000). Precipitation intensity trends in Northern Italy. *International Journal of Climatology*, **20:** 1017-1031.

Brunnetti, M., Buffoni, L., Maugeri, M. and Nanni, T. (2001). Trends in daily intensity of Precipitation in Italy from 1951 to 1996. *International Journal of Climatology*, **21:** 299-316.

Chang-Seng, D. (2007). Climate Variability and Climate Change Assessment for the Seychelles. Second National Communication (SNC) under the United Nations Framework Convention on Climate Change (UNFCCC), Ministry of Environment and Natural Resources, Republic of Seychelles, 56 pp.

Currie, R.G. and O'Brien, D.P. (1988). Periodic 18.6-year and cyclic 10 to 11-year signals in Northeastern United States precipitation data. *International Journal of Climatology*, **8(3):** 255-281.

Dai, A. and Trenberth, K.E. (1998). Global variations in droughts and wet spells: 1900-1995. *Geophysical Research Letters*, **25:** 3367-3370.

Gemmer, M., Becker, S. and Jiang, T. (2004). Observed monthly precipitation trends in China 1951-2002. *Theoretical and Applied Climatology*, **77:** 39-45.

Houghton, J.T., Meira Filho, L.G., Callander, B.A., Harris, N., Kattenberg, A. and Maskell, K. (editors) (1996). Climate Change 1995: The Science of Climate Change. Cambridge University Press, Cambridge, 572 pp.

Iloeje, N.P. (2001). A New Geography of Nigeria. New Revised Edition. Longman Press, Nigeria.

Jenkins, G.M. and Watts, D.G. (1968). Spectral Analysis and Its Applications. Holden-Day, Oakland, CA, 525 pp.

Karl, T.R., Knight, R.W. and Plummer, N. (1995). Trends in high-frequency climate variability in the twentieth century. *Nature*, **377:** 217-220.

Kleinbaum, D.G., Kupper, L.L., Nizam, A. and Muller, K.E. (2007). Applied Regression Analysis and Multivariable Methods. 4th Edition, Duxbury Press, Pacific Grove, USA, 928 pp.

Kunkel, K.E., Pielker, R.A. and Changnon, S.A. (1999). Temporal fluctuation in winter and climate extremes that cause economic and human health impact: A review. *International Journal of Climatology*, **19:** 1077-1098.

L'Hóte, Y., Mahe, G., Some, B. and Triboulet, J.P. (2002). Analysis of a Sahelian annual rainfall index from 1896 to 2000; the drought continues. *Hydrological Sciences Journal*, **47(4):** 563-572.

Lamb, P.J. (1982). Persistence of Subsaharan drought. *Nature*, **299:** 46-47.

Liu, Q., Yang, Z. and Cui, B. (2008). Spatial and temporal variability of annual precipitation during 1961–2006 in Yellow River Basin, China. *Journal of Hydrology*, **361:** 330-338.

Makarau, A. and Jury, M. (1997). Predictability of Zimbabwe summer rainfall. *International Journal of Climatology*, **17:** 1421-1432.

Mason, S.J., Waylen, P.R., Mimmack, G.M., Rajaratnam, B. and Harrison, J.M. (1999). Changes in extreme rainfall events in South Africa. *Climate Change*, **41:** 249-257.

Mitchell, J.M., Dzerdzeevskii, B., Flohn, H., Hofmeyr, W.L., Lamb, H.H., Rao, K.N. and Wallen, C.C. (1966). Climatic change. Report of a working group of the

Commission for Climatology. World Meteorological Organization (WMO) Technical note No. 79. Geneva, 79 pp.

Mitchell, T.D. and Jones, P.D. (2005). An improved method of constructing a database of monthly climate observations and associated high-resolution grids. *International Journal of Climatology*, **25:** 693-712.

Montgomery, D.C., Peck, E.A. and Vining, G.G. (2006). Introduction to Linear Regression Analysis. 4th Edition, John Wiley & Sons, UK, 640 pp.

New, M., Hulme, M. and Jones, P.D. (1999). Representing twentieth century space-time climate variability. Part I: Development of a 1961-90 mean monthly terrestrial climatology. *Journal of Climate*, **12:** 829-856.

Nicholson, S.E., Some, B. and Kone, B. (2000). An analysis of recent rainfall conditions in West Africa, including the rainy seasons of the 1997 El Nino and the 1998 La Nina years. *Journal of Climate*, **13(14):** 2628-2640.

Ogolo, E.O. and Adeyemi, B. (2009). Variations and trends of some meteorological parameters at Ibadan, Nigeria. *The Pacific Journal of Science and Technology*, **10(2):** 981-987.

Oguntunde, P.G., Friesen, J., van de Giesen, N. and Savenije, H.H.G. (2006). Hydroclimatology of the Volta River Basin in West Africa: Trends and variability from 1901 to 2002. *Physics and Chemistry of the Earth*, **31:** 1180-1188.

Oguntunde, P.G., Abiodun, B.J., Olukunle, O.J. and Olufayo, A.A. (2011). Climate change – Trends and variability in pan evaporation and other climatic variables at Ibadan, Nigeria, 1973-2008. *Meteorological Applications* (in press).

Ojo, O. (1987). Rainfall trends in West Africa, 1901-1985. The influence of climate change and climatic variability on the hydrologic regime and water resources. Proceedings of the Vancouver Symposium, August 1987. IAHS Publication No. 168, IAHS Press, Wallingford, pp. 37-43.

Olaniran, O.J. (2002). Rainfall Anomalies in Nigeria: The Contemporary Understanding. Inaugural Lecture Series, University of Ilorin, Nigeria, 55 pp.

Osborn, T.J., Hulme, M., Jones, P.D. and Basnett, T.A. (2000). Observed trends in the daily intensity of United Kingdom precipitation. *International Journal of Climatology*, **20:** 347-364.

Ozer, P., Erpicum, M., Demaree, M. and Vandiepenbeeck, M. (2003). Discussion of "Analysis of a Sahelian annual rainfall index from 1896 to 2000; the drought continues". *Hydrological Sciences Journal*, **48(3):** 489-496.

Phillips, C.L., Parr, J.M. and Riskin, E.A. (2008). Signals, Systems and Transforms. 4th Edition, Prentice-Hall, New Jersey, 779 pp.

Plummer, N.J., Salinger, A., Nicholls, N., Suppiah, R., Hennessy, K., Leighton, R.M., Trewin, B., Page, C.M. and Lough, J.M. (1999). Changes in climate extremes over the Australian region and New Zealand during the twentieth century. *Climate Change*, **42:** 183-202.

Reed, R.J., Campbell, W.J., Rasmussen, L.A. and Rogers, R.G. (1961). Evidence of downward propagating annual wind reversal in the equatorial stratosphere. *Journal of Geophysical Research*, **66:** 813-818.

Salmi, T., Maatta, A., Anttila, P., Ruoho-Airola, T. and Amnell, T. (2002). Detecting trends of annual values of atmospheric pollutants by the Mann-Kendall test and Sen's slope estimates. Publications on Air Quality, No. 31. Helsinki, Finland, 41 pp.

Sankarasubramanian, A. and Vogel, R.M. (2003). Hydroclimatology of the continental United States. *Geophysical Research Letters*, **30(7)**, 1363, doi:10.1029/2002GL015937.

Suppiah, R. and Hennessey, K.J. (1998). Trends in total rainfall, heavy rain events and numbers of dry days in Australia. *International Journal of Climatology*, **18:** 1141-1164.

Szilagyi, J. (2001). Modeled areal evaporation trends over the conterminous United States. *Journal of Irrigation and Drainage Engineering, ASCE*, **127(4):** 196-200.

Tarhule, A. and Woo, M. (1998). Changes in rainfall characteristics in northern Nigeria. *International Journal of Climatology*, **18:** 1261-1271.

Tosic, I. and Unkasevic, M. (2005). Analysis of precipitation series for Belgrade. *Theoretical and Applied Climatology*, **80(1):** 67-77.

Trenberth, K.E. (1998). Atmospheric moisture residence times and cycling: Implications for rainfall rates with climate change. *Climate Change*, **39:** 667-694.

Vörösmarty, C.J. and Sahagian, D. (2000). Anthropogenic disturbance of the terrestrial water cycle. *BioScience*, **50:** 753-765.

Wei, W.W. (1989). Time Series Analysis: Univariate and Multivariate Methods. Addison-Wesley Publishing Company, Inc., Redwood City, California, 478 pp.

Zhang, R. and Delworth, T.L. (2006). Impact of Atlantic multidecadal oscillations on India/Sahel rainfall and Atlantic hurricanes. *Geophysical Research Letters*, **33**, L17712, doi:10.1029/2006GL026267.

Appendices

Appendix A1

Standard Normal Distribution (Left side)

z_p	0	0.01	0.02	0.03	0.04	0.05	0.06	0.07	0.08	0.09
-0.0	0.5000	0.4960	0.4920	0.4880	0.4840	0.4801	0.4761	0.4721	0.4681	0.4641
-0.1	0.4602	0.4562	0.4522	0.4483	0.4443	0.4404	0.4364	0.4325	0.4286	0.4247
-0.2	0.4207	0.4168	0.4129	0.4090	0.4052	0.4013	0.3974	0.3936	0.3897	0.3859
-0.3	0.3821	0.3783	0.3745	0.3707	0.3669	0.3632	0.3594	0.3557	0.3520	0.3483
-0.4	0.3446	0.3409	0.3372	0.3336	0.3300	0.3264	0.3228	0.3192	0.3156	0.3121
-0.5	0.3085	0.3050	0.3015	0.2981	0.2946	0.2912	0.2877	0.2843	0.2810	0.2776
-0.6	0.2743	0.2709	0.2676	0.2643	0.2611	0.2578	0.2546	0.2614	0.2483	0.2451
-0.7	0.2420	0.2389	0.2358	0.2327	0.2296	0.2266	0.2236	0.2206	0.2177	0.2148
-0.8	0.2119	0.2090	0.2061	0.2033	0.2005	0.1977	0.1949	0.1922	0.1894	0.1867
-0.9	0.1841	0.1814	0.1788	0.1762	0.1736	0.1711	0.1685	0.1660	0.1635	0.1611
-1.0	0.1587	0.1562	0.1539	0.1515	0.1492	0.1469	0.1446	0.1423	0.1401	0.1379
-1.1	0.1357	0.1335	0.1314	0.1292	0.1217	0.1251	0.1230	0.1210	0.1190	0.1170
-1.2	0.1151	0.1131	0.1112	0.1093	0.1075	0.1056	0.1038	0.1020	0.1003	0.0985
-1.3	0.0968	0.0951	0.0934	0.0918	0.0901	0.0885	0.0869	0.0853	0.0838	0.0823
-1.4	0.0808	0.0793	0.0778	0.0764	0.0749	0.0735	0.0721	0.0708	0.0694	0.0681
-1.5	0.0668	0.0655	0.0643	0.0630	0.0618	0.0606	0.0594	0.0582	0.0571	0.0559
-1.6	0.0548	0.0537	0.0526	0.0516	0.0505	0.0495	0.0485	0.0475	0.0465	0.0455
-1.7	0.0446	0.0436	0.0427	0.0418	0.0409	0.0401	0.0392	0.0384	0.0375	0.0367
-1.8	0.0359	0.0351	0.0344	0.0336	0.0329	0.0322	0.0314	0.0307	0.0301	0.0294
-1.9	0.0287	0.0281	0.0274	0.0268	0.0262	0.0256	0.0250	0.0244	0.0239	0.0233
-2.0	0.0228	0.0222	0.0217	0.0212	0.0207	0.0202	0.0197	0.0192	0.0188	0.0183
-2.1	0.0179	0.0174	0.0170	0.0166	0.0162	0.0158	0.0154	0.0150	0.0146	0.0143
-2.2	0.0139	0.0136	0.0132	0.0129	0.0125	0.0122	0.0119	0.0116	0.0113	0.0110
-2.3	0.0107	0.0104	0.0102	0.0099	0.0096	0.0094	0.0091	0.0089	0.0087	0.0084
-2.4	0.0082	0.0080	0.0078	0.0075	0.0073	0.0071	0.0069	0.0068	0.0066	0.0064
-2.5	0.0062	0.0060	0.0059	0.0057	0.0055	0.0054	0.0052	0.0051	0.0049	0.0048
-2.6	0.0047	0.0045	0.0044	0.0043	0.0041	0.0040	0.0039	0.0038	0.0037	0.0036
-2.7	0.0035	0.0034	0.0033	0.0032	0.0031	0.0030	0.0029	0.0028	0.0027	0.0026
-2.8	0.0026	0.0025	0.0024	0.0023	0.0023	0.0022	0.0021	0.0021	0.0020	0.0019
-2.9	0.0019	0.0018	0.0018	0.0017	0.0016	0.0016	0.0015	0.0015	0.0014	0.0014
-3.0	0.0013	0.0013	0.0013	0.0012	0.0012	0.0011	0.0011	0.0011	0.0010	0.0010
-3.1	0.0010	0.0009	0.0009	0.0009	0.0008	0.0008	0.0008	0.0008	0.0007	0.0007
-3.2	0.0007	0.0007	0.0006	0.0006	0.0006	0.0006	0.0006	0.0005	0.0005	0.0005
-3.3	0.0005	0.0005	0.0005	0.0004	0.0004	0.0004	0.0004	0.0004	0.0004	0.0003
-3.4	0.0003	0.0003	0.0003	0.0003	0.0003	0.0003	0.0003	0.0003	0.0003	0.0002

Note: z_p = standard normal variate.

Source: Kanji, G.K. (2001). 100 Statistical Tests. Sage Publication, New Delhi, India.

Appendix A2

Standard Normal Distribution (Right side)

z_p	0	0.01	0.02	0.03	0.04	0.05	0.06	0.07	0.08	0.09
0.0	0.5000	0.5040	0.5080	0.5120	0.5160	0.5199	0.5239	0.5279	0.5319	0.5359
0.1	0.5398	0.5438	0.5478	0.5517	0.5557	0.5596	0.5636	0.5675	0.5714	0.5753
0.2	0.5793	0.5832	0.5871	0.5910	0.5948	0.5987	0.6026	0.6064	0.6103	0.6141
0.3	0.6179	0.6247	0.6255	0.6293	0.6331	0.6368	0.6406	0.6443	0.6480	0.6517
0.4	0.6554	0.6591	0.6628	0.6664	0.6700	0.6736	0.6772	0.6808	0.6844	0.6879
0.5	0.6915	0.6950	0.6985	0.7019	0.7054	0.7088	0.7123	0.7157	0.7190	0.7224
0.6	0.7257	0.7291	0.7324	0.7357	0.7389	0.7422	0.7454	0.7486	0.7517	0.7549
0.7	0.7580	0.7611	0.7642	0.7673	0.7704	0.7734	0.7764	0.7794	0.7823	0.7852
0.8	0.7881	0.7910	0.7939	0.7967	0.7995	0.8023	0.8051	0.8078	0.8106	0.8133
0.9	0.8159	0.8186	0.8212	0.8238	0.8264	0.8289	0.8315	0.8340	0.8365	0.8389
1.0	0.8413	0.8438	0.8461	0.8485	0.8508	0.8531	0.8554	0.8577	0.8599	0.8621
1.1	0.8643	0.8665	0.8686	0.8708	0.8729	0.8749	0.8770	0.8790	0.8810	0.8830
1.2	0.8849	0.8869	0.8888	0.8907	0.8925	0.8944	0.8962	0.8980	0.8987	0.9015
1.3	0.9032	0.9049	0.9066	0.9082	0.9099	0.9115	0.9131	0.9147	0.9162	0.9177
1.4	0.9192	0.9207	0.9222	0.9236	0.9251	0.9265	0.9279	0.9292	0.9306	0.9319
1.5	0.9332	0.9345	0.9357	0.9370	0.9382	0.9394	0.9406	0.9418	0.9429	0.9441
1.6	0.9452	0.9463	0.9474	0.9484	0.9495	0.9505	0.9515	0.9525	0.9535	0.9545
1.7	0.9554	0.9564	0.9573	0.9582	0.9591	0.9599	0.9608	0.9616	0.9625	0.9633
1.8	0.9641	0.9649	0.9656	0.9664	0.9671	0.9678	0.9686	0.9693	0.9699	0.9706
1.9	0.9713	0.9719	0.9726	0.9732	0.9738	0.9744	0.9750	0.9756	0.9761	0.9767
2.0	0.9772	0.9778	0.9783	0.9788	0.9793	0.9798	0.9803	0.9808	0.9812	0.9817
2.1	0.9821	0.9826	0.9830	0.9834	0.9838	0.9842	0.9846	0.9850	0.9854	0.9857
2.2	0.9861	0.9864	0.9868	0.9871	0.9875	0.9878	0.9881	0.9884	0.9889	0.9890
2.3	0.9893	0.9896	0.9898	0.9901	0.9904	0.9906	0.9909	0.9911	0.9137	0.9916
2.4	0.9918	0.9920	0.9922	0.9925	0.9927	0.9929	0.9931	0.9932	0.9934	0.9936
2.5	0.9938	0.9940	0.9941	0.9943	0.9945	0.9946	0.9948	0.9949	0.9951	0.9952
2.6	0.9953	0.9955	0.9956	0.9957	0.9959	0.9960	0.9961	0.9962	0.9963	0.9964
2.7	0.9965	0.9966	0.9967	0.9968	0.9969	0.9970	0.9971	0.9972	0.9973	0.9974
2.8	0.9974	0.9975	0.9976	0.9977	0.9977	0.9978	0.9979	0.9979	0.9980	0.9981
2.9	0.9981	0.9982	0.9982	0.9983	0.9984	0.9984	0.9985	0.9985	0.9986	0.9986
3.0	0.9987	0.9987	0.9987	0.9988	0.9988	0.9989	0.9989	0.9989	0.9990	0.9990
3.1	0.9990	0.9991	0.9991	0.9991	0.9992	0.9992	0.9992	0.9992	0.9993	0.9993
3.2	0.9993	0.9993	0.9994	0.9994	0.9994	0.9994	0.9994	0.9995	0.9995	0.9995
3.3	0.9995	0.9995	0.9995	0.9996	0.9996	0.9996	0.9996	0.9996	0.9996	0.9997
3.4	0.9997	0.9997	0.9997	0.9997	0.9997	0.9997	0.9997	0.9997	0.9997	0.9998

Note: z_p = standard normal variate.

Source: Kanji, G.K. (2001). 100 Statistical Tests. Sage Publication, New Delhi, India.

Appendix A3

Critical Values of Student's *t*-Distribution

Degrees of freedom	1−α							
	0.60	0.75	0.90	0.95	0.975	0.99	0.995	0.999
1	0.325	1.000	3.078	6.314	12.706	31.821	63.657	318.315
2	0.289	0.816	1.886	2.920	4.303	6.965	9.925	22.327
3	0.277	0.765	1.638	2.353	3.182	4.541	5.841	10.215
4	0.271	0.741	1.533	2.132	2.776	3.747	4.604	7.173
5	0.267	0.727	1.476	2.015	2.571	3.365	4.032	5.893
6	0.265	0.718	1.440	1.943	2.447	3.143	3.707	5.208
7	0.263	0.711	1.415	1.895	2.365	2.998	3.499	4.785
8	0.262	0.706	1.397	1.860	2.306	2.896	3.355	4.501
9	0.261	0.703	1.383	1.833	2.262	2.821	3.250	4.297
10	0.260	0.700	1.372	1.812	2.228	2.764	3.169	4.144
11	0.260	0.697	1.363	1.796	2.202	2.718	3.106	4.025
12	0.259	0.695	1.356	1.782	2.179	2.681	3.055	3.930
13	0.259	0.694	1.350	1.771	2.160	2.650	3.012	3.852
14	0.258	0.692	1.345	1.761	2.145	2.624	2.977	3.787
15	0.258	0.691	1.340	1.753	2.131	2.602	2.947	3.733
16	0.258	0.690	1.337	1.746	1.120	2.583	2.921	3.686
17	0.257	0.689	1.333	1.740	2.110	2.567	2.898	3.646
18	0.257	0.688	1.330	1.734	2.101	2.552	2.878	3.610
19	0.257	0.688	1.328	1.729	2.093	2.539	2.861	3.579
20	0.257	0.687	1.325	1.725	2.086	2.528	2.845	3.552
21	0.257	0.686	1.323	1.721	2.080	2.518	2.831	3.527
22	0.257	0.686	1.321	1.717	2.074	2.508	2.819	3.505
23	0.257	0.685	1.319	1.714	2.069	2.500	2.807	3.485
24	0.257	0.685	1.318	1.711	2.064	2.492	2.797	3.467
25	0.257	0.684	1.316	1.708	2.060	2.485	2.787	3.450
26	0.257	0.684	1.315	1.706	2.056	2.479	2.779	3.435
27	0.257	0.684	1.314	1.703	2.052	2.473	2.771	3.421
28	0.257	0.683	1.313	1.701	2.048	2.467	2.763	3.408
29	0.257	0.683	1.311	1.699	2.045	2.462	2.756	3.396
30	0.257	0.683	1.310	1.697	2.042	2.457	2.750	3.385
40	0.255	0.681	1.303	1.684	2.021	2.423	2.704	3.307
50	0.255	0.679	1.299	1.676	2.009	2.403	2.678	3.261
60	0.254	0.679	1.296	1.671	2.000	2.390	2.660	3.232

(*Contd.*)

Appendix A3 (*Contd.*)

Degrees of freedom	1–α							
	0.60	0.75	0.90	0.95	0.975	0.99	0.995	0.999
70	0.254	0.678	1.294	1.667	1.994	2.381	2.648	3.211
80	0.254	0.678	1.292	1.664	1.990	2.374	2.639	3.195
90	0.254	0.677	1.291	1.662	1.987	2.368	2.632	3.183
100	0.254	0.677	1.290	1.660	1.984	2.364	2.626	3.174
120	0.254	0.677	1.289	1.658	1.980	2.358	2.617	3.160
150	0.254	0.676	1.287	1.655	1.976	2.351	2.609	3.145
∞	0.253	0.674	1.282	1.645	1.960	2.326	2.576	3.090

Note: α = significance level.

Source: Shahin, M., Van Oorschot, H.J.L. and De Lange, S.J. (1993). Statistical Analysis in Water Resources Engineering. A.A. Balkema, Rotterdam, the Netherlands.

Appendix A4

Critical Test-Statistic Values of Cumulative Deviations Test

Sample size n	$Q/\sqrt{n}$			$R/\sqrt{n}$		
	90%	95%	99%	90%	95%	99%
10	1.05	1.14	1.29	1.21	1.28	1.38
20	1.10	1.22	1.42	1.34	1.43	1.60
30	1.12	1.24	1.46	1.40	1.50	1.70
40	1.13	1.26	1.50	1.42	1.53	1.74
50	1.14	1.27	1.52	1.44	1.55	1.78
100	1.17	1.29	1.55	1.50	1.62	1.86
∞	1.22	1.36	1.63	1.62	1.75	2.00

Note: Q = sensitivity to departures from homogeneity; R = range.

Source: Buishand, T.A. (1982). Some methods for testing the homogeneity of rainfall records. *Journal of Hydrology*, **58:** 11-27.

Appendix A5

Critical Test-Statistic Values of Bayesian Test

Sample size *n*	*U*			*A*		
	90%	95%	99%	90%	95%	99%
10	0.336	0.414	0.575	1.90	2.31	3.14
20	0.343	0.447	0.662	1.93	2.44	3.50
30	0.344	0.444	0.691	1.92	2.42	3.70
40	0.341	0.448	0.693	1.91	2.44	3.66
50	0.342	0.452	0.718	1.92	2.48	3.78
100	0.341	0.457	0.712	1.92	2.48	3.82
∞	0.347	0.461	0.743	1.93	2.49	3.86

Note: *U* and *A* = Bayesian Test-statistics.

Source: Buishand, T.A. (1982). Some methods for testing the homogeneity of rainfall records. *Journal of Hydrology*, **58:** 11-27.

Appendix A6

Critical Test-Statistic Values for Studentized Range of Tukey Test at 5% Significance Level

K	2	3	4	5	6	7	8	9	10	11	12	13	14	15	16	17	18	19	20
ν																			
1	17.969	26.98	32.82	37.08	40.41	43.12	45.40	47.36	49.07	50.59	51.96	53.20	54.33	55.36	56.32	57.22	58.04	58.83	59.56
2	6.085	8.33	9.80	10.88	11.74	12.44	13.03	13.54	13.99	14.39	14.75	15.08	15.38	15.65	15.91	16.14	16.37	16.57	16.77
3	4.501	5.91	6.82	7.50	8.04	8.48	8.85	9.18	9.46	9.72	9.95	10.15	10.35	10.52	10.69	10.84	10.98	11.11	11.24
4	3.926	5.04	5.76	6.29	6.71	7.05	7.35	7.60	7.83	8.03	8.21	8.37	8.52	8.66	8.79	8.91	9.03	9.13	9.23
5	3.635	4.60	5.22	5.67	6.03	6.33	6.58	6.80	6.99	7.17	7.32	7.47	7.60	7.72	7.83	7.93	8.03	8.12	8.21
6	3.460	4.34	4.90	5.30	5.63	5.90	6.12	6.32	6.49	6.65	6.79	6.92	7.03	7.14	7.24	7.34	7.43	7.51	7.59
7	3.344	4.16	4.68	5.06	5.36	5.61	5.82	6.00	6.16	6.30	6.43	6.55	6.66	6.76	6.85	6.94	7.02	7.10	7.17
8	3.261	4.04	4.53	4.89	5.17	5.40	5.60	5.77	5.92	6.05	6.18	6.29	6.39	6.48	6.57	6.64	6.73	6.80	6.87
9	3.199	3.95	4.41	4.76	5.02	5.24	5.43	5.59	5.74	5.87	5.98	6.09	6.19	6.28	6.36	6.44	6.51	6.58	6.64
10	3.151	3.88	4.33	4.65	4.91	5.12	5.30	5.46	5.60	5.72	5.83	5.93	6.03	6.11	6.19	6.27	6.34	6.40	6.47
11	3.113	3.82	4.26	4.57	4.82	5.03	5.20	5.35	5.49	5.61	5.71	5.81	5.90	5.98	6.06	6.13	6.20	6.27	6.33
12	3.081	3.77	4.20	4.51	4.75	4.95	5.12	5.27	5.39	5.51	5.61	5.71	5.80	5.88	5.95	6.02	6.09	6.15	6.21
13	3.055	3.73	4.15	4.45	4.69	4.88	5.05	5.19	5.32	5.43	5.53	5.63	5.71	5.79	5.85	5.93	5.99	6.05	6.11
14	3.033	3.70	4.11	4.41	4.64	4.83	4.99	5.13	5.25	5.36	5.46	5.55	5.64	5.71	5.79	5.85	5.91	5.97	6.03
15	3.014	3.67	4.08	4.37	4.59	4.78	4.94	5.08	5.20	5.31	5.40	5.49	5.57	5.65	5.72	5.78	5.85	5.90	5.96
16	2.998	3.65	4.05	4.33	4.56	4.74	4.90	5.03	5.15	5.26	5.35	5.44	5.52	5.59	5.66	5.73	5.79	5.84	5.90
17	2.984	3.63	4.02	4.30	4.52	4.70	4.86	4.99	5.11	5.21	5.31	5.39	5.47	5.54	5.61	5.67	5.73	5.79	5.84
18	2.971	3.61	4.00	4.28	4.49	4.67	4.82	4.96	5.07	5.17	5.27	5.35	5.43	5.50	5.57	5.63	5.69	5.74	5.79
19	2.960	3.59	3.98	4.25	4.47	4.65	4.79	4.92	5.04	5.14	5.23	5.31	5.39	5.46	5.53	5.59	5.65	5.70	5.75
20	2.950	3.58	3.96	4.23	4.45	4.62	4.77	4.90	5.01	5.11	5.20	5.28	5.36	5.43	5.49	5.55	5.61	5.66	5.71

(Contd.)

Appendix A6 (*Contd.*)

K	2	3	4	5	6	7	8	9	10	11	12	13	14	15	16	17	18	19	20
21	2.941	3.56	3.94	4.21	4.43	4.60	4.74	4.87	4.98	5.08	5.17	5.25	5.33	5.40	5.46	5.52	5.58	5.62	5.67
22	2.933	3.55	3.93	4.20	4.41	4.58	4.72	4.85	4.96	5.05	5.15	5.23	5.30	5.37	5.43	5.49	5.55	5.59	5.64
23	2.926	3.54	3.91	4.18	4.39	4.56	4.70	4.83	4.94	5.03	5.12	5.20	5.27	5.34	5.40	5.46	5.52	5.57	5.62
24	2.919	3.53	3.90	4.17	4.37	4.54	4.68	4.81	4.92	5.01	5.10	5.18	5.25	5.32	5.38	5.44	5.49	5.55	5.59
25	2.913	3.52	3.89	4.16	4.36	4.52	4.66	4.79	4.90	4.99	5.08	5.16	5.23	5.30	5.36	5.42	5.48	5.52	5.57
26	2.907	3.51	3.88	4.14	4.34	4.51	4.65	4.78	4.89	4.97	5.06	5.14	5.21	5.28	5.34	5.40	5.46	5.50	5.55
27	2.902	3.51	3.87	4.13	4.33	4.50	4.63	4.76	4.87	4.96	5.04	5.12	5.19	5.26	5.32	5.38	5.43	5.48	5.53
28	2.897	3.50	3.86	4.12	4.32	4.48	4.62	4.75	4.86	4.94	5.03	5.11	5.18	5.24	5.30	5.36	5.42	5.46	5.51
29	2.892	3.49	3.85	4.11	4.31	4.47	4.61	4.73	4.84	4.93	5.01	5.09	5.16	5.23	5.29	5.35	5.40	5.44	5.49
30	2.888	3.49	3.85	4.10	4.30	4.46	4.60	4.72	4.82	4.92	5.00	5.08	5.15	5.21	5.27	5.33	5.38	5.43	5.47
31	2.884	3.48	3.83	4.09	4.29	4.45	4.59	4.71	4.82	4.91	4.99	5.07	5.14	5.20	5.26	5.32	5.37	5.41	5.46
32	2.881	3.48	3.83	4.09	4.28	4.44	4.58	4.70	4.81	4.89	4.98	5.06	5.13	5.19	5.24	5.30	5.35	5.40	5.45
33	2.877	3.47	3.82	4.08	4.27	4.44	4.57	4.69	4.80	4.88	4.97	5.04	5.11	5.17	5.23	5.29	5.34	5.39	5.44
34	2.874	3.47	3.82	4.07	4.27	4.43	4.56	4.68	4.79	4.87	4.96	5.03	5.10	5.16	5.22	5.28	5.33	5.37	5.42
35	2.871	3.46	3.81	4.07	4.26	4.42	4.55	4.67	4.78	4.86	4.95	5.02	5.09	5.15	5.21	5.27	5.32	5.36	5.41
36	2.868	3.46	3.81	4.06	4.25	4.41	4.55	4.66	4.77	4.85	4.94	5.01	5.08	5.14	5.20	5.26	5.31	5.35	5.40
37	2.865	3.45	3.80	4.05	4.25	4.41	4.54	4.65	4.76	4.84	4.93	5.00	5.08	5.14	5.19	5.25	5.30	5.34	5.39
38	2.863	3.45	3.80	4.05	4.24	4.40	4.53	4.64	4.75	4.84	4.92	5.00	5.07	5.13	5.18	5.24	5.29	5.33	5.38
39	2.861	3.44	3.79	4.04	4.24	4.40	4.53	4.64	4.75	4.83	4.92	4.99	5.06	5.12	5.17	5.23	5.28	5.32	5.37
40	2.858	3.44	3.79	4.04	4.23	4.39	4.52	4.63	4.73	4.82	4.90	4.98	5.04	5.11	5.16	5.22	5.27	5.22	5.36
50	2.841	3.41	3.76	4.00	4.19	4.34	4.47	4.58	4.69	4.76	4.85	4.92	4.99	5.05	5.10	5.15	5.20	5.24	5.29
60	2.829	3.40	3.74	3.98	4.16	4.31	4.44	4.55	4.65	4.73	4.81	4.88	4.94	5.00	5.06	5.11	5.15	5.20	5.24
120	2.800	3.36	3.68	3.92	4.10	4.24	4.36	4.47	4.56	4.64	4.71	4.78	4.84	4.90	4.95	5.00	5.04	5.09	5.13
∞	2.772	3.31	3.63	3.86	4.03	4.17	4.29	4.39	4.47	4.55	4.62	4.68	4.74	4.80	4.85	4.89	4.93	4.97	5.01

Note: K = total number of subseries; ν = degrees of freedom.

Source: Kanji, G.K. (2001). 100 Statistical Tests. Sage Publication, New Delhi, India.

Appendix A7

Critical Test-Statistic Values for Studentized Range of Tukey Test at 1% Significance Level

K	2	3	4	5	6	7	8	9	10	11	12	13	14	15	16	17	18	19	20
ν																			
1	90.025	135.0	164.3	185.6	202.2	215.8	227.2	237.0	245.6	253.2	260.0	266.2	271.8	277.0	281.8	296.3	290.4	294.3	298.0
2	14.036	19.02	22.29	24.72	26.63	28.20	29.53	30.68	31.69	32.59	33.40	34.13	34.81	35.43	36.00	36.53	37.03	37.50	37.95
3	8.260	10.62	12.17	13.33	14.24	15.00	15.64	16.20	16.69	17.13	17.53	17.89	18.22	18.52	18.81	19.07	19.32	19.55	19.77
4	6.511	8.12	9.17	9.96	10.58	11.10	11.55	11.93	12.27	12.57	12.84	13.09	13.32	13.53	13.73	13.91	14.08	14.24	14.40
5	5.702	6.98	7.80	8.42	8.91	9.32	9.67	9.97	10.24	10.48	10.70	10.89	11.08	11.24	11.40	11.55	11.68	11.81	11.93
6	5.243	6.33	7.08	7.56	7.97	8.32	8.61	8.87	9.10	9.30	9.48	9.65	9.81	9.95	10.08	10.21	10.32	10.43	10.54
7	4.949	5.92	6.54	7.01	7.37	7.68	7.94	8.17	8.37	8.55	8.71	8.86	9.00	9.12	9.24	9.35	9.46	9.55	9.65
8	4.745	5.64	6.20	6.62	6.96	7.24	7.47	7.68	7.86	8.03	8.18	8.31	8.44	8.55	8.66	8.76	8.85	8.94	9.03
9	4.596	5.43	5.96	6.35	6.66	6.91	7.13	7.33	7.49	7.65	7.78	7.91	8.03	8.13	8.23	8.33	8.41	8.49	8.57
10	4.482	5.27	5.77	6.14	6.43	6.67	6.87	7.05	7.21	7.36	7.49	7.60	7.71	7.81	7.91	7.99	8.08	8.15	8.23
11	4.392	5.15	5.62	5.97	6.25	6.48	6.67	6.84	6.99	7.13	7.25	7.36	7.46	7.56	7.65	7.73	7.81	7.88	7.95
12	4.320	5.05	5.50	5.84	6.10	6.32	6.51	6.67	6.81	6.94	7.06	7.17	7.26	7.36	7.44	7.52	7.59	7.66	7.73
13	4.260	4.96	5.40	5.73	5.98	6.19	6.37	6.53	6.67	6.79	6.90	7.01	7.10	7.19	7.27	7.35	7.42	7.48	7.55
14	4.210	4.89	5.32	5.63	5.88	6.08	6.26	6.41	6.54	6.66	6.77	6.87	6.96	7.05	7.13	7.20	7.27	7.33	7.39
15	4.167	4.84	5.25	5.56	5.80	5.99	6.16	6.31	6.44	6.55	6.66	6.76	6.84	6.93	7.00	7.07	7.14	7.20	7.26
16	4.131	4.79	5.19	5.49	5.72	5.92	6.08	6.22	6.35	6.46	6.56	6.66	6.74	6.82	6.90	6.97	7.03	7.09	7.15
17	4.099	4.74	5.14	5.43	5.66	5.85	6.01	6.15	6.27	6.38	6.48	6.57	6.66	6.73	6.81	6.87	6.94	7.00	7.05
18	4.071	4.70	5.09	5.38	5.60	5.79	5.94	6.08	6.20	6.31	6.41	6.50	6.58	6.65	6.73	6.79	6.85	6.91	6.97
19	4.045	4.67	5.05	5.33	5.55	5.73	5.89	6.02	6.14	6.25	6.34	6.43	6.51	6.58	6.65	6.72	6.78	6.84	6.89
20	4.024	4.64	5.02	5.29	5.51	5.69	5.84	5.97	6.09	6.19	6.28	6.37	6.45	6.52	6.59	6.65	6.71	6.77	6.82

(Contd.)

Appendix A7 (*Contd.*)

K	2	3	4	5	6	7	8	9	10	11	12	13	14	15	16	17	18	19	20
21	4.004	4.61	4.99	5.26	5.47	5.65	5.80	5.92	6.04	6.14	6.24	6.32	6.39	6.47	6.53	6.59	6.65	6.70	6.76
22	3.986	4.58	4.96	5.22	5.43	5.61	5.76	5.88	6.00	6.10	6.19	6.27	6.35	6.42	6.48	6.54	6.60	6.65	6.70
23	3.970	4.56	4.93	5.20	5.40	5.57	5.72	5.84	5.96	6.06	6.15	6.23	6.30	6.37	6.43	6.49	6.55	6.60	6.65
24	3.955	4.55	4.91	5.17	5.37	5.54	5.69	5.81	5.92	6.02	6.11	6.19	6.26	6.33	6.39	6.45	6.51	6.56	6.61
25	3.942	4.52	4.89	5.15	5.34	5.51	5.66	5.78	5.89	5.99	6.07	6.15	6.22	6.29	6.35	6.41	6.47	6.52	6.57
26	3.930	4.50	4.87	5.12	5.32	5.49	5.63	5.75	5.86	5.95	6.04	6.12	6.19	6.26	6.32	6.38	6.43	6.48	6.53
27	3.918	4.49	4.85	5.10	5.30	5.46	5.61	5.72	5.83	5.93	6.01	6.09	6.16	6.22	6.28	6.34	6.40	6.45	6.50
28	3.908	4.47	4.83	5.08	5.28	5.44	5.58	5.70	5.80	5.90	5.98	6.06	6.13	6.19	6.25	6.31	6.37	6.42	6.47
29	3.889	4.45	4.80	5.05	5.24	5.40	5.54	5.65	5.76	5.85	5.93	6.01	6.08	6.14	6.20	6.26	6.31	6.36	6.41
30	3.889	4.45	4.80	5.05	5.24	5.40	5.54	5.65	5.76	5.85	5.93	6.01	6.08	6.14	6.20	6.26	6.31	6.36	6.41
31	3.881	4.44	4.79	5.03	5.22	5.38	5.52	5.63	5.74	5.83	5.91	5.99	6.06	6.12	6.18	6.23	6.29	6.34	6.38
32	3.873	4.43	4.78	5.02	5.21	5.37	5.50	5.61	5.72	5.81	5.89	5.97	6.03	6.09	6.16	6.21	6.26	6.31	6.36
33	3.865	4.42	4.76	5.01	5.19	5.35	5.48	5.59	5.70	5.79	5.87	5.95	6.01	6.07	6.13	6.19	6.24	6.29	6.34
34	3.859	4.41	4.75	4.99	5.18	5.34	5.47	5.58	5.68	5.77	5.86	5.93	5.99	6.05	6.12	6.17	6.22	6.27	6.31
35	3.852	4.41	4.74	4.98	5.16	5.33	5.45	5.56	5.67	5.76	5.84	5.91	5.98	6.04	6.10	6.15	6.20	6.25	6.29
36	3.846	4.40	4.73	4.97	5.15	5.31	5.44	5.55	5.65	5.74	5.82	5.90	5.96	6.02	6.08	6.13	6.18	6.23	6.28
37	3.841	4.39	4.72	4.96	5.14	5.30	5.43	5.54	5.64	5.73	5.81	5.88	5.94	6.00	6.06	6.12	6.17	6.22	6.26
38	3.835	4.38	4.72	4.95	5.13	5.29	5.41	5.52	5.62	5.72	5.80	5.87	5.93	5.99	6.05	6.10	6.15	6.20	6.24
39	3.830	4.38	4.71	4.94	5.12	5.28	5.40	5.51	5.62	5.70	5.78	5.85	5.91	5.97	6.03	6.08	6.13	6.18	6.23
40	3.825	4.37	4.70	4.93	5.11	5.26	5.39	5.50	5.60	5.69	5.76	5.83	5.90	5.96	6.02	6.07	6.12	6.16	6.21
50	3.787	4.32	4.64	4.86	5.04	5.19	5.30	5.41	5.51	5.59	5.67	5.74	5.80	5.86	5.91	5.96	6.01	6.06	6.09
60	3.762	4.28	4.59	4.82	4.99	5.13	5.25	5.36	5.45	5.53	5.60	5.67	5.73	5.78	5.84	5.89	5.93	5.97	6.01
120	3.702	4.20	4.50	4.71	4.87	5.01	5.12	5.21	5.30	5.37	5.44	5.50	5.56	5.61	5.66	5.71	5.75	5.79	5.83
∞	3.643	4.12	4.40	4.60	4.76	4.88	4.99	5.08	5.16	5.23	5.29	5.35	5.40	5.45	5.49	5.54	5.57	5.61	5.65

Note: K = total number of subseries; ν = degrees of freedom.

Source: Kanji, G.K. (2001). 100 Statistical Tests. Sage Publication, New Delhi, India.

Appendix A8

Critical Test-Statistic Values for Link-Wallace Test at 5% Significance Level

K	2	3	4	5	6	7	8	9	10	11	12	13	14	15	16	17	18	19	20	30	40	50
n																						
2	3.43	2.35	1.74	1.39	1.15	0.99	0.87	0.77	0.70	0.63	0.58	0.54	0.50	0.47	0.443	0.418	0.396	0.376	0.358	0.245	0.187	0.151
3	1.90	1.44	1.14	0.94	0.80	0.70	0.62	0.56	0.51	0.47	0.43	0.40	0.38	0.35	0.335	0.317	0.301	0.287	0.274	0.189	0.146	0.119
4	1.62	1.25	1.01	0.84	0.72	0.63	0.57	0.51	0.47	0.43	0.40	0.37	0.35	0.33	0.310	0.294	0.279	0.266	0.254	0.177	0.136	0.112
5	1.53	1.19	0.96	0.81	0.70	0.61	0.55	0.50	0.45	0.42	0.39	0.36	0.34	0.32	0.303	0.287	0.273	0.260	0.249	0.173	0.134	0.110
6	1.50	1.17	0.95	0.80	0.69	0.61	0.55	0.49	0.45	0.42	0.39	0.36	0.34	0.32	0.302	0.287	0.273	0.260	0.249	0.174	0.135	0.110
7	1.49	1.17	0.95	0.80	0.69	0.61	0.55	0.50	0.45	0.42	0.39	0.36	0.34	0.32	0.304	0.289	0.275	0.262	0.251	0.175	0.136	0.111
8	1.49	1.18	0.96	0.81	0.70	0.62	0.55	0.50	0.46	0.42	0.39	0.37	0.35	0.33	0.308	0.292	0.278	0.265	0.254	0.178	0.138	0.113
9	1.50	1.19	0.97	0.82	0.71	0.62	0.56	0.51	0.47	0.43	0.40	0.37	0.35	0.33	0.312	0.297	0.282	0.269	0.258	0.180	0.140	0.115
10	1.52	1.20	0.98	0.83	0.72	0.63	0.57	0.52	0.47	0.44	0.41	0.38	0.36	0.34	0.317	0.301	0.287	0.274	0.262	0.183	0.142	0.117
11	1.54	1.22	0.99	0.84	0.73	0.64	0.58	0.52	0.48	0.44	0.41	0.38	0.36	0.34	0.322	0.306	0.291	0.278	0.266	0.186	0.145	0.119
12	1.56	1.23	1.01	0.85	0.74	0.65	0.58	0.53	0.49	0.45	0.42	0.39	0.37	0.35	0.327	0.311	0.296	0.282	0.270	0.189	0.147	0.121
13	1.58	1.25	1.02	0.86	0.75	0.66	0.59	0.54	0.49	0.46	0.42	0.40	0.37	0.35	0.332	0.316	0.300	0.287	0.274	0.192	0.149	0.122
14	1.60	1.26	1.03	0.87	0.76	0.67	0.60	0.55	0.50	0.46	0.43	0.40	0.38	0.36	0.337	0.320	0.305	0.291	0.279	0.195	0.152	0.124
15	1.62	1.28	1.05	0.89	0.77	0.68	0.61	0.55	0.51	0.47	0.44	0.41	0.38	0.36	0.342	0.325	0.310	0.295	0.283	0.198	0.154	0.126
16	1.64	1.30	1.06	0.90	0.78	0.69	0.62	0.56	0.52	0.48	0.44	0.41	0.39	0.37	0.348	0.330	0.314	0.300	0.287	0.201	0.156	0.128
17	1.66	1.32	1.08	0.91	0.79	0.70	0.63	0.57	0.52	0.48	0.45	0.42	0.39	0.37	0.352	0.335	0.319	0.304	0.291	0.204	0.158	0.130
18	1.68	1.33	1.09	0.92	0.80	0.71	0.64	0.58	0.53	0.49	0.46	0.43	0.40	0.38	0.357	0.339	0.323	0.308	0.295	0.207	0.161	0.132

(Contd.)

Appendix A8 (*Contd.*)

K	2	3	4	5	6	7	8	9	10	11	12	13	14	15	16	17	18	19	20	30	40	50
19	1.70	1.35	1.10	0.93	0.81	0.72	0.64	0.59	0.54	0.50	0.46	0.43	0.41	0.38	0.362	0.344	0.327	0.312	0.299	0.210	0.163	0.134
20	1.72	1.36	1.12	0.95	0.82	0.73	0.65	0.59	0.54	0.50	0.47	0.44	0.41	0.39	0.367	0.348	0.332	0.317	0.303	0.212	0.165	0.135
30	1.92	1.52	1.24	1.05	0.91	0.81	0.73	0.66	0.60	0.56	0.52	0.49	0.46	0.43	0.408	0.387	0.369	0.352	0.337	0.237	0.184	0.151
40	2.08	1.66	1.35	1.14	0.99	0.88	0.79	0.72	0.66	0.61	0.57	0.53	0.50	0.47	0.444	0.422	0.402	0.384	0.367	0.258	0.201	0.165
50	2.23	1.77	1.45	1.22	1.06	0.94	0.85	0.77	0.71	0.65	0.61	0.57	0.53	0.50	0.476	0.453	0.431	0.412	0.394	0.277	0.216	0.177
100	2.81	2.23	1.83	1.55	1.34	1.19	1.07	0.97	0.89	0.83	0.77	0.72	0.67	0.64	0.60	0.573	0.546	0.521	0.499	0.351	0.273	0.224
200	3.61	2.88	2.35	1.99	1.73	1.53	1.38	1.25	1.15	1.06	0.99	0.93	0.87	0.82	0.78	0.74	0.70	0.67	0.64	0.454	0.353	0.290
500	5.15	4.10	3.35	2.84	2.47	2.19	1.97	1.79	1.64	1.52	1.42	1.32	1.24	1.17	1.11	1.06	1.01	0.96	0.92	0.65	0.504	0.414
1000	6.81	5.43	4.44	3.77	3.28	2.90	2.61	2.37	2.18	2.22	1.88	1.76	1.65	1.56	1.47	1.40	1.33	1.27	1.22	0.86	0.669	0.549

Note: K = total number of subseries; n = sample size.

Source: Kanji, G.K. (2001). 100 Statistical Tests. Sage Publication, New Delhi, India.

Appendix A9

Critical Test-Statistic Values for Link-Wallace Test at 1% Significance Level

K	2	3	4	5	6	7	8	9	10	11	12	13	14	15	16	17	18	19	20	30	40	50
n																						
2	7.92	4.32	2.84	2.10	1.66	1.38	1.17	1.02	0.91	0.82	0.74	0.68	0.63	0.58	0.54	0.51	0.480	0.454	0.430	0.285	0.214	0.172
3	3.14	2.12	1.57	1.25	1.04	0.89	0.78	0.69	0.62	0.57	0.52	0.48	0.45	0.42	0.39	0.37	0.352	0.334	0.318	0.217	0.165	0.134
4	2.48	1.74	1.33	1.08	0.91	0.78	0.69	0.62	0.56	0.51	0.47	0.44	0.41	0.38	0.36	0.34	0.323	0.307	0.293	0.200	0.153	0.125
5	2.24	1.60	1.24	1.02	0.86	0.75	0.66	0.59	0.54	0.49	0.46	0.42	0.40	0.37	0.35	0.33	0.314	0.299	0.285	0.196	0.151	0.123
6	2.14	1.55	1.21	0.99	0.85	0.74	0.65	0.59	0.53	0.49	0.45	0.42	0.39	0.37	0.35	0.33	0.313	0.298	0.284	0.196	0.151	0.123
7	2.10	1.53	1.20	0.99	0.84	0.73	0.65	0.59	0.53	0.49	0.45	0.42	0.39	0.37	0.35	0.33	0.314	0.299	0.286	0.198	0.152	0.124
8	2.09	1.53	1.20	0.99	0.85	0.74	0.66	0.59	0.54	0.49	0.46	0.43	0.40	0.37	0.35	0.33	0.318	0.303	0.289	0.200	0.154	0.126
9	2.09	1.54	1.21	1.00	0.85	0.75	0.66	0.60	0.54	0.50	0.46	0.43	0.40	0.38	0.36	0.34	0.322	0.307	0.293	0.200	0.156	0.127
10	2.10	1.55	1.22	1.01	0.86	0.76	0.67	0.61	0.55	0.51	0.47	0.44	0.41	0.38	0.36	0.34	0.327	0.311	0.297	0.206	0.159	0.129
11	2.11	1.56	1.23	1.02	0.87	0.76	0.68	0.61	0.56	0.51	0.48	0.44	0.42	0.39	0.37	0.35	0.332	0.316	0.302	0.209	0.161	0.132
12	2.13	1.58	1.25	1.04	0.89	0.78	0.69	0.62	0.57	0.52	0.48	0.45	0.42	0.40	0.37	0.35	0.337	0.321	0.306	0.213	0.164	0.134
13	2.15	1.60	1.26	1.05	0.90	0.79	0.70	0.63	0.58	0.53	0.49	0.46	0.43	0.40	0.38	0.36	0.342	0.326	0.311	0.216	0.166	0.136
14	2.18	1.62	1.28	1.06	0.91	0.80	0.71	0.64	0.58	0.54	0.50	0.46	0.43	0.41	0.39	0.36	0.347	0.330	0.316	0.219	0.169	0.138
15	2.20	1.63	1.30	1.08	0.92	0.81	0.72	0.65	0.59	0.54	0.50	0.47	0.44	0.41	0.39	0.37	0.352	0.335	0.320	0.222	0.171	0.140
16	2.22	1.65	1.31	1.09	0.93	0.82	0.73	0.66	0.60	0.55	0.51	0.48	0.45	0.42	0.40	0.38	0.357	0.340	0.325	0.226	0.174	0.142
17	2.25	1.67	1.33	1.10	0.95	0.83	0.74	0.67	0.61	0.56	0.52	0.48	0.45	0.43	0.40	0.38	0.362	0.345	0.329	0.229	0.176	0.144
18	2.27	1.69	1.34	1.12	0.96	0.84	0.75	0.68	0.62	0.57	0.53	0.49	0.46	0.43	0.41	0.39	0.367	0.350	0.334	0.232	0.179	0.146

(Contd.)

Appendix A9 (*Contd.*)

K	2	3	4	5	6	7	8	9	10	11	12	13	14	15	16	17	18	19	20	30	40	50
19	2.30	1.71	1.36	1.13	0.97	0.85	0.76	0.68	0.62	0.57	0.53	0.50	0.46	0.44	0.41	0.39	0.372	0.354	0.338	0.235	0.181	0.148
20	2.32	1.73	1.38	1.14	0.98	0.86	0.77	0.69	0.63	0.58	0.54	0.50	0.47	0.44	0.42	0.40	0.376	0.359	0.343	0.238	0.184	0.150
30	2.59	1.95	1.54	1.27	1.09	0.96	0.85	0.77	0.70	0.65	0.60	0.56	0.52	0.49	0.46	0.44	0.419	0.399	0.381	0.266	0.205	0.168
40	2.80	2.11	1.66	1.38	1.18	1.04	0.93	0.84	0.76	0.70	0.65	0.61	0.57	0.54	0.51	0.48	0.456	0.435	0.415	0.289	0.223	0.183
50	2.99	2.25	1.78	1.48	1.27	1.11	0.99	0.90	0.82	0.75	0.70	0.65	0.61	0.57	0.54	0.51	0.489	0.466	0.446	0.310	0.240	0.196
100	3.74	2.83	2.24	1.86	1.60	1.40	1.25	1.13	1.03	0.95	0.88	0.82	0.77	0.73	0.69	0.65	0.62	0.590	0.564	0.393	0.304	0.248
200	4.79	3.63	2.88	2.39	2.06	1.81	1.61	1.46	1.33	1.23	1.14	1.06	0.99	0.94	0.88	0.84	0.80	0.76	0.73	0.507	0.392	0.320
500	6.81	5.16	4.10	3.41	2.93	2.58	2.30	2.08	1.90	1.75	1.62	1.52	1.42	1.34	1.26	1.20	1.14	1.09	1.04	0.73	0.560	0.458
1000	9.01	6.83	5.42	4.52	3.88	3.41	3.05	2.76	2.52	2.32	2.15	2.01	1.88	1.77	1.68	1.59	1.51	1.44	1.38	0.96	0.743	0.608

Note: K = total number of subseries; n = sample size.

Source: Kanji, G.K. (2001). 100 Statistical Tests. Sage Publication, New Delhi, India.

Appendix A10

Critical Test-Statistic Values for Dunnett Test

K	2	3	4	5	6	7	8	9
			Level of Significance = 5%					
ν								
5	2.44	2.68	2.85	2.98	3.08	3.16	3.24	3.30
6	2.34	2.56	2.71	2.83	2.92	3.00	3.07	3.12
7	2.27	2.48	2.62	2.73	2.82	2.89	2.95	3.01
8	2.22	2.42	2.55	2.66	2.74	2.81	2.87	2.92
9	2.18	2.37	2.50	2.60	2.68	2.75	2.81	2.86
10	2.15	2.34	2.47	2.56	2.64	2.70	2.76	2.81
11	2.13	2.31	2.44	2.53	2.60	2.67	2.72	2.77
12	2.11	2.29	2.41	2.50	2.58	2.64	2.69	2.74
13	2.09	2.27	2.39	2.48	2.55	2.61	2.66	2.71
14	2.08	2.25	2.37	2.46	2.53	2.59	2.64	2.69
15	2.07	2.24	2.36	2.44	2.51	2.57	2.62	2.67
16	2.06	2.23	2.34	2.43	2.50	2.56	2.61	2.65
17	2.05	2.22	2.33	2.42	2.49	2.54	2.59	2.64
18	2.04	2.21	2.32	2.41	2.48	2.53	2.58	2.62
20	2.03	2.19	2.30	2.39	2.46	2.51	2.56	2.60
24	2.01	2.17	2.28	2.36	2.43	2.48	2.53	2.57
30	1.99	2.15	2.25	2.33	2.40	2.45	2.50	2.54
40	1.97	2.13	2.23	2.31	2.37	2.42	2.47	2.51
60	1.95	2.10	2.21	2.28	2.35	2.39	2.44	2.48
120	1.93	2.08	2.18	2.26	2.32	2.37	2.41	2.45
∞	1.92	2.06	2.16	2.23	2.29	2.34	2.38	2.46
			Level of Significance = 1%					
ν								
5	3.90	4.21	4.43	4.60	4.73	4.85	4.94	5.03
6	3.61	3.88	4.07	4.21	4.33	4.43	4.51	4.59
7	3.42	3.66	3.83	3.96	4.07	4.15	4.23	4.30
8	3.29	3.51	3.67	3.79	3.88	3.96	4.03	4.09
9	3.19	3.40	3.55	3.66	3.75	3.82	3.89	3.94
10	3.11	3.31	3.45	3.56	3.64	3.71	3.78	3.83
11	3.06	3.25	3.38	3.48	3.56	3.63	3.69	3.74
12	3.01	3.19	3.32	3.42	3.50	3.56	3.62	3.67

(Contd.)

Appendix A10 (*Contd.*)

K	2	3	4	5	6	7	8	9
				Level of Significance = 5%				
13	2.97	3.15	3.27	3.37	3.44	3.51	3.56	3.61
14	2.94	3.11	3.23	3.32	3.40	3.46	3.51	3.56
15	2.91	3.08	3.20	3.29	3.36	3.42	3.47	3.52
16	2.88	3.05	3.17	3.26	3.33	3.39	3.44	3.48
17	2.86	3.03	3.14	3.23	3.30	3.36	3.41	3.43
18	2.84	3.01	3.12	3.21	3.27	3.33	3.38	3.42
20	2.81	2.97	3.08	3.17	3.23	3.29	3.34	3.38
24	2.77	2.92	3.03	3.11	3.17	3.22	3.27	3.31
30	2.72	2.87	2.97	3.05	3.11	3.16	3.21	3.24
40	2.68	2.82	2.92	2.99	3.05	3.10	3.14	3.18
60	2.64	2.78	2.87	2.94	3.00	3.04	3.08	3.12
120	2.60	2.73	2.82	2.89	2.94	2.99	3.03	3.06
∞	2.56	2.68	2.77	2.84	2.89	2.93	2.97	3.00

Note: *K* = total number of subseries; ν = degrees of freedom.

Source: Kanji, G.K. (2001). 100 Statistical Tests. Sage Publication, New Delhi, India.

Appendix A11

Critical Test-Statistic (M) Values for Bartlett Test at 5% Significance Level

C	0.0	0.5	1.0	1.5	2.0	2.5	3.0	3.5	4.0	4.5	5.0	6.0	7.0	8.0	9.0	10.0
K																
3	5.99	6.47	5.89	7.20	7.38	7.39	7.22	–	–	–	–	–	–	–	–	–
4	7.81	8.24	8.63	8.96	9.21	9.38	9.43	9.37	9.18	–	–	–	–	–	–	–
5	9.49	9.88	10.24	10.57	10.86	11.08	11.24	11.32	11.31	11.21	11.02	–	–	–	–	–
6	11.07	11.43	11.78	12.11	12.40	12.65	12.86	13.01	13.11	13.14	13.10	12.78	–	–	–	–
7	12.59	12.94	13.27	13.59	13.88	14.15	14.38	14.58	14.73	14.83	14.88	14.81	14.49	–	–	–
8	14.07	14.40	14.72	15.03	15.32	15.60	15.84	16.06	16.25	16.40	16.51	16.60	16.49	16.16	–	–
9	15.51	15.83	16.14	16.44	16.73	17.01	17.26	17.49	17.70	17.88	18.03	16.22	18.26	18.12	17.79	–
10	16.92	17.23	17.54	17.83	18.12	18.39	18.65	18.89	19.11	19.31	19.48	19.75	19.89	19.89	19.73	19.40
11	18.31	18.61	18.91	19.20	19.48	19.76	20.02	20.26	20.49	20.70	20.89	21.21	21.42	21.52	21.49	21.32
12	19.68	19.97	20.26	20.55	20.83	21.10	21.36	21.61	21.84	22.06	22.27	22.62	22.88	23.06	23.12	23.07
13	21.03	21.32	21.60	21.89	22.16	22.43	22.69	22.94	23.18	23.40	23.62	23.99	24.30	24.53	24.66	24.70
14	22.36	22.65	22.93	23.21	23.48	23.75	24.01	24.26	24.50	24.73	24.95	25.34	25.68	25.95	26.14	26.25
15	23.68	23.97	24.24	24.52	24.79	25.05	25.31	25.56	25.80	26.04	26.26	26.67	27.03	27.33	27.56	27.73

Note: C = bias correction factor; K = total number of subseries.

Source: Kanji, G.K. (2001). 100 Statistical Tests. Sage Publication, New Delhi, India.

Appendix A12

Critical Test-Statistic (*M*) Values for Bartlett Test at 1% Significance Level

C	0.0	0.5	1.0	1.5	2.0	2.5	3.0	3.5	4.0	4.5	5.0	6.0	7.0	8.0	9.0	10.0
K																
3	9.21	9.92	10.47	10.78	10.81	10.50	9.83	–	–	–	–	–	–	–	–	–
4	11.34	11.95	12.46	12.86	13.11	13.18	13.03	12.65	12.03	–	–	–	–	–	–	–
5	13.28	13.81	14.30	14.71	15.03	15.25	15.34	15.28	15.06	14.66	14.07	–	–	–	–	–
6	15.09	15.58	16.03	16.44	16.79	17.07	17.27	17.37	17.37	17.24	16.98	16.03	–	–	–	–
7	16.81	17.27	17.70	18.10	18.46	18.77	19.02	19.21	19.32	19.35	19.28	18.84	17.92	–	–	–
8	18.48	18.91	19.32	19.71	20.07	20.39	20.67	20.90	21.08	21.20	21.35	21.13	20.64	19.76	–	–
9	20.09	20.50	20.90	21.28	21.64	21.97	22.26	22.52	22.74	22.91	23.03	23.10	22.91	22.41	21.56	–
10	21.67	22.06	22.45	22.82	23.17	23.50	23.80	24.08	24.32	24.52	24.69	24.90	24.90	24.66	24.15	23.33
11	23.21	23.59	23.97	24.33	24.67	25.00	25.31	25.59	25.85	26.08	26.28	26.57	26.70	26.65	26.38	25.86
12	24.72	25.10	25.46	25.81	26.15	26.48	26.79	27.08	27.35	27.59	27.81	28.16	28.39	28.46	28.37	28.07
13	26.22	26.58	26.93	27.28	27.62	27.94	28.25	28.54	28.81	29.07	29.30	29.70	29.99	30.16	30.19	30.06
14	27.69	28.04	28.39	28.73	29.06	29.38	29.69	29.98	30.26	30.52	30.77	31.19	31.53	31.77	31.89	31.88
15	29.14	29.39	29.83	30.16	30.49	30.80	30.11	31.40	31.68	31.95	32.20	32.66	33.03	33.32	33.51	33.59

Note: C = bias correction factor; K = total number of subseries.

Source: Kanji, G.K. (2001). 100 Statistical Tests. Sage Publication, New Delhi, India.

Appendix A13

Critical Test-Statistic (F_{max}) Values for Hartley Test (Right-Sided)

Level of Significance = 5%											
K / *n*–1	2	3	4	5	6	7	8	9	10	11	12
2	39.0	87.5	142	202	266	333	403	475	550	626	704
3	15.4	27.8	39.2	50.7	62.0	72.9	83.5	93.9	104	114	124
4	9.60	15.5	20.6	25.2	29.5	33.6	37.5	41.1	44.6	48.0	51.4
5	7.15	10.8	13.7	16.3	18.7	20.8	22.9	24.7	26.5	28.2	29.9
6	5.82	8.38	10.4	12.1	13.7	15.0	16.3	17.5	18.6	19.7	20.7
7	4.99	6.94	8.44	9.70	10.8	11.8	12.7	13.5	14.3	15.1	15.8
8	4.43	6.00	7.18	8.12	9.03	9.78	10.5	11.1	11.7	12.2	12.7
9	4.03	5.34	6.31	7.11	7.80	8.41	8.95	9.45	9.91	10.3	10.7
10	3.72	4.85	5.67	6.34	6.92	7.42	7.87	8.28	8.66	9.01	9.34
12	3.28	4.16	4.79	5.30	5.72	6.09	6.42	6.72	7.00	7.25	7.48
15	2.86	3.54	4.01	4.37	4.68	4.95	5.19	5.40	5.59	5.77	5.93
20	2.46	2.95	3.29	3.54	3.76	3.94	4.10	4.24	4.37	4.49	4.59
30	2.07	2.40	2.61	2.78	2.91	3.02	3.12	3.21	3.29	3.36	3.39
60	1.67	1.85	1.96	2.04	2.11	2.17	2.22	2.26	2.30	2.33	2.36
∞	1.00	1.00	1.00	1.00	1.00	1.00	1.00	1.00	1.00	1.00	1.00
Level of Significance = 1%											
2	199	448	729	1036	1362	1705	2063	2432	2813	3204	3605
3	47.5	85	120	151	184	216	249	281	310	337	361
4	23.2	37	49	59	69	79	89	97	106	113	120
5	14.9	22	28	33	38	42	46	50	54	57	60
6	11.1	15.5	19.1	22	25	27	30	32	34	36	37
7	8.89	12.1	14.5	16.5	18.4	20	22	23	24	26	27
8	7.50	9.9	11.7	13.2	14.5	15.8	16.9	17.9	18.9	19.8	21
9	6.54	8.5	9.9	11.1	12.1	13.1	13.9	14.7	15.3	16.0	16.6
10	5.85	7.4	8.6	9.6	10.4	11.1	11.8	12.4	12.9	13.4	13.9
12	4.91	6.1	6.9	7.6	8.2	8.7	9.1	9.5	9.9	10.2	10.6
15	4.07	4.9	5.5	6.0	6.4	6.7	7.1	7.3	7.5	7.8	8.0
20	3.32	3.8	4.3	4.6	4.9	5.1	5.3	5.5	5.6	5.8	5.9
30	2.63	3.0	3.3	3.4	3.6	3.7	3.8	3.9	4.0	4.1	4.2
60	1.96	2.2	2.3	2.4	2.4	2.5	2.5	2.6	2.6	2.7	2.7
∞	1.0	1.0	1.0	1.0	1.0	1.0	1.0	1.0	1.0	1.0	1.0

Note: K = total number of subseries; n = sample size.

Source: Kanji, G.K. (2001). 100 Statistical Tests. Sage Publication, New Delhi, India.

Appendix B1

Books on Advanced Statistics and Time Series Analysis

Bails, D.G. and Peppers, L.C. (1982). Business Fluctuations: Forecasting Techniques and Applications. Englewood Cliffs, Prentice-Hall, NJ, USA.

Barnett, V. and Lewis, T. (1978). Outliers in Statistical Data. John Wiley, New York.

Berthouex, P.M. and Brown, L.C. (1994). Statistics for Environmental Engineers. Lewis, Boca Raton, FL, USA.

Bethea, R.M. and Rhinehart, R.R. (1991). Applied Engineering Statistics. Marcel Dekker, Inc., New York, USA.

Box, G.E.P. and Jenkins, G.M. (1976). Time Series Analysis: Forecasting and Control. Holden-Day, San Francisco, USA.

Bras, R.L. and Rodriguez-Iturbe, I. (1985). Random Functions and Hydrology. Addison-Wesley, Reading, MA, USA.

Brockwell, P.J. and Davis, R.A. (1991). Time Series: Theory and Methods. 2nd edition, Springer Series in Statistics, Springer, New York, USA.

Brockwell, P.J. and Davis, R.A. (1996). Introduction to Time Series and Forecasting. Springer-Verlag, Inc., New York, USA.

Chatfield, C. (1980). The Analysis of Time Series: An Introduction. 2nd edition. Chapman and Hall, London, New York, USA.

Clarke, R.T. (1998). Stochastic Processes for Water Scientists: Development and Applications. John Wiley and Sons, New York, USA.

Conover, W.J. (1980). Practical Nonparametric Statistics. 2nd edition. John Wiley, New York.

Cryer, J.D. (1986). Time Series Analysis. PWS Publishers, Duxbury Press, Boston, MA.

D'Agostino, R.B. and Stephens, M.A. (editors) (1986). Goodness-of-Fit Techniques. Marcel Dekker, New York, USA.

Dahmen, E.R. and Hall, M.J. (1990). Screening of hydrologic data: Tests for stationarity and relative consistency. ILRI Publication No. 49, Wageningen.

Dixon, W.J. and Massey Jr., F.J. (1983). Introduction to Statistical Analysis. 4th edition. McGraw-Hill, New York, USA.

Gibbons, R.D. (1994). Statistical Methods for Groundwater Monitoring. John Wiley and Sons, New York, USA.

Gibbons, R.D. and Coleman, D.E. (2001). Statistical Methods for Detection and Quantification of Environmental Contamination. John Wiley and Sons, New York, USA.

Gilbert, R.O. (1987). Statistical Methods for Environmental Pollution Monitoring. Van Nostrand Reinhold, New York, USA.

Ginevan, M.E. and Splitstone, D.E. (2004). Statistical Tools for Environmental Quality Measurement. Chapman and Hall, Boca Raton, FL, USA.

Haan, C.T. (1977). Statistical Methods in Hydrology. Iowa State University Press, Iowa, USA.

Haan, C.T. (2002). Statistical Methods in Hydrology. 2nd edition. Iowa State University Press, Iowa, USA.
Hawkins, D.M. (1980). Identification of Outliers. Chapman and Hall, New York, USA.
Helsel, D.R. and Hirsch, R.M. (2002). Statistical Methods in Water Resources. Chapter A3, Book 4. Hydrologic Analysis and Interpretation, Techniques of Water-Resources Investigations of the United States Geological Survey (USGS), USGS, Reston, VA.
Himmelblau, D.M. (1969). Process Analysis by Statistical Methods. John Wiley and Sons, New York, USA.
Hipel, K.W. and McLeod, A.I. (1994). Time Series Modeling of Water Resources and Environmental Systems. Elsevier, Amsterdam, The Netherlands.
Hoel, P.G. (1954). Introduction to Mathematical Statistics. 2nd edition. John Wiley & Sons, New York, USA.
Hoff, J.C. (1983). A Practical Guide to Box-Jenkins Forecasting. Lifetime Learning Publications, London, U.K.
Kanji, G.K. (2001). 100 Statistical Tests. Sage Publication, New Delhi, India.
Kendall, M.G. (1948). The Advanced Theory of Statistics 2. Charles Griffin & Co. Ltd., London, U.K.
Kendall, M.G. (1973). Time Series. Charles Griffin & Co. Ltd., London, U.K.
Kendall, M.G. (1975). Rank Correlation Methods. Charles Griffin & Co. Ltd., London, U.K.
Kenney, J.F., and Keeping, E.S. (1954). Mathematics of Statistics: Part One. D. Van Nostrand Company, Inc., New York, USA.
Kiesiel, C.C. (1969). Time Series Analysis of Hydrologic Data. *In:* V.T. Chow (editor), Advances in Hydroscience, 5.
Kottegoda, N.T. (1980). Stochastic Water Resources Technology. Macmillan & Co. Ltd., London, U.K.
Lehmann, E.L. (1975). Nonparametrics: Statistical Methods Based on Ranks. Holden-Day, Inc., San Francisco, CA, USA.
Lehmann, E.L. (1999). Elements of Large Sample Theory. Springer, New York, USA.
Lin, L.M., Hudak, G.B., Box, E., Muller, M. and Tiao, G. (1994). Forecasting and Time Series Analysis Using the SCA Statistical System. Scientific Computing Associates Corp., Oak Brook, IL, USA.
Madansky, A. (1988). Prescriptions for Working Statisticians. Springer-Verlag, New York.
Makridakis, S., Wheelwright, S.C. and Hyndman, R.J. (1998). Forecasting: Methods and Applications. 3rd edition, John Wiley & Sons, New York, USA.
McCleary, R. and Hay, R.A. (1980). Applied Time Series Analysis for the Social Sciences. Sage Publications, Beverly Hills, CA.
McCuen, R.H. (2003). Modeling Hydrologic Change: Statistical Methods. Lewis Publishers, CRC Press LLC, Florida, USA.
McDowall, D., McCleary, R., Meidinger, E.E. and Hay, R.A. (1980). Interrupted Time Series Analysis. Sage Publications, Beverly Hills, CA, USA.
McGhee, J.W. (1985). Introductory Statistics. West Publishing Co., New York, USA.
Montgomery, D.C., Johnson, L.A. and Gardiner, J.S. (1990). Forecasting and Time Series Analysis. 2nd edition. McGraw-Hill, New York, USA.
Natrella, M.G. (1963). Experimental Statistics. National Bureau of Standards Handbook 91, Washington, DC.
Neville, A.M. and Kennedy, J.B. (1964). Basic Statistical Methods for Engineers and Scientists. Intertext Books Co., London, U.K.

O'Connel, P.E. (1977). ARIMA models in synthetic hydrology. *In:* T.A. Ciriani, U. Maione and J.R. Wallis (editors), Mathematical Models in Surface Water Hydrology. John Wiley & Sons, New York, USA.

Ostrom, C.W. (1978). Time Series Analysis: Regression Techniques. 2nd edition, Sage University Papers Series on Quantitative Applications in the Social Sciences, Number 07-009, Thousand Oaks, CA.

Ott, W.R. (1995). Environmental Statistics and Data Analysis. Lewis, Boca Raton, FL.

Owen, D.B. (1962). Handbook of Statistical Tables. Addison-Wesley, Reading, Mass.

Pankratz, A. (1983). Forecasting with Univariate Box-Jenkins Models: Concepts and Cases. Wiley, New York, USA.

Rao, A.R., Hamed, K.H. and Chen, H.-L. (2003). Nonstationarities in Hydrologic and Environmental Time Series. Water Science and Technology Library, 45.

Sachs, L. (1972). Statistische Auswertungsmethoden. 3rd edition. Springer-Verlag, Berlin.

Salas, J.D. (1993). Analysis and Modeling of Hydrologic Time Series. *In*: D.R. Maidment (editor-in-chief), Handbook of Hydrology. McGraw-Hill, Inc., USA.

Salas, J.D., Delleur, J.W., Yevjevich, V. and Lane, W.L. (1980). Applied Modeling of Hydrologic Time Series. Water Resources Publications, Littleton, CO, USA.

Shahin, M., Van Oorschot, H.J.L. and De Lange, S.J. (1993). Statistical Analysis in Water Resources Engineering. A.A. Balkema, Rotterdam, The Netherlands.

Shapiro, S. (1980). How to Test Normality and Other Distributional Assumptions. Volume 3, The ASQC Basic References in Quality Control: Statistical Techniques, American Society for Quality Control, Milwaukee, WI.

Snedecor, G.W. and Cochran, W.G. (1980). Statistical Methods. The Iowa State University Press, Ames, Iowa, USA.

Spiegel, M.R. and Stephens, L.J. (2000). Schaum's Outlines Statistics. 3rd edition. Tata McGraw-Hill Publishing Company Ltd., New Delhi, India.

Thode, H.C. (2002). Testing for Normality. Marcel Dekker, New York.

Tukey, J.W. (1977). Exploratory Data Analysis. Addison-Wesley, MA.

USEPA (1992). Guidance Document on the Statistical Analysis of Ground-Water Monitoring Data at RCRA Facilities. EPA/530/R-93/003, United States Environmental Protection Agency (USEPA), Office of Solid Waste, Washington DC.

USEPA (1996). Guidance for Data Quality Assessment: Practical Methods for Data Analysis. EPA QA/G-9, EPA QA/G-9, United States Environmental Protection Agency (USEPA), Office of Research and Development, Washington, DC.

USEPA (2006). Data Quality Assessment: Statistical Methods for Practitioners. Guidance Document EPA QA/G-9S, United States Environmental Protection Agency (USEPA), Office of Environmental Information, Washington DC.

Vandaele, W. (1983). Applied Time Series and Box-Jenkins Models. Academic Press, New York, USA.

von Storch, H. and Navarra, N. (editors) (1995). Analysis of Climate Variability: Applications of Statistical Techniques. Springer-Verlag, Berlin, Germany.

Walpole, R. and Myers, R. (1985). Probability and Statistics for Engineers and Scientists. 3rd edition, Macmillan, New York, USA.

Wei, W.S. (1990). Time Series Analysis. 2nd edition. Addison Wesley, Menlo Park, CA.

Yevjevich, V.M. (1972). Stochastic Processes in Hydrology. Water Resources Publications, Fort Collins, CO, USA.

Appendix B2

Web Resources on Time Series Analysis

- www.itl.nist.gov/div898/handbook/eda/section3/eda35g.htm.
  (NIST/SEMATECH (2007): e-Handbook of Statistical Methods)
- http://mathworld.wolfram.com/TimeSeriesAnalysis.html
  (Wolfram MathWorld: the web's most expensive mathematics resource)
- http://sites.google.com/site/bookenglishlang/Nonlinear-Time-Series-Analysis
  [Kantz, H. and Schreiber, T. (2004). Nonlinear Time Series Analysis]
- http://www.iasri.res.in/ebook/EBADAT/5-Modeling%20and%20Forecasting%20Techniques%20in%20Agriculture/2-time_series_analysis_22-02-07_revised.pdf
  (Ramasubramaniam, V. Time Series Analysis. Lecture Notes, Indian Agricultural Statistical Research Institute, New Delhi, India)
- http://www.statistik-mathematik.uni-wuerzburg.de/fileadmin/10040800/user_upload/time_series/the_book/2011-March-01-times.pdf
  [Falk, M., Marohn, F., Michel, R., Hofmann, D., Macke, M., Spachmann, C. and Englert, A. (editors) (2011). A First Course on Time Series Analysis: Examples with SAS. Chair of Statistics, University of Würzburg]
- http://processtrends.com/toc_trend_analysis_with_excel.htm
  (Trend Analysis with Excel)
- http://www.wessa.net/tsa.wasp
  (Free Statistics and Forecasting Software)
- http://www.statisticssolutions.com/resources/directory-of-statistical-analyses
  (Directory of Statistical Analyses)
- http://stats.stackexchange.com/
  (Q&A for Statisticians, Data Analysts, Data Miners and Data Visualization Experts)
- http://www.math.yorku.ca/SCS/StatResource.html
  (Statistics and Statistical Graphics Resources)
- http://www.statistics.com/glossary/
  (Glossary of Statistical Terms)
- http://www.top4download.com/free-statistical-analysis-software/
  (Download Statistical Analysis Software)
- http://statpages.org/
  (Web Pages that Perform Statistical Calculations)
- http://statpages.org/javasta2.html#Freebies
  (Free Statistical Software)
- http://statpages.org/javasta3.html#Textbooks
  (Statistical Books, Manuals and Journals)
- http://statpages.org/javasta5.html#OtherSites
  (Links to Other Statistical Web Sites)

- http://www.abs.gov.au/websitedbs/d3310114.nsf/4a256353001af3ed 4b2562bb-00121564/b81ecff00cd36415ca256ce10017de2f!OpenDocument
  (Time Series Basics, Australian Bureau of Statistics)
- http://www.scribd.com/doc/7194005/Handbook-for-Statistical-Analysis-of-Environmental-Background-Data
  (Handbook for Statistical Analysis of Environmental Background Data)

Appendix B3

Software for Time Series Analysis

STATISTICA (www.statsoft.com)

DataPlot (www.itl.nist.gov/div898/software/dataplot/)

SPSS (http://spss.co.in/)

SYSTAT (http://www.systat.com/)

SAS Software (http://www.sas.com/technologies/analytics/statistics/stat/)

DataQUEST – Data Quality Evaluation Statistical Toolbox (US Environmental Protection Agency, Report No. EPA QA/G-9D)

NCSS – Statistical and Power Analysis Software (http://www.ncss.com/)

DTREG – Software For Predictive Modeling and Forecasting (http://www.dtreg.com/)

Estima – Econometrics and Time Series Analysis Software Package (http://www.estima.com/)

Catterpillar SSA – Time Series Analysis and Forecasting Software (http://www.gistatgroup.com/cat/)

GMDH Shell (www.gmdhshell.com)

Tableau Software (http://www.tableausoftware.com/)

Modified WEKA – Waikato Environment for Knowledge Analysis (http://davis.wpi.edu/~xmdv/weka/)

STEM – Short Time-series Expression Miner (http://gene.ml.cmu.edu/stem/)

STSA – Statistical Time Series Analysis Toolbox (http://www.omatrix.com/stsa.html)

HYDROSPECT – Software for Detecting Changes in Hydrological Data (http://water.usgs.gov/osw/wcp-water/detecting-trend.pdf)

TSM – Time Series Modeling (http://www.timeseriesmodelling.com/)

Index